Sickness, Suffering, and the Sword

C&C

CAMPAIGNS & COMMANDERS

GREGORY J. W. URWIN, SERIES EDITOR

Sickness, Suffering, and the Sword

The British Regiment on Campaign, 1808–1815

Andrew Bamford

Foreword by Donald E. Graves

University of Oklahoma Press : Norman

This book is published with the generous assistance of
The Kerr Foundation, Inc.

Library of Congress Cataloging-in-Publication Data

Bamford, Andrew, 1981–
  Sickness, suffering, and the sword : the British regiment on campaign,
1808–1815 / Andrew Bamford ; foreword by Donald E. Graves.
     pages cm. — (Campaigns and commanders ; volume 37)
  Includes bibliographical references and index.
  ISBN 978-0-8061-4343-9 (hardcover : alk. paper) 1. Great Britain. Army—
History—Peninsular War, 1807–1814. 2. Great Britain. Army—Organization—
History—19th century. 3. Peninsular War, 1807–1814—Psychological aspects.
4. Military morale—Great Britain—History—19th century. 5. Wellington,
Arthur Wellesley, Duke of, 1769–1852—Military leadership. I. Title.
  DC232.B336 2013
  940.2'7420941—dc23
                                                                2012044494

Sickness, Suffering, and the Sword: The British Regiment on Campaign, 1808–1815
is Volume 37 in the Campaigns and Commanders series.

Contents

List of Illustrations   vii

List of Tables   viii

Foreword, by Donald E. Graves   ix

Preface and Acknowledgments   xiii

Introduction   xvii

List of Abbreviations   xxiii

   1. The British Army and Its Campaigns   3

   2. Regimental Identity and Leadership   44

   3. The Regimental System in Practice   86

   4. The Limits of the System   127

   5. Beyond the Regiment   177

   6. Strategic Consumption   217

   7. Beasts of Burden   260

Conclusion: A System Reassessed   286

Appendix 1. Details of Data Sample   303

Appendix 2. Men Returned as Sick after the Corunna
   Campaign   305

Appendix 3. Sample of Returned Deserters in the Peninsula   307

Notes   309

Bibliography   337

Index   349

Illustrations

1. Campaign Manpower in the 1/40th   93
2. Campaign Manpower in the 29th   97
3. Campaign Manpower in the 2/83rd   101
4. Campaign Manpower in the 1/88th   105
5. Campaign Manpower in the 16th Light Dragoons   111
6. Campaign Manpower in the 2/87th   123
7. Deaths on Active Service, 1808–1815   221
8. Sickness and Death Rates in the Peninsula   233
9. Peninsular Horse Strength by Month and Arm of Service   265
10. Peninsular Horse Deaths by Month and Arm of Service   267

Tables

1. Total Manpower Commitments, 1808–1815 41

2. The Battalions of Detachments 132

3. The Corps of Embodied Detachments 136

4. Battalions Deployed to the Peninsula, 1810–1813 143

5. Peninsular Provisional Battalions 149

6. Battalions Sent from Britain to the Netherlands, December 1813 163

7. York's Provisional Battalion Scheme 173

8. Average Unit Experience (Months) of Peninsular Divisions 198

9. Sickness and Death Rates by Theater 223

10. Average Desertion Rates by Theater 246

11. Desertions on Active Service, 1808–1815 247

12. Comparative Equine Mortality in the Peninsula 275

13. Peninsular Equine Mortality Rates by Regiment 277

Foreword

In the long and proud history of the British Army, one of the most prominent chapters is its participation in the conflict waged with revolutionary and imperial France from 1793 to 1815. Known for nearly a century afterward as the "Great War with France," this epic struggle cost the lives of nearly five million people in Europe and the Middle East. Britain's fatal casualties alone were, per capita, as severe as they were during the First World War and were only ameliorated by the fact they were suffered over a period of two decades, rather than in four years. Unchallenged on the oceans after Nelson's victory at Trafalgar in 1805, Britain's main effort in the latter years of the conflict was the campaigns of its land forces in the Iberian Peninsula, most notably those conducted by the Duke of Wellington. From 1808 to 1814, these forces were rarely defeated but, instead, won a series of victories whose lyrical names—Albuera, Badajoz, Barrosa, Bussaco, Ciudad Rodrigo, Corruna, Fuentes D'Onor, Nive, Nivelle, Orthes, the Pyrenees, Roleia, St. Sebastian, Salamanca, Talavera, Toulouse, Tarifa, Vimiera, Vittoria, to use the Battle Honours spellings—are emblazoned on the Colours of most modern British regiments.

Those nineteenth-century campaigns in Portugal and Spain have been a great source of regimental tradition, lore, and mythology that is proudly commemorated to this day. An example from the 1811 battle of Albuera—the bloodiest single engagement of the Peninsular War—will suffice. During that engagement, the Fusilier Brigade (7th and 23rd Regiments of Foot, Royal Fusiliers, and Royal Welch Fusiliers respectively) mounted an attack on Marshal Soult's army that decided the battle. This attack was immortalized in the magnificent prose of William Napier. "Nothing," Napier wrote in his *History of the War in the Peninsula,* "could stop that astonishing infantry. No sudden burst of undisciplined valour, no nervous enthusiasm, weakened the stability of their order; their flashing eyes were bent on the dark columns in their front, their measured tread shook the ground, their dreadful volleys swept away the head of every formation, their deafening shouts overpowered the dissonant cries that broke from all parts of the tumultuous crowd, as slowly and with horrid carnage, it was pushed by the incessant vigour of the attack to the farthest edge of the height." Stirring stuff, to be sure, and it will come as no surprise that the title of one of the many regimental histories of the Royal Welch Fusiliers is *That Astonishing Infantry.*

Unfortunately, the truth of that attack is somewhat more prosaic as I found when I wrote my 2010 book, *Dragon Rampant,* a history of the 23rd Foot during the Great War with France. First, the battle did not take place quite as neatly as Napier described it—but then real action never does. Second, the casualties on both sides were horrendous (52 percent of the Fusilier Brigade was killed or wounded). Third, Napier does not mention that five battalions of Portuguese infantry were present to help out the fusiliers. Nonetheless, the attack of the Fusilier Brigade at Albuera in 1811 was a great feat of arms, and I so described it in *Dragon Rampant.*

But it was the only major battle that the 23rd Foot participated in between the time it landed in Lisbon in the last days of October 1810 until it went into winter quarters during the first days of November 1811. During that year, this unit, or parts of it, only came under enemy fire on ten days, yet the records reveal that it lost nearly five hundred men from all causes—"sickness, suffering, and the sword," in the words of Captain Moyle Sherer of the 34th Foot, which the author has aptly taken as the title of the book that follows. It required constant efforts to keep regiments like the 23rd Foot up to strength in the peninsula. In *Sickness, Suffering, and the Sword: The British Regiment on Campaign, 1808–1815,* Andrew Bamford has set himself the rather difficult task of explaining how the regimental system of the British Army functioned during the latter years of the Great War with France.

In my opinion the author has very successfully accomplished that task. Bamford begins with a detailed description of the evolution and organization of the regimental system. He analyzes the important changes made to the system when the Duke of York became commander in chief of the British Army in the 1790s and how this influenced its operation in 1808–15, his period of emphasis. He then discusses the important matter of unit identity and cohesion, stressing that leadership was all-important as many a bad unit was transformed by a good commander while many a fine unit was damaged by a poor one. In that respect, Lt. General Sir Brian Horrocks perhaps said it best when he commented that "one good commanding officer is worth a ton of tradition with mould on it."

Bamford continues with a detailed discussion of the regimental system in practice, its strengths, weaknesses, problems, and extemporized solutions, the continual draining away of strength by noncombat losses. He does not, however, neglect higher organizations and presents a history of the divisional system within Wellington's army, which is extremely useful. Personally, I found the chapter on "Beasts of Burden" to be fascinating as, in no uncertain terms and with a great deal of factual data, it demonstrates the importance of the grain-fuelled prime mover of the time to military operations. "Data" is the operative word here because Bamford backs up all his analyses and statements with statistical information from War Office and other departmental records that he has mined with impressive industry.

The result is that *Sickness, Suffering, and the Sword* is a historical study that not only reflects a very high level of scholarship but constitutes a most welcome and useful addition to the literature on the British Army in the latter years of the Napoleonic period. I only wish that I had had it on my desk when I was working on my history of a peninsular regiment, and I am certain that it is destined to become a standard source for all serious students of not only the British Army of 1808–15 in particular, but Napoleonic armies in general.

Donald E. Graves
"Maple Cottage"
Valley of the Mississippi, Upper Canada
Victoria Day, 2012

Preface and Acknowledgments

This work began life some years ago under a different name and in a rather different form. Having returned to academia as a postgraduate, I became increasingly interested in how Napoleonic armies functioned on campaign—that is to say, not in combat but during the weeks and months of marching, countermarching, winter quarters, and minor skirmishes that made up the bulk of early nineteenth-century warfare. First, I was able to explore these ideas through a master's dissertation looking at the British Army in the first six months of the Peninsular War. This study helped me develop a statistical methodology, which, after long hours in the National Archives at Kew, enabled me to do the same for the period of 1808–15, forming the focus first of my doctoral thesis and now of this book.

My initial title was "The British Army on Campaign," and it was as such that my thesis saw the light of day. But it strikes me now that such a focus was misleading. It is easy enough to talk of the British Army as a single entity, but even today it remains in many ways a collection of regiments. This situation was all the more apparent in the eighteenth and nineteenth centuries, and one only has to consult some of the excellent new wave of regimental histories covering the period to see how varied the regimental experience of the Napoleonic Wars could be. As I began to develop my research, it seemed increasingly apparent that what I was really addressing—what much of my thesis had in any case been arguing for—was the effectiveness not of the British Army but of Britain's regimental system. It is, therefore, the question of how well that system stood up to the Napoleonic Wars that forms the backbone of this book.

The organization of the British Army of the Napoleonic era was complex, as was the nomenclature used to differentiate its units, outlined in more detail in chapter 1. A few explanatory notes are necessary, however. Regiments were generally numbered, although the Royal Horse Guards were not, and the second regiment of Foot Guards is never referred to as such—since this would imply subordination to the 1st Foot Guards—having always been the Coldstream Guards. Most infantry regiments also had titles, either geographic or linking them to the royal family, and cavalry regiments were further distinguished by type: dragoon guards (with their own numbering sequence), dragoons, light dragoons, and hussars. Thus, being formal, we might speak of the 3rd (Prince of Wales's) Dragoon Guards, the 6th (Inniskilling) Dragoons, or the 84th (York and Lancaster) Regiment of Foot. In general, I have used the numbers only. Infantry regiments frequently had multiple battalions—as many as eight in one special case—and in the text these are given before the regimental number so that, for example, 2/44th signifies the second battalion of the 44th Foot and 4/1st signifies the fourth battalion of the 1st Foot. Where regiments fielded only a single battalion, I have used the number alone: thus 20th, 51st, 104th.

On the subject of names, two points need clarifying. Firstly, this narrative is thematic rather than chronological, whereas names and ranks changed over time as individuals rose through the military hierarchy and also, for the few who reached high rank, into and through the peerage as well. I have used the titles appropriate to the period being discussed, thus the narrative alternates between Wellesley and Wellington as we move back and forth across time. Similarly, John Spencer Cooper is variously private, lance corporal, and sergeant; William Tomkinson variously lieutenant and captain. The other variable concerns place names; in many cases the present name and the name used in 1808 are not the same, whilst the attempts made by British writers of the time frequently differ both from established forms and from each other. When a place name appears in a direct quote, I have retained the original spelling but otherwise have used the modern place names: thus Vitoria rather than Vittoria, Merksem rather than Mercxem. The only exception is those places where we British, in our ignorance or arrogance, have long since rechristened a major town or city to suit our tongues: thus, Lisbon rather than Lisboa; Oporto rather than Porto; Brussels rather than Bruxelles. Like all compromises, this will doubtless offend someone somewhere, but that was not the intention.

I must finally thank the many individuals whose involvement, interest, and support have helped make this book possible. Since it began life as an academic project, I must begin with my supervisors at the University of Leeds, Kevin Linch and Edward Spiers, as well as John Childs, John Gooch, and Alan Forrest, who were all involved in the examination of my thesis and many of whose suggestions have subsequently been incorporated into this revised work. The other great source of scholarly support has been the Napoleon Series website and the many contributors to its discussion forum; I must in particular name Bob Burnham, David Buttery, Anthony Gray, Ron McGuigan, and Howie Muir as amongst those whose contributions have been of particular help, and most of all Don Graves, who has provided continued support for this project and has most generously furnished it with a foreword.

There are countless other friends and colleagues, also deserving of a mention, who have provided help and encouragement over the last six years. Some have helped with sources, others have provided specialist advice in areas where my own research could never match theirs, and others have read and commented on sections of the manuscript. A big thank you therefore goes out to Dave Beckford (who stepped in to create the graphs when it became clear that my own IT skills were not up to the job), Dave Brown, Andy Burbidge, Ed Coss, Mick Crumplin, Carole Divall, Vic James ("You Know He Makes Tents"), Rory Muir, Ian Robertson, and John White. I must also thank Mike Galer at the 9th/12th Royal Lancers Museum; Linda, Monty, and the team at the Prince Consort's Military Library; and the staff at the National Archives. In bringing the project together, Chuck Rankin, Greg Urwin, Emily Jerman, and Emmy Ezzell at the University of Oklahoma Press treated my queries, questions, and quibbles with far more patience than I likely deserved.

Lastly, this book would never have been possible without the help, encouragement, and patience of my family. Within that, special thanks go to my father, Mick Bamford, who must surely now regret ever getting me interested in history now that he has read and reread thousands of words of manuscript as my indefatigable proofreader, and my fiancée, Lucy, whose love and support has sustained me throughout this project.

Introduction

During the winter of 1816–17, whilst on garrison duty in Ireland, Captain Moyle Sherer of the 34th Regiment of Foot sought to relieve the tedium of home service by beginning to compose his memoirs of active service during the Peninsular War. He would be one of the first of many British officers—and a rather smaller number from amongst the British Army's rank and file—to put his experiences down on paper, but, unlike many who came after him, Sherer was writing of events that were still in recent memory and was doing so alongside some of those who had gone through the same experiences with him. Sherer was then serving, as he had in Spain and Portugal, in the second battalion of his regiment; now, in the inevitable reductions of peacetime, the battalion was being run down and would soon be disbanded. Whether or not this played any part in Sherer's decision to start writing is unclear, but it certainly gave him cause to reflect on the fate of those who had gone to war with him seven years before: "One thousand and seventy bayonets, all fine sized, efficient men, then mustered under our colours. My regiment has never been very roughly handled in the field, although it has borne handsome share of honourable peril. But, alas! What between sickness, suffering, and the sword, few, very few of those men are now in existence. We had yearly supplies of men from the depôt; they too have for the most part disappeared."[1]

Sherer's is not only one of the most eloquent summaries of the experience of a battalion at war in the Napoleonic era but also one of the most representative. For all that has been written on the great battles of the Napoleonic Wars, these events typify only a few days in a campaigning experience that spanned months or years. It therefore comes as little surprise that the bulk of memoir accounts focus not on the battles but on the day-to-day experience of active service and that their focus is rarely on the army as a whole but on the regiment with which the writer served. Decision on the field of battle rested with commanders at a higher level, but it was at a regimental level that the daily burden of war—Sherer's grim trio of sickness, suffering, and the sword—was felt. Thus, ultimately, it was on the success or failure of the regimental system that the fate of Britain's campaigns rested.

This work is not a narrative history of the British Army's campaigns against Napoleon and his allies; neither is it a social history of the army during those campaigns, nor, at least in the conventional sense of providing detailed orders of battle, does it track the organization of the forces involved. Many excellent works have been produced that cover all of these topics; they are listed in the bibliography, and this work would have been impossible without them. What I have attempted here, however, is to draw on all these various approaches, alongside a systematic study of original returns and reports, in order to offer a fresh insight into how the British Army functioned in the field during the Napoleonic Wars. This approach, in turn, starts to pose new questions as to just how effective the British regimental system actually was. The question is not one of impact on battlefield performance, where the historical record largely speaks for itself, but rather one of how the system stood up to months and years of sustained campaign service in a war far more immense in scale than anything it had been designed for.

One may well argue that since the ultimate purpose of an army is to defeat the forces of the enemy on the battlefield, the effectiveness of a military force can only be judged by looking at factors bearing directly on this. Yet it must be remembered that warfare in the Napoleonic era saw very few days of actual pitched battle, and that when one consults the correspondence of senior commanders, the majority of exchanges deal not with strategy and tactics but with administration, discipline, logistics, and manpower. It is also unwise for the historian to judge an army solely on its battlefield record, since there is evidence aplenty to suggest that units that performed well in battle did not always do so well during ordinary campaigning; the indiscipline of the 88th Foot was as notorious as its combat record was exemplary, whilst the foreigners of the Chasseurs Britanniques fought with a determination that does not fit with that regiment's continual record of desertion off the battlefield. What must be considered is how individual units, and the larger formations of which they formed a part, were able to sustain an effective strength whilst on service. This ability, in turn, was directly responsible for dictating the level of strength that could be brought to bear on the battlefield.

In conventional narratives, the time spent between combats is all too frequently dispensed with as essentially a topic for social histories, best left to memoir accounts. Except in extreme cases, such as the privations suffered during the retreat from Burgos, or the Guadiana and Walcheren fever epidemics, little consideration is typically given as to how the rigors of that time between battles might affect the outcome of the next combat. For the writer of battle history, there is no pressing need to explain why, to invent a hypothetical case, such and such a battalion embarked on campaign with 800 men, and fought its first battle a month later with just 650. The assumption is seemingly that, if the 650 were enough to win the day, the fate of the remainder is unimportant. Within the strict narrative confines of battle history, this reasoning cannot be faulted. On the other hand, it also cannot be denied that considerable effort has been expended in attempting to explain these sorts of cases for other armies of the period, suggesting that a similar treatment of the British Army is needed to fill the gap.

Such methods date back, at the very least, to Charles Joseph Minard's 1861 graphical representation of the strength of the Grande Armée of 1812.[2] More recently, Scott Bowden's application of the concept of strategic consumption has been utilized for analysis of the reraised Grande Armée of 1813 in a work that has inspired much of the methodology behind this study.[3] In order to achieve something similar here, much time was spent in the National Archives at Kew, transcribing the monthly returns covering the majority of the campaigns in which the British Army was engaged between 1808 and 1815, and it is an analysis of these figures that underpins much of the content of this book.

I shall return to the contents of these tables, and their uses, later, but it is first necessary to say something about the scope and remit of this study. Firstly, the time frame has, with a few minor deviations, been maintained at 1808–15. This was in part for practical reasons relating to the amount of numerical data that could legitimately be processed during the course of what started life as a three-year doctoral project, but also, and ultimately more importantly, because those years encapsulate a distinct and self-contained narrative. The period in question marks one of the few occasions—only the First World War, and perhaps the War of the Spanish Succession, can be fairly compared—in which the British Army took a sustained leading role in warfare on the European continent. Field armies were deployed in more than one theater simultaneously, and over a five-year period a force was maintained in the Iberian Peninsula that won some of the most decisive victories of the Napoleonic Wars. That this great achievement was possible seems all the more surprising when one considers that the British Army as a whole had barely completed—indeed, to some extent was still completing—one of the greatest internal overhauls in its history, simultaneously with a swing in strategic policy that exchanged a doctrine of short-term deployments, with specific limited goals, for one of sustained conflict directed at the defeat of the enemy's troops in the field and the liberation of occupied allied territory. Thus, the reforms instigated by the Duke of York during the late 1790s and early 1800s form the background to this narrative since it was these, along with unit-level traditions that were far older, that created the systems by which the British Army was organized in 1808.

In order to judge just how effective those systems were when subjected to the test of sustained active service, in the Iberian Peninsula, in North America, and in northern Europe, I have deliberately avoided any detailed consideration of the British Army's role in domestic or colonial affairs where different circumstances applied. Nevertheless, the British Army was controlled from headquarters at Horse Guards in London, and drew its recruits from all corners of the British Isles. This being so, it is impossible to ignore the domestic scene completely, for the circumstances of their home service and recruiting did much to affect the subsequent performance of the various battalions and regiments once they deployed overseas. A vital resource here are the reports and returns that form the lasting record of the biannual inspections carried out, at least in theory, on all regiments, corps, and detachments of the British Army. Because few such records have survived from inspections carried out on active service—particularly in the years before 1812—this is a patchy record at best, and one that therefore requires supplementing with reports made in the months prior to a unit's departure overseas. By supplementing these, where possible, with those reports from overseas that have survived, as well as with the correspondence of commanders in the field and the commander in chief at Horse Guards, it is generally possible to fill in most of the gaps in the story of a particular regiment and thereby follow it from deployment, through the rigors of campaign service, to its eventual return home.

Alongside this written paper trail, there is also a numerical one in the shape of the monthly returns already mentioned. Whilst these documents represent a resource comprising vast numbers of tabulated data, there has, as yet, been no large-scale systematic study of them. They have, it is true, been extensively mined for information on specific regiments, or to establish the strength of the forces deployed at a given battle, but they have not previously been used to underpin a study as extensive as this one. Thanks to modern spreadsheet software, it has been possible to transcribe over 300 tables giving monthly snapshots of data from various theaters. From these, over 250 further, more sophisticated, tables have been created, analyzing specific units and tracking change over time. In this way it is possible to compare the various trends within a single unit across a period of time, or to compare a number of units present simultaneously in the same theater. This method has been used, for example, to analyze the performance of provisional and detachment battalions during the Peninsular War, or to pick up on different sickness and death ratios—for both men and horses—in the different arms of service.

My hope and intention was that once this statistical analysis had been used to pinpoint areas of interest, investigation by more conventional means would reveal the reasons behind the trends and anomalies identified, and on the whole this has proved to be the case. However, some of the results produced by this methodology have proved to be so unexpected that they present substantial challenges to long-standing beliefs that have gone unchallenged for many years. This, in turn, has questioned the validity of some of the memoir sources on which these beliefs are based, and, although I have used firsthand accounts wherever possible throughout this book, there are places—usually, in fairness, when the author is writing long after the fact, or about events not personally witnessed—where conflicting evidence renders the memoir palpably inaccurate and a hindrance rather than a help to understanding. Revisionism for its own sake is a dangerous thing, but there are nevertheless several instances where the statistical data leaves no other option than a substantial reassessment of the assumptions, and some of the sources, on which much of the analysis of the British Army of this period has been based.

Abbreviations

AGO   Additional General Order
GO   General Order
KGL   King's German Legion
NCO   Non Commissioned Officer
RHA   Royal Horse Artillery
TNA   The National Archives
WO   War Office

Sickness, Suffering, and the Sword

CHAPTER 1

The British Army and Its Campaigns

The British Army of 1808, despite having undergone an extensive program of reform over the previous decade, was still very much a mixture of the old and the new. Thanks to the measures instituted under the aegis of the Duke of York following his appointment as commander in chief in 1795, the British Army was a far more professional body than heretofore. During this thirteen-year period, its command structure, both military and civil, was reorganized, leading to the concentration of responsibility into a relatively small number of posts. This restructuring enabled Viscount Castlereagh, then occupying the senior civil post of secretary of state for war and the colonies, to complement York's military reforms with a shake-up of the way in which the British Army was recruited and its manpower systems organized. It was the resulting increase in manpower available for active service, at the expense of reducing the numbers of men assigned to home defense, that enabled Britain to mount the campaigns covered by this work.[1]

Yet although York could influence how a regiment was trained and led, and although Castlereagh's reforms gave it better access to fresh manpower, the regiment nevertheless remained a decidedly independent entity. In theory at least, a great deal remained in the hands of the regimental colonels, who, position notwithstanding, tended to be fairly senior general officers. On a more day-to-day level, much depended on the seniority and reputation of a regiment, on the character and experience of the field officers commanding it, on its station and service, and on its success or otherwise in attracting recruits. But whilst this book is ultimately about successes and failures at a regimental level, that focus would make little sense without an understanding of how the higher levels of command functioned. Such an overview is all the more essential since it also serves to place the statistical sources used in this study in their historical context. The modern historian may have the advantage of software-aided analysis, but the returns and reports that underpin the conclusions presented in this work were just as keenly pursued at the time, at all levels of the military and civil hierarchy, and, at least when the system was working efficiently, decisions of strategic import were taken as a result.[2]

Indeed, one could even go so far as to assert that strategy was at times directly dictated by the rise and fall of the totals at the bottom of every column of every return. In 1809, for example, Britain found that it simply did not have the manpower available to mount successful campaigns in two different theaters of war, whilst in 1814 there were three different theaters competing for a decidedly finite pool of resources. It then, of course, fell to the government to decide on its priorities, and for the staff at Horse Guards to allocate manpower accordingly. The same was true on an operational level. Not only did the availability of manpower dictate the course of events on a local scale, determining whether troops could be spared to reinforce a garrison here, or to mount an expedition there, but also in a larger sense, regulating whether or not a force could operate in a particular area at all. During the winter of 1809–10, for example, Wellington was torn between political demands that called him to maintain his army on Spanish soil, and the practicalities of the fact that the soil in question was the malarial marshes of the Guadiana valley around Badajoz. In the end, the practical outweighed the political, and he was compelled to make a substantial redeployment of his troops for the simple reason that many of them would otherwise have succumbed to illness. Without an understanding of the course of the campaigns, therefore, one cannot fully appreciate the nature of the demands that they placed on Britain's military systems; without an understanding of those systems, it is impossible to fully appreciate why these same campaigns took the form that they did.

Military Organization: The High Command

The professional head of the British Army was the commander in chief, who was always London-based in this period although the Duke of York had taken the field in the Netherlands in 1799 and entertained some hopes of doing so again when it was first mooted to send an army to Portugal.[3] York, who had first been appointed to the post in 1795, remained in it until forced to resign in March 1809, as a result of the Mary-Ann Clarke scandal, in which his former mistress was implicated in alleged corruption relative to taking payments in return for using her influence with the Duke. General Sir David Dundas filled the post for the next two years until York, his reputation rehabilitated, was able to return in June 1811, serving thereafter until his death in 1827. The other two senior home-based staff appointments were those of adjutant general, filled throughout this period by Lt. General Sir Harry Calvert, and quartermaster general, held by Lt. General Sir Robert Brownrigg until 1811, and by Lt. General Sir James Willoughby Gordon thereafter. The latter post was concerned with arranging the movements and postings of regiments, whereas the former dealt primarily with matters of drill and discipline. Additionally, York raised the profile of the military secretary, previously a civilian post, by appointing Gordon to the role in 1809 to shoulder a portion of the commander in chief's administrative tasks, replacing him with Colonel Henry Torrens when Gordon stepped up to become quartermaster general.

On the civil side—although the military-civil distinction was blurred by the appointment of serving officers to some civil posts—the major roles were those of the secretary at war, responsible for financial matters, and the secretary of state for war and the colonies, a senior Cabinet minister with overall responsibility for coordinating the war effort. The former post was successively held by General Sir James Murray-Pulteney, Lord Granville Leveson-Gower, and Viscount Palmerston; the latter by Viscount Castlereagh until September 1809, then by Lord Liverpool until his becoming prime minister in June 1812, and finally by the Earl of Bathurst until after the peace. Also worth mentioning at this juncture, since his name repeatedly crops up in correspondence throughout the period, is Colonel Henry Bunbury, an experienced staff officer and would-be strategist who served as undersecretary of state for war and the colonies—that is to say, the junior minister assisting the secretary of state—from November 1809.[4] Finally, the roles of head of the ordnance services and de facto inhouse military advisor to the Cabinet were combined in the office of master general of the ordnance, which was filled by Lt. General the Earl of Chatham until 1810, and by General the Earl of Mulgrave thereafter.[5]

The resulting system was complex, with the civil element at times duplicating military functions: something that can be seen as stemming, at least in part, from deep-seated fears of military absolutism dating back to the seventeenth century. Most strikingly, the secretary of state for war and the colonies had considerable control—albeit by indirect means—over matters that for all intents and purposes were purely military issues. Bathurst in particular elected at times to meddle in matters that did not, by rights, lie within the remit of his office. Yet, despite its unnecessary complexities, the system did work effectively enough when put to the test, but the relationships between the various posts need to be understood correctly if the successes and failures of the period are to be understood and a degree of credit and blame apportioned. The key point to be grasped is the comparatively subordinate nature of the commander in chief and his staff, who were, in effect, there merely to put into practice the military policies decided upon by the Cabinet. That body relied on the master general and the secretary of state for military advice, supplemented, where necessary, by input from the first lord of the admiralty, the foreign secretary, and the prime minister himself. In practice, however, as well as the always-possible wild card of royal intervention, the commander in chief could exercise a considerable amount of restraint on the government, with York successfully quashing proposals for a descent on Boulogne in 1808 by making it clear that the troops required were simply not available.[6]

It was for this reason that the maintaining of an up-to-date appreciation of the strength and fitness of the various regiments and battalions became a matter of vital importance, and, by keeping himself up to date, the commander in chief was far more effectively positioned to turn the demands of government into practical military operations. A prime example of military statistics being used to this end is in the exhaustive scouring of manpower to put together the Walcheren expedition of 1809, for which Dundas was quickly able to call upon a detailed and accurate account of the state of the troops then in Britain. Conversely, as we shall see with regard to the years 1813 and 1814 in particular, misunderstanding or misinterpreting the information contained in these returns could lead to troops being sent overseas in a manifestly unfit state. The equal significance of these reports, states, and returns for a commander in the field are obvious. Wellington utilized them to keep an extremely close account of the state of his forces in the peninsula, and used Warren Peacocke, commandant at Lisbon, to carry out additional inspections of newly arrived units so as to instantly gauge their fitness.[7]

Commanders needed to know two primary things about any given unit—its strength and its quality. The first was in many ways the most important—not least because it provided a partial answer to the second—and each regiment, battalion, or detachment on active service was required to provide a complete return once a month. Until June 1809 this return was to be completed on the first of the month, but this was then changed to the 25th.[8] From that date, the regulations required these returns to show "[T]he exact state of the Corps, in which every Officer, Non-commissioned Officer, and Private Soldier, belonging to the Corps, is to be accounted for. . . . The casualties which have occurred from the 25th day of each Month to the 24th day of the Month following, both days inclusive, must be accurately inserted in the respective Columns."[9] Although the regulations went on to state that copies of this return be sent to the "General Officer under whose Command the Regiments may be serving," as well as to the adjutant general back in London, the latter practice does not seem to have been much followed; certainly, few examples have survived to be incorporated into the National Archives. However, barring the need for occasional reminders to jar recalcitrant units into getting their paperwork in, and some confusion in the case of detached commands, such as Cadiz, as to which general officer the returns should be sent to, theater commanders seem to have generally been able to obtain regular monthly returns for units under their command.[10] This was of course vital for their own understanding of their forces, but also necessary in that it gave them the data to provide, as they were required by regulation to do "A Return, *as soon as it can be made up after the 25th of each month*, of the Troops, and of the General and Staff Officers employed at each Station."[11] This return was to be sent not only to the adjutant general but also to the secretary of state for war and the colonies and to the secretary at war, although the copy for the adjutant general required, in addition, details of all regimental officers present and absent.

The exact manner in which these returns were made out differed from theater to theater, and their format also changed over time with, for example, an initial distinction made between those ill but still present with their unit, and those in hospital, later being replaced by a single total of the unit's sick. At no point, it should be stressed, was any distinction made between the ill and the wounded—all came under the heading of "Sick." Another innovation, first seen in June 1810, was the listing of men sent home, alongside those lost through death or desertions, whilst Canadian-raised units serving in that country, being able to send men on leave, made considerable use of the "Furlough" column that was, necessarily, generally left blank in returns from other theaters. The main columns, however, which remain constant throughout and which have been the most heavily used in the preparation of this work, are those giving the rank-and-file strength of each unit, broken down into "Effective," "Sick," and "On Command," plus the running totals of deaths and desertions; for mounted units, or those employing draught animals, the total number of horses and the number dead since the last return must also be added. The meaning of these various headings should be self-explanatory, with the exception of "On Command," which may be best understood as a catchall term for men actively employed in the service but away from their parent unit. In practice, this category could variously encompass anything from a handful of men overseeing some small task under the supervision of an NCO to several companies detached on an expedition, as well as most things in between. With such a level of detail, one can glean a great deal as to the effectiveness of the unit in question from these returns, including the general state of its health and, if stationed at home, the progress of its recruiting. It is data from these returns, their coverage detailed in appendix 1, that form the statistical backbone of this study.

Nevertheless, it is apparent now as it was then that only so much could be gleaned from statistical data alone. Accordingly, the *General Regulations* for the British Army also contained the requirement that all units, at home or overseas, be subjected to a biannual inspection by a general officer.[12] In theory, these inspections were to take place in May and October, although the dictates of active service meant that the latter was frequently delayed until the end of active campaigning. At home or on active service, an inspection was an occasion of some importance, and even those memoirs that otherwise touch little on military matters tend to mention them in some detail. One can imagine that the preparations, no doubt entailing much cleaning and polishing, and punctuated by the curses of harassed adjutants and angry sergeants, made quite an impression on those concerned. John Green remarked in his account of life in the ranks of the 68th that when the battalion was due to be inspected whilst still on home service, the extra drills required to perfect the battalion's evolutions began some three weeks before the appointed day.[13] The day itself was also evidently a lengthy one; the men of the 2/34th were under arms from five in the morning for their inspection by Brigadier General Catlin Craufurd at Lisbon in 1809, although this may well have represented an unusually early start due to the need to complete proceedings before the hottest hours of the day.[14]

The reports of these inspections were to be forwarded to the commander in chief via the general officer commanding the district or station in which the inspected unit was serving, and were required to be wide-ranging in their scope. The unit commander, field and company officers, staff officers, and NCOs were all to be reported upon, with particular attention to their fitness for their roles. With respect to the rank and file, attention was also to be paid to health, cleanliness, and general bearing, as well as to the arrangements for their messing and provision of clothing and other necessaries. Details of recruiting on the one hand, and of men discharged on the other, were also to be provided, and particular attention was to be given to the standard of drill and field exercise. For mounted units, detailed information was also to be provided for the horses. Lastly, record was to be made of all courts-martial held since the previous inspection and an opportunity given for any complaints to be noted. These two latter elements form a valuable resource, even if the historian is frequently left wishing for more detail on some of the more unusual cases, where the dry reportage keeps the human element tantalizingly beyond reach. The written report was required to be accompanied by a detailed return, of a similar nature to those already provided on a monthly basis but with the addition of greater detail covering such things as the age, origins, and terms of enlistment of the rank and file.[15]

Had all reports been completed to the degree of detail required by regulations—or, in some cases, completed at all—the historian would have a truly invaluable resource; as it is, many inspections were either never carried out, or, if they were, the records have not survived. Understandably, the greatest discrepancies and omissions are of those units with which we are most interested here: those engaged in active service. However, there are also many cases where although a report has survived, the level of detail required by the regulations is absent. Typically in these cases, all that is to be found is a repetition of the same stock phrases— that the men are well-drilled and healthy, or that such and such an officer has conducted himself in accordance with regulations— and it is not uncommon for a single letter to contain reports on three or four units from a particular brigade or garrison with the descriptions of each couched in broadly the same terms and thus effectively useless as an aid to understanding the state of the units concerned. On the other hand, many inspecting officers did carry out their duties in closer accord with the spirit of the regulations, and their reports, as we shall see, frequently provide detailed and thoughtful comments on the units in question. Some inspecting officers even received subsequent admonishment for using their reports to make suggestions that fell outside the remit of what they were required to comment on—Lt. General Sir Colin Campbell, commanding at Gibraltar, seems to have found repeated fault with his subordinates in this regard, and their reports are frequently accompanied by cover letters to the adjutant general containing Campbell's own, alternative, opinion of units in question.[16]

Military Organization: The Regiments

In 1808, Britain was able to field a total of 197 regular battalions of infantry, divided amongst 103 regiments of the line. In addition to these, there were also three regiments of Foot Guards totaling seven battalions, ten battalions of the King's German Legion recruited from Hanoverian refugees, and five other assorted foreign line regiments. To this may be added several irregular or sedentary foreign infantry regiments, nine West India regiments, twelve Royal Veteran battalions and eight garrison battalions.[17] During this period there was little change, but for the addition of a further regular regiment when the New Brunswick Fencibles were taken into line as the 104th Foot, and the creation of a 13th Royal Veteran Battalion and the 1st Foreign Veteran Battalion. In 1810, the troops of the exiled Duke of Brunswick were also taken into British service, forming an infantry battalion and a hussar regiment. Conversely, the number of garrison battalions was steadily reduced and one West India regiment disbanded, whilst the peace of 1814 saw widespread cuts in the number of veteran and foreign units, and of the second and third battalions of line regiments. Although infantry regiments ideally maintained two battalions, some, generally those of good reputation, raised a larger number. The 1st Foot had four in 1808, and the 1st Foot Guards and the 27th Foot three each, whilst the 14th and 56th Foot, along with the 95th Rifles, would each also raise a third during the period covered by this book. Conversely, some thirty-eight regiments fielded only one battalion in 1808, although several subsequently increased this to two. There was also the special case of the 60th Foot—still nominally, and at one time actually, the Royal Americans—which had by this time taken on something of the character of a foreign legion. It had in 1808 six battalions, of which four were line troops earmarked for colonial service and the remainder light infantry or rifles; two more would be added by the end of the period.

Traditional British practice in all wars up to and including that against revolutionary France had been to augment its army through the raising of new regiments rather than augmenting old ones. Accordingly, prior to 1803 very few infantry regiments had numbered more than a single battalion. On the one previous occasion that a widespread creation of second battalions had occurred, during the Seven Years' War, these units, once established, had rapidly been re-designated as regiments in their own right.[18] Indeed, this practice had continued on a smaller scale up until 1802, when the second battalion of the 52nd Foot, raised four years previously, became the new 96th Foot.[19] The advantages of creating new regiments rather than augmenting existing ones were considerable in terms of the potential for patronage, creating lucrative colonelcies and allowing for the rapid promotion of officers who brought in men for the new regiments. Yet whilst the number of marching regiments was pushed as high as 135 during the French Revolutionary Wars, the bulk of these units served only to absorb valuable manpower without ever firing a shot in anger. Accordingly, the bulk of them were disbanded in one of the Duke of York's first reforms upon becoming commander in chief, and their manpower drafted into those regiments that remained.[20] Upon the resumption of hostilities in 1803, rather than repeat this error, the sensible decision was therefore taken to do away with the old methods, and increase the forces by means of adding additional battalions to the ninety-six infantry regiments already in existence.

This move was at the heart of Castlereagh's reform of the manpower systems, being based on an accurate appreciation that too many men had previously been tied up in home defense roles, as part of the Militia, Volunteers, and Fencibles, to the detriment of the British Army itself. When hostilities recommenced in 1803, only the 1st and 60th Foot had more than a single battalion; now, by using manpower from the Militia and Army of Reserve, second battalions were authorized for the fifty infantry regiments then stationed in the British Isles.[21] Whilst invasion remained a possibility, these battalions would continue to fulfill the same home defense roles as the units from whence their manpower had come, but if the threat were to diminish they would be available for other duties. This policy was further extended as time went on, although initially not to the exclusion of forming some new regiments. However, after the raising of the 101st Foot in 1806, all new battalions formed part of preexisting regiments; the addition of the 102nd through 104th Foot represented existing unnumbered battalions being taken into the line rather than the creation of new units.[22] Usually commanded by a lt. colonel, a battalion was ordinarily established with an organization of ten companies of which two—light and grenadier—were designated as elite "flank companies," and the remainder, distinguished by number, were collectively known as center companies. Whilst the number of companies remained constant, ensuring that there was always a cadre to expand upon, their established size could range from 33 to 120 rank and file depending on the role and station in which the battalion was intended to serve.[23]

The primary result of Castlereagh's reforms was to free up sufficient troops to create a disposable force that could be sent on active operations overseas without unduly weakening Britain's defenses at home. Furthermore, greater utility was obtained from the Militia by making it not only the primary defense against threats—domestic and foreign—within the British Isles but also by using it as the first stage for a soldier's progression through the military hierarchy. The hope, which to some extent was realized, was that many of those called up by ballot for the Militia would develop a taste for soldiering and subsequently be willing to volunteer for a line regiment. In theory at least, time in the Militia would have already prepared a man for service in the regular forces should he be persuaded to transfer. Thereafter, service in the second battalion of a line regiment would complete his training and ultimately fit him for active duty overseas with his new regiment's first battalion.

Junior battalions were not, in the first instance, envisaged as being sent on active service overseas at all. The domestic duties of the home station were to have formed their main employment while they simultaneously acted as reserves of manpower to keep the senior battalions up to strength. Recruits would be trained in the second battalion before being posted to the first, and newly promoted officers would, in theory, be posted to the second battalion to obtain experience in their new rank before ultimately moving back to active service in the first battalion.[24] Unfortunately, although an excellent concept in theory, this was not always the most practical system when the demands of war caused second battalions to go overseas too, and all the more so if the battalions of a regiment were left widely dispersed as a result. An extreme case is exemplified by an episode from Moyle Sherer's service with the 34th Foot. In late 1811 he reached sufficient seniority as a lieu-tenant as to warrant a posting from the 2/34th, in Portugal, to the 1/34th in India. Accordingly, he returned to England in Decem-ber 1811 to await a passage to his new battalion, but, before this could be arranged, he was promoted to captain; thus, being the regiment's junior officer in that grade, Sherer was again posted to the 2/34th, which he rejoined in August 1812. Sherer doubt-less enjoyed his leave, and was grateful that he had "escaped my banishment to India," but his regiment had been deprived of his services for nine months to no useful end.[25] In like fashion, when multiple battalions of a regiment were overseas, the first battalion had priority for drafts of recruits, although in practice an element of common sense seems to have been employed here with regard to occasions when the senior battalion was in some distant garri-son and the second or third on active service and quite patently a priority. The 3/1st and 3/27th both benefited from this approach and were accordingly able to serve lengthy stints in the peninsula, with the former going on to fight at Waterloo as well.

A similar flexibility was incorporated in the *General Regulations* to enable newly promoted officers to remain with their former battalions in their new rank if that battalion had fewer officers than its establishment. In practice, however, this loophole seems to have been used to the advantage of first battalions at the ex-pense of their junior counterparts, particularly if the former were on active service and the latter in garrison.[26] Not surprisingly, many officers preferred the potential for distinction with an active unit, with the newly promoted Lt. William Grattan making strin-gent, and for a long time successful, efforts to use this regulation to enable him to stay in the peninsula with the 1/88th rather than return to his rightful posting in the second battalion. As a result, although Grattan became a lieutenant on April 12, 1812, he did not join the 2/88th until a year later when the first battalion had enough of its own lieutenants present for duty.[27] This system was all very well for the senior battalions, which were, after all, supposed to bear the brunt of active service, but meant that junior battalions were frequently left extremely short of officers, to the extent that there were too few for some units even to be effective in a garrison role. In an extreme case, the 2/89th at Gibraltar in 1811 lacked any company officers of its own at all, and had to borrow some from other regiments in the garrison, but second battalions in Britain were also left short at times, or with too many young and inexperienced officers, to the detriment of their training and state of readiness.[28]

This policy of creating second battalions worked because, since recruiting was carried out at a regimental level, the only way in which to utilize surplus manpower within a given regiment was to create an additional battalion. The system also had the advantage of creating employment for deserving half-pay officers that enabled them to return to duty and draw full pay. However, whilst it was not uncommon for a regiment's total strength to exceed the establishment of one battalion, it was rather more rare for it to total enough for it to be able to field two at full strength. As a result, although they were generally established at a lower strength to begin with, second battalions frequently remained weak, greatly limiting their utility. Although junior battalions were meant to remain at home, it was frequently the case that such units, particularly later in the war, were too strong for the depot role, yet too weak for overseas service—on the face of things, a waste of manpower. There are, however, two caveats to be inserted with regard to this apparent inefficiency. Firstly, although the threat of invasion largely disappeared after 1805, it was still necessary to keep at least some reasonably effective battalions on the home stations for the maintenance of law and order, particularly in Ireland, and for local defense in the Channel Islands. A weak second battalion of three or four hundred men represented a good solution to these problems, but deploying such a unit in this role limited how many drafts could be sent to the first battalion without rendering the second unfit for purpose at home.[29] Secondly, the distinction has to be made between the original crop of second battalions created between 1803 and 1808, which generally had a high establishment of around 800 rank and file, and those raised afterward, which frequently had a far lower established strength. Battalions in the former category were strong enough to play an active role during the first half of our period, but those in the latter were generally made of poorer stuff. Particularly poor were the new second and third battalions created during 1813 and 1814, which were, on the whole, weak, ineffective, and largely unfit for service—to the detriment, as we shall see, of Britain's attempts to mount a campaign in the Netherlands during the closing months of the war.

Generally speaking, the senior battalion of a regiment would, upon embarking on campaign, exchange its ineffective personnel into the junior battalion, and be brought up to strength by drafts of effectives from that unit. By extension, the third battalion of a three-battalion regiment would have two active senior battalions to support, and the unfortunate 4/1st, three. In this way, at least in theory, a senior battalion would depart with an active strength of around 1,000 rank and file, and could, again in theory, be kept at a high strength over a prolonged campaign by further drafts. This process can be seen in action during the preparations for Wellesley's 1808 expedition to Portugal, where the bulk of the battalions needed to be topped up in order to embark at full establishment strength. The largest influx on this occasion would seem to have been the 261 men sent to reinforce the 1/91st, who thereby ended up forming roughly a quarter of that battalion's total rank and file.[30] On occasion, however, men might not be available, or at least available in time. Whilst most of Wellesley's battalions in 1808 could make arrangements to receive drafts from their second battalions, the 1/36th and 1/45th, having been on garrison duty in Ireland and reassigned to Portugal at the last minute, had to be brought up to strength by drafts from the Irish Militia, receiving 300 and 365 men respectively.[31] This use of the Militia as a source from which active battalions could be directly augmented continued throughout the war, often at similarly short notice. Thus it was that elements of Wellesley's army at Talavera a year later were remarked upon as still having their old Militia distinctions on their equipment, and that many of the men who defended Hougoumont did so wearing the Militia uniforms of their former regiments, rather than those of the Foot Guards regiments into which they had newly volunteered.[32]

However, the system did not entirely debar second battalions from service, and if a two-battalion regiment had 1,600 or more effective rank and file on its strength, its second battalion would also become eligible for overseas deployment. Due to this caveat in the regulations, in 1808 the 9th Foot took 1,508 men to Portugal between its two battalions; whilst the combined strength apparently falls short of the required total, these figures make no account of additional manpower left in Britain.[33] This deployment of both battalions was possible because the regiment was able to obtain an unusually high total of 833 volunteers from the Militia in 1807, which, sensibly, were distributed between both battalions in roughly equal proportions—a more typical distribution would have been to place them all in the second battalion and concentrate the experienced men in the first. Even so, the presence of the 2/9th on active service is still impressive, not least because it had given over a draft of one hundred men to the first battalion only a month before its own embarkation.[34]

Lastly with respect to infantry, the British Army contained a number of foreign units, raised from various sources and differing greatly in their attributes and utility. Whilst foreign troops were employed in considerable numbers throughout the conflict, the last years of the Napoleonic Wars saw both an increase in the numbers actively deployed but also in their numbers tout court. Although the King's German Legion was run down from 1812, after peaking at an all-arms strength of over 14,000,[35] other foreign auxiliary formations continued to thrive, as with the Calabrian Freecorps and Italian Levy serving in Sicily and eastern Spain, and also the mercenary regiments of Dillon, De Roll, Watteville, and De Meuron, which served both in the Mediterranean and in North America. There was also the last vestige of the émigré formations of the 1790s in the shape of the single-battalion Chasseurs Britanniques. Recruiting for these regiments, however, had largely shifted from genuine volunteers to the employment of the steady stream of deserters from the Grande Armée, along with prisoners of war who preferred turning their coats to a life—and possible death—in the hulks. The practice of enlisting such men was hardly new, but subsequently became increasingly formalized with the KGL getting the pick of any non-French and the remainder going into one or another of the colonial corps. Exception would seem to have been made for "French" soldiers who were in fact of German or Flemish origins, allowing them to also join the KGL.[36] This distribution of secondhand manpower is reflected in a return of July 1810, which shows that of 560 deserters enlisted—since when is not, unfortunately, made clear—147 men had gone to the KGL infantry and 58 to its cavalry, whilst the remaining 355 found themselves drafted into the York Light Infantry Volunteers and shipped off to the West Indies.[37] Only in the closing months of the war does this system seem to have relaxed somewhat, and less reliable men enlisted into the KGL from which many of them subsequently deserted.[38]

The organizational situation for the cavalry was rather simpler than for the infantry, since although each regiment was composed of several squadrons these did not have the same level of autonomous existence as did battalions of an infantry regiment. Indeed, it was almost unheard of for the active squadrons of a regiment to be split up. The differences between the regiments themselves, however, were confused by an overcomplex system of nomenclature that went through further changes during this period. In 1808 there were thirty-four regiments of British cavalry, household and line, and a further five belonging to the KGL. Officially, the mounted arm composed two regiments of Life Guards, and one of Royal Horse Guards—between them forming the Household Brigade, although strictly speaking the latter only formally became a household unit in 1820—plus seven regiments of dragoon guards and twenty-four regiments of dragoons numbered 1st through 4th and 6th through 25th. The number five was vacant due to the 5th Dragoons having been disbanded following implication in the 1798 Irish Rebellion. In terms of role, however, there were only two distinctions—heavy and light. The Household Cavalry, dragoon guards, and the five senior regiments of dragoons made up the heavies, whilst the junior regiments—7th through 25th—were designated light dragoons. To confuse matters further, however, the 7th, 10th, 15th, and 18th Light Dragoons were uniformed as hussars, and ultimately adopted this title officially; other than a certain self-assumed elite status, this made no difference to their organization. The KGL cavalry, meanwhile, initially comprised two regiments of dragoons and three of light dragoons. The latter had always been uniformed as hussars, and were generally referred to as such; in 1813 this designation was made official, whilst the two heavy regiments were simultaneously converted to light dragoons. Although there had been a proliferation of émigré cavalry units during the Revolutionary Wars, these had mostly disappeared by the turn of the century, and, aside from the KGL cavalry, the only other foreign mounted unit in our period was the Brunswick Oels Hussars, which saw service in eastern Spain.[39]

Lastly, though not part of the regimental system or even under Horse Guards control, the troops of the ordinance nevertheless formed a vital—though usually small—part of all the campaigns with which this work is concerned. Although their corps was organized into battalions for administrative purposes, the men of the Royal Artillery were generally employed to form batteries manning six field guns. Such formations, combining a Royal Artillery company and a detachment of Royal Artillery Drivers, were confusingly known as brigades. Their mounted equivalent was the Royal Horse Artillery troop, which included its own drivers as an integral part of the permanent organization.[40] To give some idea of the manpower requirements of these units, Captain Hew Ross's A Troop, RHA, deployed to the peninsula in 1809 with 149 rank and file and 162 horses.[41] Naturally for a horse artillery unit, this figure included riding horses for the gunners as well as draught animals, so the equine requirements of a brigade of foot artillery would be proportionally less. The Royal Engineers were represented by a handful of specialist officers, with manpower provided by the Royal Military Artificers, renamed Royal Sappers and Miners as of 1813. Because this small pool failed to produce sufficient manpower, 1798 had seen the creation of a parallel institution under Horse Guards control in the shape of the Royal Staff Corps. Equally small, this formation is not to be confused with the Staff Corps of Cavalry created by Wellington in the peninsula and Netherlands, who were orderlies-cum-policemen attached to army headquarters.[42] Lastly in this rundown of manpower comes the commissary services, the bulk of which were effectively civilian in makeup, in the sense of being British clerks or indigenous hired muleteers and carters, but the men of the Irish Wagon Train and Royal Wagon Train also make appearances in the various returns, in numbers sufficiently small as to make it clear why their supplementation from outside the military system was necessary.[43]

Active Service, 1808–1815

Whilst an overview of the main campaigns during the period helps understand the shifting pressures placed on the British Army, it is important to appreciate that there were also significant permanent overseas commitments to be met, whose demands diverted manpower away from the active theaters of war. Although garrison forces were cut back to a bare minimum as demands for active manpower became more pressing, substantial numbers of troops had to be provided for stations across the British Empire. Firstly, troops were needed simply to maintain a military presence in the various colonial outposts: the West Indies; South Africa; Malta; right down to the convict guard in New South Wales. North America, at least until 1812 when the situation was transformed, may also be included in this category. These, by and large, were stations where there was no immediate threat, but which nevertheless could not be left completely undefended due to their strategic and/or commercial importance. Regiments often remained for long periods in these posts, particularly on the smaller stations, but there was greater mobility in the larger commands, and troops could be drawn from these for operations elsewhere as in 1809 when troops from North America formed part of the force assembled to capture Martinique.

Of a rather more significant nature, both in terms of commitment of troops and relevance to the developing combat abilities of the units and commanders involved, were the two foreign stations that absorbed the greatest numbers of troops. These were India, with 21,534 British troops in January 1808, exclusive of the forces of the Honourable East India Company (HEIC), and Sicily. The latter was unusually short of men in the same month, the garrison being down to only 9,343, but there was a 5,194-strong reinforcement on the way that would bring it back to its usual level.[44] India in 1808 was a less active station than it had been in the years immediately prior to our period, the Maratha Wars having been brought to a successful conclusion in 1806. Nevertheless, there remained an ongoing and drawn-out backcountry war in Ceylon, and troops were still required on minor policing duties, as well as to keep an eye on the largely native-recruited HEIC forces, elements of which had mutinied in 1807.[45] King's regiments serving in India tended to do so for some length of time—the future Duke of Wellington took the 33rd Foot out in 1796, for example, and the regiment did not return until 1811—and soldiers unwilling to leave the Indies, East or West, would often volunteer into regiments remaining on the station, thus leaving only a cadre to come home to be rebuilt.[46] It is for this reason, as well as a very different campaigning environment, that India has been left out of this study, although commanders who had served in India, Wellington not least amongst them, were able to successfully apply some of the lessons learned there to European warfare. The presence on Sicily, meanwhile, dated back to 1805 and served a twofold purpose: firstly, to hold the island and keep an eye on its unreliable ruling dynasty; secondly, to provide forces for offensive operations elsewhere in the Mediterranean.[47] Unfortunately these two requirements were largely mutually exclusive, with the result that it was never possible to dispatch sufficient forces from the island to make an appreciable difference elsewhere. Sir John Stuart's 1806 descent on Calabria achieved success on the battlefield of Maida, which did great things for morale but achieved no lasting results, and troops from the Sicilian garrison also participated in the futile 1807 descent on Egypt.[48]

With the major permanent garrisons accounted for, and leaving aside units stationed in Britain either recruiting or retained to supplement the Militia and Volunteers in the increasingly unlikely occurrence of a French invasion or Irish rebellion, the remainder of the British Army was available for active operations. To lead any such expeditions, a "disposable" or expeditionary force was kept at readiness to embark on transports specially set aside for its carriage. Whilst this was an excellent concept, there was some initial uncertainty as to how best to deploy these troops, newly available thanks to Castlereagh's manpower reforms, and for the most part they were initially dispersed in penny packets. Nevertheless, it was from this force that troops were variously sent to Copenhagen, South America, and the Baltic during 1806 and 1807, and it was with an eye toward service in South America again that troops were being assembled at Cork when developments in Portugal and Spain opened up an entirely new arena of operations.[49]

Britain's initial commitment to the war in the Iberian Peninsula represented the same sort of opportunistic strategy that had characterized its previous expeditions ever since the recommencement of hostilities in 1803. Only after 1809 would it become apparent that there was no quick solution to be had, and that a prolonged commitment to continental warfare was required to achieve the sort of gains needed to bring France to heel. This realization would accordingly lead to a reassessment of priorities, as the focus of the war effort became increasingly centered on one theater at the expense of the standing commitments already outlined. By following the shifts in strategic priorities on a year-by-year basis, it is easy to see how Britain's limited military resources were placed under increasing pressure. When this pressure could be withstood, and a balance between the requirements of the various active and inactive theaters of war achieved, the British Army was able to cope well enough; when that balance was upset, military overstretch ensued and, with it, the potential for disaster.

In 1808, however, all this was in the future, and although there was some initial hesitation, as well as mutual difficulties so far as working with former enemies was concerned, the success of the disposable force concept allowed for the speedy dispatch of significant forces to Portugal to help evict Général de Division Andoche Junot's invading French. Five thousand men under Major General Brent Spencer were diverted from a planned reinforcement for Sicily, whilst the bulk of Lt. General Sir Arthur Wellesley's corps was already assembling, as we have seen, with a view to operations in South America. In like fashion, Lt. General Sir John Moore's 12,000-strong command had been engaged on a futile expedition to the Baltic, but could also be rapidly redeployed thanks to the ready availability of the necessary transport ships. To create the force of 29,000 men deemed necessary by Horse Guards, it was therefore necessary only to add a couple of extra battalions to Wellesley's command before it left Ireland, bringing it up to a strength of around 8,500, and to prepare two additional brigades to embark from ports in England. With the system apparently working perfectly, more trouble was created by the need to find a suitable commander than to assemble an army for him to command; Wellesley and Moore both lacked political credibility, and the former was also too junior. Accordingly, once York had been dissuaded from assuming command in person, Lt. General Sir Hew Dalrymple, the commander in chief at Gibraltar, was given the post with Lt. General Sir Harry Burrard as his second. By the time the two senior generals arrived, Wellesley had already defeated the advanced guard of Junot's army at Roliça on August 17 with only the six brigades of his own and Spencer's commands. Four days later, when the French commander took to the field in person and gathered up a more respectable force to attack the British around Vimeiro, the only additional troops on hand were Wroth Acland's and Robert Anstruther's brigades newly arrived from Britain. Even this small reinforcement was sufficient to afford Wellesley a considerable numerical advantage, which, along with Junot's unwise division of his forces, was enough to ensure a crushing French defeat. Only with the fighting at Vimeiro coming to a close did Burrard arrive on the scene, with Dalrymple finally taking command a day later and finding that Wellesley had largely done his work for him.[50]

One is inclined to suspect, with the example of Lt. General Sir John Whitelocke at Buenos Aires still recent, that Dalrymple now preferred to secure Wellesley's victory by opening negotiations rather than risk it all, and his career with it, by further fighting. The infamous Convention of Cintra was accordingly concluded on September 30, by which Junot's troops were to be evacuated to France, in British ships, with their arms, baggage, and personal effects: this latter category being flexibly interpreted by the French to include the loot of their occupation.[51] Once the news of Cintra was out, both Dalrymple and Wellesley found themselves rapidly recalled, with Burrard soon to follow, thus leaving Moore in command with the unenviable task of cooperating with the armies of the Spanish insurgent juntas. In order to compensate for the garrison to be left in Portugal, Moore was to be reinforced with a corps of 12,000 men under Lt. General Sir David Baird, which was shipped to Corunna before marching overland to rendezvous with the main body. This gave Moore a field army exceeding 30,000 men, but this counted little against the 244,000 that Napoleon was leading into Spain.[52] Moore was able to distract the French for a time by his operations in the north of Spain, with his cavalry doing well in two minor actions, but retreat became an inevitability if the army was to be saved, and only Moore's posthumous battlefield victory before the walls of Corunna restored some luster to an otherwise failed campaign.[53]

The initial reports reaching Britain concerning Moore's progress were largely positive; indeed, in its overenthusiasm, the *Times* at first credited Lord Paget's cavalry with the capture of a French marshal.[54] Once the news of the retreat was out, however, the tone changed, and the ragged state of many of the survivors as they came ashore in Britain did not help matters. Parliamentary questions were asked concerning the number of casualties, although Castlereagh's brother, Brigadier General Charles Stewart, was quickly on his feet to stifle rumors that over 8,000 men had been lost since the army entered Spain.[55] Moore's troops had also returned without much of their equipment and supplies, abandoned during the retreat or destroyed at Corunna, and this loss too became subject to parliamentary investigation. Amongst other equipment losses, the commissary in chief had to report the loss of 248,523 pounds of salt meat, 192,191 pounds of biscuit, 5,387 1/2 gallons of rum, 8,872 blankets, and 4,382 pairs of shoes, all of which would need replacing before a new campaign.[56]

As a result of the return of Moore's army to Britain, a sizeable disposable force was again available for service for the campaigning season of 1809. With Austria again preparing for war, two strategic options were now open: either a return to the peninsula to support the continuing Spanish resistance, or a new campaign in northern Europe where Britain could potentially further its own designs in the Low Countries whilst providing a diversion in favor of Austria. Although attempts were still being made to persuade the Spanish to admit a British garrison to Cadiz, Moore's campaign had drawn the bulk of French forces into the north and west of Spain, and under these circumstances the reorganized Spaniards were able for the time being to secure Andalusia unaided. Therefore, British operations would have to be based on Lisbon, where the garrison under Lt. General Sir John Cradock could form the nucleus for a new army. However, Cradock had less than 10,000 men, largely units that had been deemed too sickly or inefficient to march into Spain with Moore.[57] Before his death, Moore had claimed that Portugal could not be held, but Wellesley, with his Indian experience of sustained operations in conjunction with indigenous forces, had the foresight to see that, assuming the cooperation of a remodeled Portuguese army, Britain could feasibly maintain its presence in the country and ultimately use it as a base for further operations.[58] With his reappointment to the peninsular command, which he would hold for the rest of the war, Wellesley won the argument. However, for 1809 the peninsula was not to be Britain's main theater of operations. Instead, the lure of objectives closer to home spawned the "Grand Expedition" to the Low Countries under the military command of Lt. General the Earl of Chatham, which, though successful in capturing Flushing, was unable to push on to the main objective of Antwerp; all this did nothing to help the Austrians who were already defeated by the time the expedition sailed. Meanwhile, the British troops were stricken with fever in the swamps of Walcheren Island, and suffered considerable losses before the last of them were finally withdrawn in December.[59]

Although they ultimately achieved little, the 39,000 men who sailed to Walcheren in July 1809 represented the largest single expeditionary force that Britain sent overseas during the whole of the conflict. Its assembly, however, was achieved only after a rigorous scouring of the British Isles for available troops, since it had also been necessary to send reinforcements out with Wellesley when he returned to Portugal in April. Even though the peninsula was a secondary priority for 1809, the total number of British troops serving under Wellesley had risen to some 26,000 men by the time that the Walcheren expedition sailed, with a further 9,000 on their way as reinforcements.[60] This commitment of such significant forces in two different theaters placed a considerable strain on the British Army—quite simply, to attempt a maximum effort at this stage in the war was asking too much too soon and led to the first of two periods of strategic overstretch.

The primary symptom of this overstretch was the commitment to active service of infantry battalions that would not, under ordinary circumstances, have been earmarked for this role. During 1808 the infantry serving in the peninsula eventually amounted to forty-nine battalions. Thirteen of these had remained in Portugal, and thirty-six had returned from Corunna in varying states of disorganization. The bulk of those that returned became the nucleus for the Walcheren expedition; only the first battalions of the 43rd and 52nd Light Infantry and 95th Rifles were exempt, instead forming a light brigade under Brigadier General Robert Craufurd to join Wellesley in Portugal. Whilst there were still some sixty-three battalions of the line remaining in Britain in January 1809 that had not taken part in the previous year's campaigning, the greater part of these were weak and fit only for home service. Nevertheless, ten of these—including two excellent battalions of Foot Guards but otherwise mostly second battalions of line regiments—would ultimately go to Portugal during the course of the year, and a further ten would go to Walcheren. The situation regarding mounted troops was rather better, since fewer regiments had been engaged in 1808 and additional horsemen had been readied for service as a potential reinforcement for Moore. The year 1809 saw seven cavalry regiments go out to the peninsula whilst a further five were earmarked for the Walcheren expedition although not all sailed.[61]

Although Wellesley did not receive priority so far as reinforcements from home were concerned, he was able to obtain some additional troops from the Mediterranean garrisons and also benefited from the belated arrival of some units that had initially been intended to reinforce Moore. With these forces, he was able to clear Portugal of the French for a second time, launching a devastating surprise attack on Marshal Jean-de-Dieu Soult at Oporto on May 12 and then harrying the retreating French through the mountains of the Trás-os-Montes.[62] Scarcely had the victors recovered from the rigors of the pursuit than they were on the move again, returning to the valley of the Tagus and then heading east into Spain where they were to cooperate with the Spanish in a combined offensive designed to liberate Madrid. Poor interallied cooperation thwarted this goal, and although the fighting around Talavera on the 27th and 28th of June represented a tactical success, the action was by no means free from error. Furthermore, many troops began the battle on empty stomachs as the commissary system began to fail under the strain. Continuing supply shortages, and the massing of French reinforcements against his lines of communication, forced Wellesley—created Lord Wellington after Talavera—to abandon the campaign and order a withdrawal that took the army first to the unhealthy Guadiana and then back to Portugal.[63]

Although 1809 had represented a huge effort by the British Army, the year had been a strategic failure, with Austria's defeat ending any hopes of maintaining the Fourth Coalition against France. In the peninsula, too, the year had ended in disaster for the allies as the French drove south into Andalusia and laid siege to Cadiz. Although some help was sent to the Spanish, Britain's priorities for the next two years were focused on the defense of Portugal. Financial support, and the endeavors of Lt. General Sir William Beresford and his team of seconded British officers, ensured that the Portuguese armed forces would function under British command, and from 1810 onward Wellington would command a unified Anglo-Portuguese field army. In effect, therefore, the events of 1809 forced a return to a more traditional grand strategy, whereby British gold took much of the strain away from British manpower.[64] The active campaigning of 1810 was dominated by the defense of Portugal against Marshal André Masséna's 65,000-strong French Armée de Portugal. By bringing the Portuguese into the field, Wellington was able to double the size of his regular forces, but this still only gave a field army of 52,000. However, by combining delaying actions with a scorched earth policy, Wellington hoped to wear the French down and thus negate Masséna's numerical advantage.[65] Having captured the border fortresses of Ciudad Rodrigo and Almeida, Masséna finally advanced into Portugal during September 1810, meeting Wellington in battle for the first time on the Serra do Busaço on the 27th. Although tactically victorious, the allies continued to fall back toward Lisbon and on October 10th entered the newly completed Lines of Torres Vedras. The secrecy with which these fortifications had been constructed led to their existence being a complete surprise to the French, but the repulse of several probing attacks soon made it clear that they could not be taken by force. After six weeks, Masséna accepted the inevitable and ordered a withdrawal to Santarém on the Tagus.[66]

Meanwhile, Wellington had been receiving a steady trickle of reinforcements as the survivors of the Walcheren campaign began to become available for a return to service. Along with units reassigned from overseas garrisons, these new arrivals saw Wellington commanding some 48,000 British troops, once the last major reinforcements had arrived in mid-March, along with half that number of Portuguese.[67] Although this gave Wellington an edge over Masséna, the French capture of Badajoz forced the dispatch of troops to the southern theater of war, leading to a division both of focus and of force that would continue throughout 1811 and 1812.[68] Wellington was able to harry Masséna's forces out of Portugal, but supply problems prevented a further advance, and the detachment of Beresford with three divisions to the southern theater left Wellington at a disadvantage when the French commander regrouped and launched a counterattack.[69] The resulting Battle of Fuentes de Oñoro, fought between the 3rd and 5th of May 1811, was the narrowest and least creditable of Wellington's victories, forcing the admission that "If Boney had been there, we should have been beaten."[70] Further embarrassment would follow, when a series of blunders allowed the French garrison of Almeida to effect its escape and reach Masséna's lines, an incident that revealed a startling lack of professionalism amongst certain of the senior officers involved.[71]

Meanwhile, events in the south had led to another general action, even bloodier than Fuentes de Oñoro. This took place at Albuera on May 16, where a multinational army under Beresford—commanding by virtue of his commission as marshal of Portugal—attempted to defeat Soult's attempt to relieve the French garrison in Badajoz. Soult managed to turn the allied right flank, and the battle degenerated into a series of fierce combats as successive allied formations moved up to stem the French tide.

During the early stages of the action, Lt. Colonel John Colborne's brigade was caught in line by French cavalry and suffered severe casualties, and it was only after a prolonged infantry firefight that the French were finally driven back. The number of British casualties—4,156—made Albuera Britain's bloodiest peninsular battle; coming on top of the 1,493 casualties at Fuentes de Oñoro, this represented a sizable reduction of British effective strength in the theater, and necessitated a wholesale reorganization to rebuild the battered Second and Fourth Divisions.[72]

The bulk of the next twelve months saw a concentration on warfare by maneuver, with Wellington seeking to obtain an advantageous position that would allow him to successfully besiege either Ciudad Rodrigo or Badajoz. In the face of three separate French armies, this was a hard task, since although he possessed an individual advantage over Soult's Armée du Midi, covering Badajoz, or Général de Division Jean Dorsenne's Armée du Nord covering Ciudad Rodrigo, either force could potentially be reinforced by the Armée de Portugal, which was now under the command of Marshal Auguste Marmont and based around Salamanca. Obtaining a local advantage necessitated the rapid movement of substantial forces, but with each north-south movement representing 150 miles over indifferent roads, this inevitably had a deleterious effect on sick lists that were already swollen by the wounded from May's battles. Throughout 1811 Wellington found his moves thwarted, and it was only in the January of 1812, after a surprise winter campaign, that he was able to capture Ciudad Rodrigo, albeit at the cost of 1,111 casualties.[73]

Although the defense of Portugal had been the primary focus of Britain's commitment to the peninsula during 1810 and 1811, there had also been an increase in activity along the Mediterranean coast of Spain. The French advance into Andalusia during 1809 finally persuaded the Spanish to accept British aid in defending Cadiz, which in February 1810 welcomed a British brigade as reinforcement for its garrison. Thereafter, the British presence was rapidly built up to a peak of 7,070 rank and file, plus a further 1,259 Portuguese under British command, in July of that year. A reduction in French pressure once it became apparent that Cadiz could not be captured by assault, combined with a growing need for reinforcements in Portugal, saw the British contingent reduced to the region of 4,000 to 5,000 men, plus another thousand or so Portuguese, at which level it remained until 1812.[74] From March 1810 command was invested in Lt. General Sir Thomas Graham, who later led an Anglo-Portuguese division in the operations of March 1811 that culminated in the Battle of Barrosa.[75] As well as troops from Cadiz, this campaign drew on troops from the Gibraltar garrison, which had previously also contributed forces to an ill-advised and ill-fated amphibious descent on Fuengirola in October 1810.[76] More successful was the occupation of Tarifa, held by a British garrison from September 1810 onward, which withstood a French siege between December 1811 and January 1812.[77] In British hands, the fortress provided a useful sally point for operations in the interior, including those leading to Barrosa. Although relatively small in scale, representing a return to the old strategy of using the British Army as a means of extending the reach of the Royal Navy, these operations did help keep substantial numbers of Soult's forces tied down in the south of Spain, and this would be of great utility to Wellington as he began planning for the campaign of 1812.

Even in 1811, the dispersal of Soult's Armée du Midi had allowed Lt. General Rowland Hill, commanding the Anglo-Portuguese forces in the southern theater, to launch a surprise attack against one of Soult's divisions at Arroyomolinos on October 28 and inflict one of the most one-sided defeats suffered by the French in the peninsula.[78] Now, when Wellington moved south in force during March 1812 to take Badajoz, Soult was unable to muster a creditable relief force. Nevertheless, the possibility that Marmont might yet reinforce Soult led to the siege operations being rushed, leading to heavy casualties during the successful storm of April 5, 1812. Allied casualties totaled 4,670 for the siege as a whole, which, combined with the losses at Ciudad Rodrigo, left the army appreciably weakened even before the main business of the year's campaigning could begin.[79]

By the time Badajoz fell, Wellington's forces had been engaged in active warfare for well over a year, with little respite. Exertion and combat had depleted the ranks to a considerable degree, and a similar depletion, through death and wounds as well as illness, had removed a substantial number of Wellington's key subordinates. Sizeable reinforcements came out during late 1811 and early 1812, amounting to some 8,000 men including two full brigades of heavy cavalry, but many of the infantry battalions that came out during this period were those that had taken part in the Walcheren expedition and still had the aftereffects of its fevers running through their ranks. Nevertheless, the new arrivals made up for the Badajoz losses in quantity if not quality, whilst the French armies in the peninsula were now weaker thanks to the drafting off of regiments for participation in Napoleon's invasion of Russia. Wellington was therefore able "to move forward into Castille, and to endeavour to bring Marmont to a general action."[80] On July 22, outside Salamanca, he did just that and won the greatest battlefield victory of his career, albeit at a cost of a further 3,129 British casualties.[81]

By this point, however, the pressures of such sustained operations were beginning to make themselves felt, and many of the regiments that had joined during the weeks prior to Salamanca were now in a poor state due to the rigors of campaigning having brought about a return of the fevers that had bedeviled them since Walcheren. Thus, although the allies were able to enter Madrid on August 12, a substantial body of men had to be left there when Wellington continued in pursuit of the Armée de Portugal, with entire battalions detached from their divisions in order to recuperate. With numbers thus reduced, and the French beginning to rally, the allied advance was checked before the walls of Burgos. Hearing that Soult had abandoned Andalusia and was marching north with 60,000 men, Wellington accordingly fell back, first to Salamanca where the army was able to regroup, and then to the Portuguese frontier around Ciudad Rodrigo.[82] Although Wellington's troops were largely unmolested by the French, whose combined forces considerably exceeded those of the allies, the latter stages of the retreat placed the army under great strain, with bad weather and logistical mismanagement adding to exhaustion to bring the whole force close to collapse. Once the last stragglers had come in, it was reckoned that some 3,000 men had been lost during the retreat, leading to an irate Wellington castigating the army as a whole for losses that he perceived to be due to indiscipline in the ranks and neglect on the part of the officers.[83] Nevertheless, the 1812 campaign as a whole had been a huge achievement, clearing Andalusia and central Spain, and allowing the reconcentration of Wellington's forces into a single field army. However, strategic planning could no longer take into account the situation in the peninsula alone. Not only did Napoleon's defeat in Russia shift the focus of grand strategy back to central Europe for the first time since 1809, but, on the other side of the Atlantic, Britain had become embroiled in a whole new conflict.

Because of the steady rotation of its most combat-ready units away from overseas garrisons in order to concentrate a worthwhile field army in Europe, Britain began the War of 1812 against the United States with only a skeleton force of a little under 10,000 regular troops in the various commands making up British North America.[84] The bulk of these forces had seen little prior active service, and included a substantial number of locally raised Fencible units. Canada also had its Militia to help take on some of the burden, and although many of its members formed part of sedentary units for local defense, this source also furnished six Embodied Militia battalions for more active duty and a further two volunteer battalions.[85] As was the case in Europe and India, good use was also made of indigenous allies, with substantial numbers of Native Americans playing an effective part in the early stages of the war thanks to successful diplomatic efforts on the part of the Indian Department.[86] A further advantage was the mobility provided by use of water transport, and this enabled troops to be rapidly shifted to threatened points. In this way, Major General Isaac Brock, commanding in Upper Canada, was able first to secure the surrender of one U.S. force at Detroit on August 16, and then, at the cost of his own life, to defeat another on the Niagara frontier at the Battle of Queenston Heights on October 13. Nevertheless, with the United States unwilling to give up the contest, it was evident that British reinforcements would be required to force a decision. Thus, whilst events in Europe finally presented the opportunity for Britain to play a meaningful role in a renewed anti–French coalition, the country was forced to do so with one eye cast over its shoulder.[87]

An immediate effect of the fact that the peninsular army was no longer Britain's sole priority was that Wellington was forced to relinquish the services of several units that had been particularly worn down during the 1812 campaign. Although these would be replaced by fresh troops, this withdrawal marked the beginning of a clash of wills that drew in both York and the Earl of Bathurst, newly appointed secretary of state for war and the colonies. Yet whilst he was husbanding every battalion, Wellington began his new campaign with a force that was more powerful than ever since some 9,000 reinforcements had joined by the time that campaigning resumed in May 1813, many of them reassigned from Cadiz and Gibraltar now that the south of Spain was free of French forces. Wellington not only began his campaign with 52,484 British and 28,792 Portuguese troops in his field army but was now also generalissimo of the Spanish armies, which, rejuvenated over the past year by British financial aid, provided a further 46,292 men for the main offensive.[88] In contrast, the French in the peninsula were weakened through having had to supply reinforcements for central Europe. Even by combining the troops of three nominally distinct armies, Joseph Bonaparte could muster only 63,000 men to Wellington's 69,000 when he stood at bay outside Vitoria, and was defeated in a manner so complete that the achievement won Wellington the baton of a field marshal.[89] In the aftermath of the battle, the French survivors crossed the Pyrenees and regrouped around Bayonne, where they were reorganized into a single Armée d'Espagne under Soult, whilst Wellington set down to besiege the last French garrisons south of the mountains. A French counteroffensive during late July achieved some initial successes in the Pyrenean passes, with Soult managing to get within sight of Pamplona before being turned back after several disjointed days of hard fighting.[90] Defeat of the French counteroffensive allowed siege operations to resume and for San Sebastián, which had withstood one assault on July 25, to finally be stormed on August 31, albeit with heavy losses during the attack and considerable disorder in the aftermath.[91]

With the French now in full flight over the Pyrenees for a second time, only eastern Spain remained host to substantial hostile forces. Between 1810 and 1812, Marshal Louis Suchet's Armée d'Aragon had methodically cleared the Spanish from much of Aragon and Valencia, and this eventually compelled the detachment of British troops to support the remaining Spanish forces. Other than a single battalion and some artillery reassigned from Cadiz, the British contingent was drawn entirely from the force based on Sicily, that station again demonstrating its utility as a central base for expeditions around the Mediterranean. However, the employment of troops from Sicily also ensured that these troops remained under the orders of the commander on Sicily, Lt. General Lord William Bentinck, and meant that operations were at times restricted by conflicting strategic priorities stemming from Bentinck's preoccupation with events in Italy. Lt. General Frederick Maitland commanded the initial expeditionary force, which landed at Alicante in August 1812; he then fell ill over the following winter and was replaced by Lt. General Sir John Murray. Murray was able to help the Spanish check French offensive moves during early 1813, but failed badly when subsequently tasked with capturing Tarragona. Bentinck then briefly assumed command in person, working with the Spanish to push Suchet back over the Ebro, but his heart was always drawn to Italy. Accordingly, he turned the Catalonian command over to Lt. General William Clinton, who continued to cooperate with the Spanish armies until the close of the Peninsular War.[92]

Although initially useful in shoring up Spanish resistance, this 12,000-strong force subsequently achieved little of value, and in April 1814 was broken up with half going to reinforce Wellington and the remainder returning to Bentinck for operations in Italy, although neither element reached its destination before the war's end.[93] In the meantime, with forces drawn directly from Sicily, Bentinck was able to assemble an Anglo-Sicilian force of 14,658 all ranks, including three battalions of regular British foot and three more from the KGL, and land them at Livorno during March 1814. Although these forces were able to help secure the eviction of the French from Genoa in the closing days of the war, their operations were limited by Bentinck's ill-advised meddling in Italian politics, and the troops saw little action.[94]

By the autumn of 1813, with Wellington successful in Spain and the continental powers of the Sixth Coalition having finally obtained the upper hand in Germany, Britain was at last in a position to reassess its strategic commitments. Whilst there were successes that could potentially be exploited, there were also new problems to be surmounted. What was more, with victory now in sight, there was the postwar world to be considered, and if Britain expected to have a say in the settlement of European affairs then it was vital that its armies were seen to be doing their utmost. Serious thought was therefore given to leaving the Spanish and Portuguese to defend the line of the Pyrenees whilst Wellington's British forces were shipped to northern Europe to join the allied advance. Wellington was able to produce arguments sufficient to demonstrate the flaws in such a movement, but found that his attempts to obtain additional manpower were increasingly fruitless, although he was more successful in resisting York's attempts to reclaim units already serving in the peninsula. Nevertheless, with attrition outstripping reinforcements, Wellington's forces did shrink slightly during the closing months of the war, with the strength of his British contingent peaking at 63,595 rank and file in November 1813.[95] North America, too, received little by way of reinforcement from home, although reassignment of units from other colonial commands allowed for a steady increase in troop numbers that was sufficient to turn back the amateurish and halfhearted American plans for an autumn advance on Montreal at the battles of Chateauguay and Crysler's Farm.[96] Only in the far west were the Americans able to achieve superiority, allowing them to crush their British and Native American opponents at the Battle of the Thames on October 5, 1813.[97]

Since it was impossible to remove manpower from either active theater, any new campaign would need to be mounted using the increasingly slender resources available from the home islands. So far as the armies of the Sixth Coalition were concerned, British gold had helped bankroll them, and British equipment had helped fit them out, thus, although a small force had gone to the Baltic under Major General Samuel Gibbs in mid-1813 to guard the ports through which this flood of supplies was sent, no major military presence was needed to give Britain continued credibility as an ally. A few British units did make it onto the battlefields of Gohrde and Leipzig, but this was as much by accident as design.[98] Matters changed, however, as the fighting shifted toward the Low Countries where revolt broke out in favor of the House of Orange. As in 1809, control of this area was deemed vital to secure Britain's maritime interests, and it was therefore politically expedient to send troops there in force. By recalling battalions from the Baltic, and scraping the depots of Britain for available men, a corps with an eventual strength of 11,812 rank and file was assembled by early 1814, but it is testament to the poor quality of many of the units concerned that, of this number, only 8,597 were actually fit and present for duty.[99] Command of this dubious army was offered to Sir Thomas Graham, who accepted only against his better judgment. In the event, Graham was successful in a minor action at Merksem on January 13, 1814, but could not take Antwerp without Prussian help, and failed in an attempt to take the smaller fortress of Bergen op Zoom by storm on the night of March 8. Only with the coming of peace was Graham able to negotiate an entry into Antwerp after the French had evacuated the place.[100]

The opening of this third front was in many ways the final straw for Britain's manpower systems, which were now stretched to a point that they could no longer be counted on to provide a continued supply of combat-worthy manpower. The nature of these failings, and the increasingly drastic measures taken to try and counter them, form the subject of a later chapter, but it became clear that a breaking point was being reached and that the last six months of hostilities, up to Napoleon's first abdication in April 1814, represent a period of overstretch more severe even than that of 1809. It is true that, on paper, there were still 51,530 regular troops stationed in Britain in January 1814, and a further 12,303 in Ireland,[101] but this is a misleading total since the bulk of these troops represented the depots of units that were already overseas. Those complete units still at home, barring those retained for internal policing duties and a handful of other exceptions such as the Channel Island garrisons, were those that had not been deemed fit to accompany Graham's expedition to the Netherlands; when one appreciates the parlous state of many of the battalions that *did* embark with Graham, the condition of those left behind should need no elaboration.

Under such circumstances, it is therefore rather disconcerting to see that Wellington, even in the closing months of the war, was still doing his best not only to hang on to every last man that he had but also to gain any potential reinforcements going.[102] This, too, when Soult was steadily being stripped of more and more of his best troops to provide reinforcements for Napoleon's own struggles in the north. Such single-mindedness on Wellington's part would be understandable had every man been needed in the front line, but once the allies had broken out of the Pyrenees and into France itself, following successes at the Nivelle, Bidasoa, and Nive, the victorious troops became increasingly dispersed and diverted by secondary tasks.[103] Indeed, the end of the war found Wellington facing Soult at Toulouse with six of his Anglo-Portuguese divisions, plus a Spanish corps; two further divisions, along with more Spaniards, invested Bayonne, and the last division—Lord Dalhousie's Seventh—was occupying Bordeaux. Admittedly, the French remained full of fight. Soult's army put up a strong though ultimately futile resistance at Orthez on February 27, and again at Toulouse on April 10, whilst the garrison at Bayonne launched a most damaging sortie at the very end of the war, taking Lt. General Sir John Hope prisoner.[104] In order to strengthen his case, Wellington continued to express doubts almost until the end of hostilities that a separate peace between France and Spain might yet free Suchet's army to unite with that of Soult. Nevertheless, it was a gross and ungrateful exaggeration for Wellington to describe the powerful field army that he commanded at the conclusion of the Peninsular War, as he did to Bathurst in his last and most unappealing complaint over the lack of reinforcements, as merely a "handful of brave men."[105]

It is hard to escape the conclusion that, by April 1814, Britain's systems of military manpower were on the verge of collapse, and that peace in April 1814 came at a most timely moment. However, rather than consolidate, Britain's politicians sought to return to a peace establishment in Europe whilst, at the same time, sending a substantial reinforcement to North America in order to bring

a swift conclusion to what was now a largely pointless conflict. At
the beginning of 1814, before the arrival of reinforcements from
Europe, regular forces in Canada totaled around 14,500 rank and
file; by the end of the year, this had almost doubled with a peak of
27,918 in January 1815. In similar fashion, forces based on Nova
Scotia grew from less than 4,000 at the end of 1813 to over 7,000 a
year later, allowing for the local commander, Lt. General Sir John
Sherbrooke, to carry out offensive operations deep into Maine.[106]
In Canada itself, fighting still retained a defensive character, since
the increase in British numbers coincided with a rise in American capabilities. The first indication of the threat posed by the
rejuvenated American forces was their occupation of much of
the Niagara peninsula during July 1814, capturing Fort Erie and
defeating a British brigade at Chippewa. Directed by the senior
officer in Upper Canada, Lt. General Sir Gordon Drummond,
subsequent fighting, at Lundy's Lane on July 25 and then during
the botched attempt to recapture Fort Erie, largely degenerated
into stalemate. It was eventually the presence of Anglo-Canadian
naval forces on Lake Ontario, threatening American supply lines,
that forced the American retreat in late autumn, bringing an end
to major operations on the Canadian front.[107]

Although the Niagara campaign had involved reinforcements
from Europe, including units drawn both from Wellington's disbanded army and from the Mediterranean, the bulk of the 1814 reinforcements were committed to offensive operations elsewhere.
As early as January 1814, Bathurst looked ahead to a future European peace and envisaged the massing of the best troops from
Wellington's army into a single powerful corps of 20,000 men by
which a knockout blow could be delivered. In practice, nothing
like this concentration of force was achieved, and by the time the
troops were available the proposal had been scaled back to 12,000
or 14,000 men only.[108] The closest thing to Bathurst's intentions
was an 11,000-strong force—roughly half of the men newly arrived
reinforcements—assembled for an advance by way of Lake Cham-
plain in a campaign that was overseen by the governor general of
British North America, Lt. General Sir George Prévost. Poor relations between Prévost and his subordinates did not make for effective cooperation, and the campaign was sent permanently into

reverse following a failed attack on the main American depot at Plattsburg on September 11, 1814.[109]

Although failure at Plattsburg owed much to Prévost's inability to coordinate land and naval forces, this was not the case elsewhere, and a more effective partnership was developed along the Atlantic Seaboard. With support from the Royal Navy, Major General Robert Ross was able to land with an expeditionary force of 4,800 men and briefly occupy Washington. A subsequent move against Baltimore was less successful, with Ross being killed by a sharpshooter during the advance on the city, and the attempt was ultimately abandoned once it became clear that the American defenses were stronger than had been anticipated.[110] Whilst burning the White House was undoubtedly an excellent coup, such operations represented another return to the tried and tested methods of using the British Army to extend the reach of the Royal Navy, and would not have been out of place in the previous Anglo-American conflict forty years previously. The same, indeed, could be said for the war's last operation. The only difference with the expedition to the Gulf of Mexico that culminated in defeat at the Battle of New Orleans on January 8, 1815, was the greater size of the forces involved: 8,869 men under Major General Edward Pakenham, although not all of these were present at the battle itself.[111]

By the time that New Orleans had been fought, the political situation in Europe was rapidly deteriorating to the extent that the former allies might end up fighting amongst themselves over the spoils of victory. These pressures contributed to the reduction in the planned reinforcement for North America, and also hastened the Anglo-American peace negotiations at Ghent, which were eventually concluded on Christmas Eve 1814—prior to the fighting at New Orleans.[112] Yet, despite the unsettled political situation, the British military establishment was already being cut back. Some of this was done with positive effect, with twenty-four weak second and third battalions being disbanded to bring their regiments' first battalions up to strength. Even so, there were substantial cutbacks, with the cavalry and artillery establishments being reduced and a wholesale disbanding of the veteran battalions and foreign regiments. The bulk of the KGL was concentrated in the

Netherlands, where the remnants of Graham's army remained in garrison, in preparation for its breaking up and incorporation into the reraised Hanoverian army, a measure which, once fully implemented, would have removed several thousand veterans from the British establishment.[113] Several of the émigré units were also broken up, and there was also a heavy cull from ranks of the various colonial corps. Although events would preserve the KGL to fight at Waterloo, the brief peace saw 37,000 Britons and a further 10,000 foreigners discharged from the British Army.[114]

It was against this background that Napoleon's return from Elba threw everything into chaos, but, with much of its best infantry still on the other side of the Atlantic, Britain was hard-pressed even to create the "infamous army" that fought the Battles of Quatre Bras and Waterloo.[115] As we have seen, Wellington was not above talking down the quality of the forces under his command if doing so might obtain him reinforcements, but in this case the criticism—which may be interpreted as much as a comment on the 52,783 Netherlanders and Germans under his command as on the 42,640-strong British contingent—has rather more validity.[116] That said, a better description might well have been an "incomplete army"; as we shall see, there were substantial differences in strength and organization between the forces that stood on the ridge at Waterloo and those that marched into Paris three weeks later. This distinction, however, serves only to highlight the narrowness of the margin by which a credible British force was assembled in the Netherlands. A French delay of a week or two would have found a much stronger British force, but, equally, bringing operations forward would have found one far weaker. Waterloo itself may well have been a near run thing on the battlefield, but the same can be said of the campaign as a whole so far as the British Army's preparations for it are concerned.

Having brought this overview of the campaigns of the British Army during the period to a close, it remains only to review the nature of the overall manpower commitment, an exercise that demonstrates both the shifting priorities between different theaters of war, and the narrow margin that at times stood between the success and failure of the military system as a whole. Table 1 breaks down manpower deployments by theater during the period.

Table 1. Total Manpower Commitments, 1808–1815

	July 1808	July 1809	July 1810	July 1811	July 1812	July 1813	July 1814	July 1815
Northern Europe	300	39,237	288	391	404	1,128	14,180	54,406
Mediterranean	22,489	23,459	27,991	30,311	35,057	38,943	23,864	17,396
Peninsula	0	36,438	40,186	56,825	63,121	59,555	5,929	0
East Indies	22,734	24,729	26,819	28,815	26,667	29,679	29,701	26,587
N. America and W. Indies	29,426	29,059	30,123	27,706	30,307	38,524	43,803	30,418
Other Overseas Stations	8,924	7,989	8,147	6,059	7,307	6,335	6,085	5,611
Home Islands	81,411	52,486	77,468	66,579	71,117	64,566	94,422	47,073
Total	165,284	213,397	211,022	216,686	233,980	238,730	217,984	181,491

Source: Data primarily derived from digest in TNA, WO17/2814, supplemented where necessary with data from elsewhere in the WO17 series.

One thing that immediately becomes apparent is the relative insignificance of the peninsula in terms of overall numbers. Although for some time the largest single overseas deployment, the peninsula was nevertheless greatly surpassed at all times by the sum total of other such deployments. For all its domination of the literature pertaining to the period, the Peninsular War did not represent the norm for the average experience of military service. Also obvious are the two periods of extreme commitment to multiple theaters, leading to potential overstretch. As noted, the first was 1809 when significant forces were simultaneously deployed to Walcheren and the peninsula, and the second, attained through a steady increase in deployments from 1810 onward, came in early 1814 when forces were committed to operations in the peninsula, North America, and the Low Countries. The last point to be kept in mind is the relatively fixed strength of the forces in the East and West Indies. In the former case, the figure represents a bare minimum garrison, although even this was stretched by the need to find troops to occupy Java and Mauritius by the end of the period. For the West Indies and America, on the other hand, the increase associated with the War of 1812 is by no means as large as might be expected. Strength in Canada increased considerably, but many of these troops came from elsewhere in Britain's North Atlantic and Caribbean possessions; only with Napoleon defeated, at least for the time being, could troops be redirected from Europe in any numbers.

This redeployment was therefore achieved at the expense of entrusting Britain's holdings in the West Indies to a very large extent to foreign and indigenous troops. This measure, however, was only part of a growing swing toward using non-British manpower, which the total figures used in table 1 do not immediately emphasize. Although the table shows an increase in the overall strength of the British Army, what it does not show is that this increase was largely composed of foreigners. In fact, the total number of British troops increased by relatively little, from 152,400 in January 1808 to a peak of 189,837 in July 1813. Thus, by early 1814 and the time of Britain's greatest military efforts, total British manpower had actually begun to decline, with foreigners making up the shortfall. Although the better foreign units—the KGL and some of the émigré regiments—were of equal value to British regulars, and prioritized for active employment alongside them, it was rare for colonial troops to be so used. The War of 1812, which saw active use of the Canadian Fencibles and men of the West India regiments, was the only major exception to this rule. On the other hand, minor garrisons such as Surinam and Curaçao came largely to be manned by foreign or colonial units, and it was also rare for the Bahamas to have more than a company's worth of British regulars. Even in India, an average garrison strength of around 20,000 British regulars becomes relatively insignificant when set against the contribution made by the troops of the HEIC, whose forces can be reckoned at approximately 6,000 European and 111,000 Indian troops as an average across the period.[117] Because of these increasing demands for service in the overseas garrisons, the number of foreigners in British service rose at a far greater rate than the number of Britons in the years before 1814: from 31,891 in January 1808 to peak at 52,416 in January 1813. This represents an increase of 61 percent as opposed to only a 25 percent growth for the British, and is clearly indicative of the need to increase recruitment from this source and use the resulting units to replace British ones. Yet, even despite these measures, Britain's total military manpower remained in decline after mid 1813.

The shortfall of manpower by 1813–14, at a time when the British Army was being pushed harder and harder by new strategic imperatives over and above its normal global commitments, meant that obtaining manpower, and keeping units at a viable combat strength, became severe problems both within individual theaters and for the British Army as a whole. Never for a moment should the reader forget that this period, whilst seeing great triumphs of arms, also saw Britain pushed to the breaking point, and one is forced to conclude that, had peace not come when it did in 1814, Britain would have been forced to choose between two, equally politically unthinkable, options: either cut back on Imperial commitments or introduce something akin to conscription. That Britain was able to finish the Napoleonic Wars as a significant power within the Sixth and Seventh Coalitions was, as we shall see, as much due to its ability to work with and around its military system, as to that system's own inherent strengths.

CHAPTER 2

Regimental Identity and Leadership

In the spring of 1814, having led the second battalion of the 73rd Foot through a year of hard and distinguished campaigning that had taken it from the Baltic to Belgium, the popular and respected Lt. Colonel William Harris took a well-earned respite and left the battalion under the temporary command of Major Dawson Kelly. As Major General Gibbs reported after inspecting the battalion a short time later, Kelly was "an officer of great respectability but having been chiefly employed on the Staff of the Army does not appear acquainted with the Drill or management of a Regiment. The corps is not in as good state of discipline as they were and very unsteady awkward and careless in manoeuvre."[1]

Lest there be any doubt, Gibbs reiterated his point when commenting on the battalion's private soldiers, explicitly stating that this decline in standards was Kelly's fault. However, Kelly's failings as a commander had had a wider effect as the rank and file, who, although "a very good body of men with a general appearance of health and cleanliness," were found to be "neither well drilled, except in the firelock exercise[,] or attentive." Some of this could be put down to hard service, but Gibbs also noted discord amongst the officers—with whom Kelly had in turn expressed dissatisfaction—and, in particular, amongst the battalion's NCOs. Thomas Morris, who served in the ranks of the 2/73rd throughout its time in Germany and the Netherlands, later identified the ready availability of cheap gin as having played a key part in the tailing off of the battalion's efficiency, but also noted that Kelly's

44

response to the deteriorating situation was a resort to increasingly heavy punishments, something that is confirmed by Gibbs's report.[2] As much as anything, Kelly—who had a good combat record with the battalion prior to taking command of it—seems simply to have been out of his depth, but there can be no denying that both the morale and effectiveness of the battalion suffered all the same.

The 2/73rd would, as we shall see, in time recover its condition and go on to fight with distinction at Waterloo, but the example of this battalion clearly highlights the importance of unit identity, morale, and esprit de corps, and the vital role that a battalion or regimental commander played in this—for good or ill. With the right individuals in the right places, the regimental system was ideally suited to facilitate the development of all these things. Officer appointments still involved purchase, but the worst abuses of that system had been abolished with set amounts of time to be served in each rank and the establishment of training schools for prospective officers. However, since officers attained command on the basis of seniority rather than competency, the system crucially failed to ensure that the right man would end up in the right job—in other words, a unit was as likely to find itself with a Major Kelly in command as a Lt. Colonel Harris, and Dawson Kelly was by no stretch of the imagination the worst sort of officer to attain such an appointment.

Thus, the regimental system as it operated during the Napoleonic era was in many ways something of a double-edged sword. On the one hand, it shifted a great deal of administrative responsibility away from Horse Guards, most notably in respect of recruitment, and it also provided the focus for a strong, and growing, sense of unit identity and esprit de corps. The differences between various regiments, particularly those aspiring to a self-defined elite status, impacted both on their internal sense of identity and—albeit not always with positive results—on their campaign and combat performance. Nevertheless, the fact that the system allowed some units to raise themselves above the majority should not obscure the fact that the same system bore much of the responsibility for the poor state of others.

The Role of Identity

What is perhaps the most interesting aspect of the issues surrounding regimental identity is the fact that in many cases it was a relatively new development. Admittedly, there were regiments that traced their ancestry back to the Restoration and even beyond, but over half of the line infantry and around a quarter of the cavalry regiments dated only from the mid-eighteenth century. Indeed, the 78th Highlanders, and all regiments of foot junior to it, had been raised as a result of the revolutionary and Napoleonic conflicts, as had the 21st–25th Light Dragoons. As a result, other than the 94th, which had previously been the old unnumbered Scotch Brigade, these regiments were completely lacking in prior history and traditions.[3] Care also needs to be taken in placing too much value on the sense of identity surrounding regimental names, for this was still an age of numbered regiments, and many of the most famous mass appeals to units use the number rather than any title. Consider Lt. Colonel William Inglis's possibly apocryphal "Die hard, 57th! Die hard!" cry at Albuera, or Sir Thomas Picton's rallying cry at Quatre Bras: "28th! Remember Egypt!"[4] The former certainly created a nickname, but neither exhortation makes any reference to an existing name, official or otherwise. In part, this usage may be attributed to the fact that regimental titles were, relatively speaking, a new invention, and still secondary to the numerical designation of a unit. For the bulk of line regiments, the existence of a geographic county name would have little resonance to a rank and file whose composition in terms of background was far more disparate.

Only in the oldest regiments with their royal or other distinguishing titles, and in some of the newest units, did the name have any potential as a unifying focus of identity. In this fashion, for example, Major General Edward Pakenham could favor a name that was far older than the regimental number to encourage the 1/7th at Aldea de Ponte by shouting, "Forward Fusiliers!"[5] Whilst geographical titles predominate throughout the list, having been allocated en masse in 1782, the patriotic language of the 1790s is apparent in the use of "Loyal" and/or "Volunteers" in the titles of units raised in that decade, indicative of the new forms and

expressions of patriotism then evident. This was the era of the loy-
alist "Counter-Revolution" posited by the historian John Cookson,
but, although its legacy survived in the titles of the 80th–82nd,
85th, and 90th Foot, it is less clear to what extent these appellations were recognized in the same way by the 1810s.[6] The award
of additional titles as a mark of favor should also be noted, with
the 87th becoming the Prince of Wales's Own Irish after its exploits at Barrosa. On the other hand, county titles were frequently
and sometimes radically reallocated on occasion, with the 39th
exchanging East Middlesex for Dorsetshire in 1807, and the 70th
making the jump from Surrey to Glasgow in 1812. Whilst these
measures may well have helped attract new recruits, the nature of
their effect on the men already in the ranks—if they had an effect
at all—is rather more questionable.[7]

Distinctive regimental names, even archaic ones that no longer
represented any actual difference from the norm, also gave certain units a justification for considering themselves a cut above
the rest. The three regiments of Foot Guards were self-evidently
distinct from the line, but there were a number of other regiments
that may be considered, or that at least considered themselves, as
being elite units. In no case however, even with the Guards, was a
regiment entirely composed of specially selected men, and many
of these self-defined elites differed from their counterparts only
in distinctions of dress and nomenclature, as with the fusiliers
and highlanders. There were three regiments of the former and
eleven of the latter in 1808, although the 72nd–75th and 91st were
ordered to cease using the highland dress and designation after
April 7, 1809, in an attempt to broaden recruiting.[8] Highlanders
and fusiliers may be considered as having successfully cultivated
a greater esprit de corps, and at least some of the highland regiments benefited in addition from a far more homogenous recruiting base than was the norm; this was the case for the 42nd, 78th,
79th, and 92nd. The 91st, although "de-kilted," managed to keep
its first battalion, if not purely highland, at least largely Scots but
ended up with all manner of mongrels in the ranks of its second.[9]
In at least some highland units, the 1/92nd under Lt. Colonel
John Cameron of Fassifern in particular, a strong element of the
old clan paternalism was also retained, although the dictates of

manpower replacement meant that this was by now a fading tendency.[10] Nevertheless, both fusiliers and highlanders were to all intents and purposes ordinary battalions of the line so far as their organization and tactical use were concerned.

The other distinctive group, where role as well as title served as a distinction from the line, were the light infantry and rifles. The initial Light Brigade, trained at Shorncliffe from 1803, comprised the 43rd, 52nd and 95th, and these regiments all subsequently raised additional battalions. Along with the light battalions of the 60th and KGL, these represented the official sum of the light infantry at the opening of 1808. More battalions were in the process of being organized under the superintendence of Major General Rottenburg, author of the 1798 *Regulations for the Exercise of Riflemen and Light Infantry*, which was used for all light infantry training. The 51st, 68th, 71st and 85th were all converted during the course of 1808 and 1809 to add a further five light battalions, although the 51st and 1/71st were on active service during 1808 and were therefore only reorganized after their return from Corunna.[11] The 90th (Perthshire Volunteers) had been raised in 1794 as a light infantry corps but was not formally recognized as such until 1815, despite serving in that role as early as the 1801 Egyptian campaign.[12] Other units that were nominally regular battalions of the line were put through the same training regime at the discretion of their commanders; this is known to have been the case for the 20th and 2/78th, which may well explain why the former was brigaded with light infantry units during the Corunna campaign.[13] The ethos of light infantry training, taken to even greater extremes in the rifles, involved a combination of independent thinking, open-order fighting, and marksmanship, which in turn demanded a better class of man, in terms of ability, both in the ranks and the officers' mess.[14] Of course, these regiments still had their bad characters; Lt. Colonel John Colborne, who commanded the crack 1/52nd in the peninsula and at Waterloo, still asserted that *any* regiment, irrespective of special circumstances and pretensions to elite status, would be lucky that did not contain some fifty or so irredeemable hard bargains.[15]

Whilst the last decade of the eighteenth century and the first decade of the nineteenth saw a proliferation in new names and designations for regiments, the name used by the rank and file to distinguish their unit would frequently be very different from that noted in the *Army List*. For the vast majority of regiments, at least one nickname can be identified as in use during the Napoleonic Wars, with several actually being coined during the period as a result of some exploit or misadventure on campaign.[16] In some cases, these names were so established as to have become semiofficial and used as much by senior officers as by rank and file—witness Rowland Hill's misassumption at Talavera that the French night attack was "the old Buffs, as usual, making some blunder"[17]—but others would seem to have had more limited circulation. Some of the more obtuse or obscure pseudonyms—"Pontius Pilate's Body Guards" for the 1st Foot, or "Kirke's Lambs" in ironic commemoration of an early colonel of the 2nd—seem more likely to have been restricted to the officers' mess. Likewise, some of the more cant or dialect-based names—"Havercakes" for the 33rd, "Jaggers" for the 60th's rifle battalions, or "Pot Hooks" for the 77th—may equally be assumed to be restricted to the rank and file.[18] Whilst a few of the names are downright insulting epithets bestowed by outsiders, the bulk are either complementary or, if mildly insulting then at least humorously so as with "Honeysuckers" for the 58th Foot and "Cherrypickers" for the 11th Light Dragoons, both of which stem from thieving escapades in the peninsula.[19]

Collectively, the widespread use of nicknames indicates a healthy sense of self-identity and regimental rivalry as existing across all ranks. It is true that very few memoir accounts use nicknames for the writer's own regiment, but many do employ them for other regiments, which reinforces rather than negates the importance of such nicknames as it still implies a regimentally based view of identity within the British Army as a whole, with other regiments being praised or denigrated as appropriate. If this conception sets up the rest of the army as an "other" to be judged and commented upon, by definition it also ascribes a collective "us" definition to the writer's own unit, suggesting a "bottom-up" model of self-identity based on the regiment rather than a "top-down" one based on the British Army or even the nation.[20]

If it was rare for even complementary nicknames to be used in reference to the writer's own unit, it cannot be denied that unit pride did exist. For a start, there is an honest pride in many memoirs in relation to those events in which the writer's unit particularly distinguished itself. Grattan, for example, explicitly titled one of the subsections in his chapter on Busaço as "Achievements of the 88th," and in it assured his readers with respect to the bayonet charge made by his battalion that "The conception of this attack, its brilliant execution, which ended in the total overthrow of Reynier's column, all belong to Colonel Alexander Wallace of the 88th Regiment" and that this was due "to the splendid state of perfection in which that corps then was."[21] Regimental pride also made it possible to reconcile difficult service with the logic that one's unit had been given the task because it was the army's best. George Gleig in later life was willing to concede that the same spirit was no doubt entertained in the ranks of every regiment going into action, but as a young lieutenant in the 85th was quite sure that the choice of his battalion to attack the village of Urrugne during the Battle of the Nivelle was a mark of the high regard in which Wellington held it. Whatever Wellington's actual views on the unit, it was the fact that the men of the battalion felt themselves to have been specially chosen that was important, and the desire to live up to the reputation helped ensure that the 85th put up a sterling performance.[22]

For those regiments with a longer history, a similar degree of pride could be taken in earlier triumphs—Emsdorf in 1760 for the 15th Hussars, for example—but more recent successes were also rapidly incorporated into regimental tradition.[23] Lieutenant Swabey of the RHA observed that the first anniversary of Albuera was alcoholically celebrated within those regiments that had fought there, a circumstance that the editor of his diary remarked upon as being not uncommon.[24] To a certain extent, this reinforcing of identity with reference to past service was aided by official and semiofficial measures. The most obvious of these was the bestowing of battle honors, but there were also other means, such as the additional Assaye Colour carried by the 74th, which can even be found being referred to in official paperwork as the "Assaye Regiment."[25] In a similar fashion, the previously unnamed 76th was for a time officially titled as the Hindoostan Regiment in reflection of its own services on the subcontinent.[26] However, distinguished service in combat was only one source of pride. John Le Couteur of the 104th was keen to record in some detail the "unexampled march" of his regiment from New Brunswick to Kingston in the depths of the winter of 1813, taking pride not only in the physical achievement but also in the fact that, "[D]uring this long march, under considerable privations and hardships, not one single robbery was committed by the men, nor was there a single report made against them by the inhabitants to the commanding officer."[27]An interesting sidelight on the standards of the time, if a clean sheet was so evidently an unusual and credit-worthy achievement!

Regimental pride and identity could be put to practical use through systems of rewards and punishments. A growing tendency can be identified toward promoting good behavior, on or off the battlefield, by medals, badges, or other distinctions, such as long service distinctions used in the 71st and 72nd, or the Valiant Stormer badges awarded by the 52nd to those who had survived a forlorn hope.[28] In another case—the 1/71st in the peninsula under the popular and well-respected Lt. Colonel Henry Cadogan—prizes were given both for sporting prowess and for skill in trade as demonstrated by competitions amongst the battalion's cobblers and other craftsmen.[29] Nevertheless, the value of such measures was not universally recognized, as may be inferred from an 1813 inspection of the 4th Dragoon Guards, in which the inspecting officer, Colonel Sir Granby Calcraft, discovered an odd inclusion in the men's books:

> On examining the charges against the men, I remarked an almost universal one of one shilling per month "to St. Patrick's Fund." As I conceive that no such charge is sanctioned by King's Regulations, I have felt it my duty to examine into the nature of that Fund. It appears to have been instituted under the sanction of the Commanding Officer in 1805, and was submitted to Lt. Genl. Stavely, Colonel of the Regiment, who not only approved of the measure but himself became a liberal subscriber. The Rules and Regulations were drawn up under the nominal title of St. Patrick's Fund, and subject to a certain Act of Parliament entitled "an act for the encouragement and relief of friendly societies."

The object of this fund appears to be, to grant pensions to NC Officers & Soldiers discharged after certain terms of service, and altho' sanctioned by such authority and continued so long, I have deem'd it my duty to remark on its existence.[30]

Evidently, the fund was a semiofficial way in which men who had served well could obtain some measure of subsequent relief from poverty. Equally evidently, Sir Granby does not entirely seem to have seen the point. The report says more about the limitations of the inspecting officer than about the utility of the fund; Calcraft was one of several colonels who were discretely posted away from the field army on reaching major general's rank, indicating that Wellington had little time for his services or his opinions.[31] In passing, it may be noted that the naming of the fund serves as a reinforcement of the unit's identity as the Royal Irish Dragoon Guards.

Alongside the carrot, there also existed the stick. Since the practice of corporal punishment served to make a public example of the victim, it also presented him as an object of shame before his comrades. For the shaming element of the punishment to be effective, there had to exist a bond of shared identity in the first place. Indeed, Joseph Donaldson of the 94th, using his memoirs to speak out against the practice of corporal punishment, insisted that the humiliation of a flogging was at least as degrading as the physical element. Speaking of a private of his regiment who was sentenced to five hundred lashes for stealing wine during the defense of Fort Matagorda at Cadiz, only to be pardoned at the last minute, Donaldson "did not think it any great act of favour to pardon him, after exposing him, *stripped* and *tied up* before his comrades. His back certainly remained whole; but his feelings must have been as much hurt, as if he had received the punishment."[32]

In a great many cases, the number of lashes sentenced was not the number issued, with the punishment being cut short or else never carried out at all. The fear of a flogging was widely acknowledged to be as bad as if not worse than the experience itself, to the extent that some men even attempted suicide rather than await punishment.[33] Under such circumstances, a heavy sentence remitted at the last minute—particularly if, as with the case mentioned by Donaldson, the remission did not take place until the man had been exposed before his comrades—could serve as a severe punishment in itself without the lash being used at all.

Thus, punishment, or the threat of punishment, relied for its effect in some part on the shared bonds of regimental comradeship that existed between rank-and-file soldiers, but also sought to manipulate those bonds in order to improve discipline. In 1811, for example, Private George Rider of the 1/7th was brought before a General Court Martial, along with several innocent comrades, on a charge of robbery from a house in Valverde. Rider, who was indisputably guilty, confessed in order to save his comrades from punishment; he, in turn was then pardoned. Whilst Wellington used the General Order detailing the case to condemn any crime against the Portuguese people, he also held up Rider's honesty before the court as a positive example.[34] Men could also be pardoned if their unit had performed well in action, and it was common usage that a man who went into action with a punishment pending, and who survived the action, earned the right to a reprieve. Nor did a good battlefield performance only benefit a unit's bad characters; deserving NCOs from units that had distinguished themselves might be promoted to commissioned rank as a collective reward and example to their regiment.[35]

Units that had distinguished themselves could also be singled out for praise, and held up as an example to the forces as a whole, thus helping foster a spirit of pride and emulation. Typical of this was Wellington's General Order after the action at El Bodón, which named the units and commanders concerned, detailed the services that had merited "his particular thanks," and expressed the recommendation of "the conduct of these troops to the particular attention of the Officers and soldiers of this army, as an example to be followed in all such circumstances."[36] However, it was not just battlefield heroics that could get a unit mentioned in General Orders. John Aitchison of the 3rd Foot Guards wrote home in October 1809, telling his father, "It is a real pride to belong to a corps so highly thought of by all ranks of the army— many compliments are paid in General Orders—the other day alluding to inaccuracies in Returns—'it is but justice to both battalions of Guards to state their returns have always been accurate in every particular, as the conduct of these excellent corps in the field has been regular and exemplary throughout.'"[37] Evidently, anything that added luster to the regimental reputation was worth crowing about, from conduct in the field to the thoroughness of the adjutant and his clerks.

However, if a sense of shared identity could be put to positive use in preventing misconduct within the regiment, it also led to attempts to minimize or conceal misdemeanors of which the efforts by the 1/3rd after Albuera, and by the 2/69th after Quatre Bras, to cover up the loss of their Colours are a somewhat extreme example. Although the Buffs were able to fudge the issue since the regiment had recovered the flags if not the poles, the loss by the 2/69th, only eighteen months after also losing a Colour at Bergen op Zoom, resulted in a doomed attempt to disguise the loss by producing a homemade replacement.[38]

If it was impossible to keep battlefield misadventure from public knowledge, it was at least possible to ensure that blame was firmly placed elsewhere. Aitchison's letters for some months after Talavera contain repeated assertions that, contrary to whatever his family might have heard, the Guards Brigade had been by no means as roughly handled during the repulse of the First Division's counterattack as newspaper accounts were implying. Furthermore, in his first full letter home after the battle he sought to make it clear that the fact that the Guards had fallen back at all was down to the misconduct of others, writing, "[U]nfortunately the infantry of the German Legion, which formed the left of our division, gave way and this made it necessary for our brigade to retire. When we faced about, the enemy that were flying rallied and opened a heavy fire and we were taken in our left flank by that part of the enemy which ought have been driven back by the Germans."[39] Since he could at this date have no conception of how the battle would be reported back at home, Aitchison was evidently making sure that he had his excuses in place first. It should be noted, however, that where scapegoats are provided for disaster it is almost invariably the case, as here, that the troops in question are foreign rather than British.

In a similar fashion, incidents of crime and indiscipline could also be reinterpreted before being presented for wider

consumption. In marked contrast to the pride taken by Le Couteur in the good conduct of the 104th, William Grattan's memoirs
contain several instances where the conduct by his men is passed
off as an amusing jape but in fact undoubtedly represented criminal acts. The story of "Ody Brophy and the Bacon," in which Lt.
John D'Arcy was also complicit in failing to punish a blatant theft,
stands as a particularly flagrant example of this tendency. Despite having apprehended Private Brophy rushing out of a "good-
looking house" near Robleda "with half a flitch of bacon under
his arm," D'Arcy feigned to believe Brophy's tall tale that he had
intended to purchase the bacon on the lieutenant's behalf, only
to be misunderstood and set upon by its owners. The result was
that D'Arcy paid off the aggrieved Spaniards, retained the bacon,
and let Brophy off scot-free. The tale is unashamedly told by Grattan as a fine example of the extent to which the men of the 88th
possessed the gift of the gab, or, as he put it, "had a *taking way* with
them."[40] Clearly, the hard truth would not have been appropriate in the context of the memoirs, and whilst Grattan would have
struggled to ever portray the Connaught Rangers as orderly and
disciplined, his characterization of the unit as roguish rather than
criminal is an inspired attempt to portray a shared corporate identity in the best possible light.

Over and above the issue of how the officers and men of the
British Army figured themselves and their comrades, unit identity
also had a tangible impact on unit performance, both on and off
the battlefield. During Moore's retreat to Corunna, his infantry
lost, through all causes, 22.8 percent of its strength. Yet an average
for the losses of those regiments that might be assumed from designation to identify themselves as a cut above the ordinary line—
Foot Guards, highlanders, light infantry, and fusiliers—comes
out notably lower, at 17.6 percent.[41] Battalions able to draw on
higher morale, better training, or both, as these units could, were
naturally more likely to hold together and maintain a more disciplined state, and this was noticed at the time. Even a cavalryman's
prejudice against those who did their soldiering on foot could
not prevent Captain Gordon of the 15th Hussars from asserting,
"The Guards bore up against all these hardships far better than
the rest of the army. Their ranks were always well closed, and the

battalions mustered strong; they lost comparatively few of their number by straggling."[42]

Gordon asserted that the same was true of the light infantry, but that straggling amongst the highlanders was bad. This latter distinction is not immediately obvious from the data, which would suggest that at least these stragglers, unlike many from the line regiments, retained both the intention and the ability to rejoin when they could; contemporary accounts would imply that this was also true of those who fell out from light infantry units.[43] Lest it be thought that this is an isolated example, similar trends can also be noted when a comparison is made of the effective strength of units at the end of the 1812 retreat from Burgos. The differing condition of units at the commencement of this retreat, and different unit experiences during it, prevent a direct comparison being drawn with Moore's troops, but when one examines the figures from 1812 and compares like for like, the best ratios of effectives in proportion to total strength are again to be found amongst the light infantry, highlanders, and Foot Guards.[44]

Whilst hard evidence can be used to demonstrate a correlation between assumed elite status and low levels of strategic consumption, the connection between this issue and that of combat effectiveness is somewhat harder to quantify. Indeed, there were occasions where pretensions to elite status led to unnecessary losses caused either by overconfidence or a desire to live up to the myth.[45] The overextension of the Foot Guards following the failure of the First Division's counterattack at Talavera is one example that may be cited in this regard, and the near loss of a significant part of the Light Division after it lingered too long on the wrong bank of the Côa during the combat of July 24, 1810, is another, although the latter case also has much to do with divisional commander Robert's Craufurd's overrating of his own abilities. The misadventures of the heavy cavalry at Waterloo may also be ascribed to a similar cause—its men were, in the words of Sergeant Johnston of the 2nd Dragoons, "cock sure of victory," and paid the price for that overconfidence when their charge left them vulnerable to counterattack.[46] Most telling of all is the case of the 1/42nd when serving with the "Highland Brigade" of the Sixth Division during the Battle of Toulouse. Having contrived to invert the order of companies within the battalion during its advance, Lt. Colonel Robert Macara insisted on reforming his command under heavy fire before going forward to join the remainder of the brigade in the assault on the Great Redoubt. Such matters of seniority and precedence were considered particularly important in Scots units, but the battalion suffered 414 casualties, compared to 214 in the 1/79th and 111 in the 1/91st that fought alongside it: an unnecessarily high price for maintaining the standards of the regiment. As the 42nd was also the senior highland corps, the impetus to do things in the correct manner was doubly important, requiring Macara to demonstrate his battalion's superiority over the two junior highland units forming the remainder of the brigade, as well as over the rest of the Sixth Division.[47]

The other area where unit identity has a tangible connection to unit performance is the stimulus it could provide to recruits, in particular volunteers from the Militia. These latter, being more conversant with the military life, were generally well aware of the advantages and disadvantages of particular units, and those regiments with a good reputation were able to recruit extensively from this source. The most extreme example of this is to be seen in the huge influx of militiamen into the 95th Rifles as a result of the greatly enhanced reputation with which it returned from the Corunna campaign, which saw the 3/95th—raised specifically to accommodate the influx—having no less than 1,004 rank and file on the books within a month of the battalion being established.[48] The combination of good reputation and the progressive attitude toward the rank and file inherent in their training made the light infantry corps consistently popular—the 52nd, 71st, and 95th seemingly the most so judging by the ease with which they were able to maintain strong battalions in the field—but other factors could of course play a part, and do so in favor of any regiment, line or light.

Militiamen following a popular officer commissioned into the line, or choosing to transfer to a regular corps that had been serving alongside their Militia unit, are common. The case of the Limerick militiamen who elected to join the regiment of whichever recruiting officer could beat them in a sporting contest is, however, a rather extreme example of the same trend.[49] Equally, some men seem to have chosen apparently elite units simply by chance—neither John Cooper of the 7th Fusiliers, nor the anonymous sergeant of the 43rd Light Infantry, make any reference to the particular glamour or attributes of their respective corps on the occasion of their joining, whilst William Wheeler, who ended up in the 51st Light Infantry, simply went with his friends in order to escape a tyrannical Militia officer.[50]

On the other hand, the preference of some men for service in highland units may have had far more to do with their ethnic composition than their elite status, although in this sense the two are linked. Highland regiments form a crossover between those units whose reputation arose primarily from assumed superiority manifested in distinctions of dress and designation, and those regiments that were able to cultivate a specifically national or regional identity. In this second category must be placed the majority of those units with either Irish or lowland Scots affiliations, as well as the 23rd as the token Welsh regiment and the handful of English units where the county designation was more than just a name.[51] The prime example of the latter would be the 14th Foot, which developed its connections with Buckinghamshire to such a degree that it was able to exchange titles with the 16th and assume the name of that county in lieu of that of Bedfordshire. That this re-titling was possible, however, was in a large part due to the fact that the 14th's colonel was Sir Harry Calvert, with all the weight of his status as adjutant general behind him, and the case is to some extent the exception to prove the rule so far as English regiments are concerned.[52] For Scots and Irish regiments, the ethnic identity of the units was far easier to maintain, not least through deliberately favoring the recruitment of men from the appropriate nationality. This, however, was something of a cyclical process, for as the cultivation of an ethnically specific identity made these units more appealing to men of a specific background, this in turn facilitated their ability to obtain the manpower necessary to maintain that character. Furthermore, whilst it was unlikely that any regiment would turn away a recruit able only to speak Gaelic, in either its Scots or Irish form, such a recruit would be far more likely to enlist in a regiment where his mother tongue was commonly spoken. Gaelic is known to have been spoken in the ranks of the 27th and 88th, to which the 87th with its Gaelic nickname of Faugh a Ballaghs ("Clear the Ways") might also reasonably be added, and some vestiges of the older tongue may well have survived in the ranks of those highland regiments exempted from the "de-kilting."[53]

Whilst Scots and Irish regiments littered the *Army List*, only the 23rd Fusiliers bore a distinctly Welsh—or, as the regiment would insist, "Welch"—title. Contrary to modern boundaries, Monmouthshire, the assigned county of the 43rd Light Infantry, was at this time considered part of England. The presence of Welsh speakers—as opposed to Welsh born—in the 23rd is impossible to confirm or deny. The year 1807 saw only 146 out of the 991 NCOs and men of its first battalion as being of Welsh birth, but thereafter a concerted effort was made to cultivate links with the Welsh Militia and thus obtain more recruits from the principality and develop the Welshness of the regiments.[54] Even if it were not always possible for practical steps to be taken to retain or increase a regiment's national character, regiments with pretensions to a regional or national identity took great care to cultivate it, and thus passed it on to those members previously having no connection with the locality with which the regiment was associated. Captain Thomas Browne recorded the celebration of Saint David's Day by the 23rd in a not dissimilar fashion to Grattan's account of Saint Patrick's Day in the Irish-dominated 88th.[55] These identities could also be extended onto the battlefield, as when Lt. Colonel Hugh Gough of the 2/87th ordered his bandsmen to strike up the Irish tune "Garryowen" as the battalion went into action at Tarifa.

The identity of a unit, or the way in which it was presented to outsiders, was also a factor in recruiting more generally, and both old fame and recent exploits were trotted out as part of the recruiting sergeant's patter. In like fashion, distinctions of dress always helped set a unit apart, be that seen in the green of the rifles, the busbies and pelisses of those light dragoon regiments that chose to restyle themselves as hussars, or the countless smaller conceits of lacing, braid, and badges that set the regiments of the line apart from each other. That said, for a regiment with little to boast about, and still more so if the active element of that regiment was away in the distant colonies, picking up recruits was inevitably far harder. This book is focused on active service, not recruiting, but, since responsibility for obtaining fresh manpower was delegated to the regimental level, unit reputation and unit manpower became linked in such a way that it is impossible to ignore the issue completely. So far as keeping up the numbers was concerned, success bred success for the lucky few regiments, whilst the bulk made the best of what was available and an unfortunate few fared decidedly badly. Not only did this in turn impact on how these units performed in the field—or, indeed, whether or not they ever got the chance to do so—but the decentralized nature of the recruiting and manpower system would prove to be the Achilles' heel of the whole regimental system when the demands of war placed it under the greatest strain, and would make responding to the resulting demands all the more complicated.

The Role of Leadership

Writing in response to Wellington's critical memorandum issued in the aftermath of the retreat from Burgos, in which the peninsular army as a whole was castigated for its disorders and what Wellington saw as unnecessary losses, Moyle Sherer not only used his memoirs to refute the most sweeping of his former commander's criticisms but also identified the reason that had contributed to some units having performed every bit as badly as Wellington's words had implied. Sherer's conclusions have a wider significance since, in seeking to disprove Wellington's implication that the whole army was equally guilty, he identified the key factor that explained why there could be such substantial differences between the campaign performance of units that, on the face of it, had much in common: "[T]here were corps, and many corps, who maintained their discipline, and whose casualties were comparatively trifling, and most satisfactorily accounted for. I believe the interior economy of British regiments, and the discipline of a British company, *in a regiment well commanded*, to be superior to that of any army in the world."[56]

The role of leadership was therefore just as significant as unit identity and esprit de corps, so far as the functioning of the regimental system is concerned. Whilst, as Sherer emphasized, leadership in this context extends down to the company or troop level to encompass the example set by NCOs and junior officers, the issue is primarily concerned with the role of the unit commander. After all, in a battalion with a good commander, the captain who neglected his men would find himself sharply reminded of his duty, but in a badly led unit even a paragon amongst captains could have little influence beyond the confines of his company. Recognizing the importance of this one individual, the General Regulations covering half-yearly inspections stressed that the "Commander-in-Chief should have, as far as possible, a personal Knowledge of the Merit and Capacity of Officers in the Command of Regiments, with the view to their being called forth on future occasions to Situations of more extensive Service."[57] The unit commander, in conjunction with the field officers in general, was not only responsible for the interior economy and field exercise of his command but was also tasked with ensuring that junior officers received sufficient instruction for them to be able to perform their duties. Accordingly, further provision was made for the inspecting officer to submit an additional special report on any officer who was unfit or incompetent to serve.[58]

In exercising command, it was imperative that even the best commander had the support and cooperation of his junior officers. To some extent, this did depend on the quality or otherwise of those juniors—particularly the other field officers—hence the requirement for an inspecting officer to report on them too, but it was for the unit commander to ensure that harmony reigned as much as was feasible within the officers' mess. Such harmony was one of the marks of a good regiment, with William Hay noting of his first twelve months in the peninsula, as a young ensign with the crack 1/52nd, that the time had been spent "as a child in a happy and well conducted family, for, at the time I write of, nothing could exceed the mutual kindliness of feeling which existed in that most estimable corps. Such a feeling as selfishness did not exist, and probably accounted for this excellent state of things; a stranger, entering with a different disposition, soon found himself so much out of place, that he did not remain long in the regiment."[59] Hay was also lucky enough to find a similar state of contentment prevailing when he was promoted into the 12th Light

Dragoons under the command of the "active and intelligent" Lt. Colonel the Hon. Frederick Ponsonby, so that he later turned down a captaincy in another regiment in order to remain a lieutenant of the 12th.[60]

On the other hand, commanders who either provoked dispute themselves, or allowed it amongst their officers, were censured. In 1812, Major General Sir George Anson noted of the 11th Light Dragoons that there existed a long-standing "difference of opinion" between Colonel Henry Cumming, commanding, and the second-in-command, Lt. Colonel James Sleigh. Since Anson added, "I am inclined to report, that Colonel Cumming is not easily pleased," it is evident where he elected to place the responsibility.[61] More serious in this regard were two cases where the impossibility of good relations between commander and officers led to wholesale replacements of those concerned. In the first instance, that of the 85th upon returning from its first short and undistinguished spell of peninsular service, both Lt. Colonel Henry Cuyler, commanding, and almost all the officers were removed, being considered equally bad. Only a single captain absent at the Royal Staff College was exempted from this transfer, which saw an influx of picked officers under whose leadership the battalion was soon restored to order and able to return to service.[62] However, whilst it was the Prince Regent's intervention that removed the officers of the 85th, it was also through the prince's friendship that Lt. Colonel George Quentin of the 10th Hussars was able to retain command of his regiment notwithstanding a breakdown of order almost on a par with that in the 85th. His officers, who had not only complained about his misconduct but accused him of cowardice, were all replaced en masse, even though—or, more likely, because—a court-martial had found Quentin guilty of one of the several charges they had brought against him.[63]

So far as this era is concerned, effective leadership seems primarily to have consisted of paternalism cut through with a streak of the benevolent despot.[64] Some feeling of this can be obtained from the memoir literature, which contains a variety of individuals who all justified, for a variety of reasons, the admiration and affection bestowed upon them by the men under their command. That said, quite how one defines a "good" commander depends to a considerable extent on who is doing the defining, and whilst the ideal commander might well tick both boxes it does not necessarily follow that a popular commanding officer would also be an effective one, nor, indeed, that an effective one would be popular. Lt. Colonel John Montague Mainwaring of the 51st received a decidedly mixed reception from his seniors at the time so far as his military judgment was concerned, and has continued to do so from historians, but was quite obviously held in high regard by the men he led.[65] On the other hand, there were martinets like William Inglis who were certainly competent, and may have won the grudging respect of their men, but were by no stretch of the imagination popular.

So far as we are concerned, then, a "good" commander may be considered as being one who was not only capable of competently commanding his unit in the field, but who was also able to maintain its effectiveness in all regards whilst on active service. Thus, the ideal officer would maintain discipline without being harsh or unfair, and oversee an efficiently run unit so as to keep it at an appropriate level of training and readiness for service. The methods by which these goals were achieved, however, differ as widely as the characters of the officers concerned. One can find little in common between the hard-boiled professionalism of William Inglis of the 1/57th, the charismatic eccentricity of Charles Donellan, killed leading the 1/48th at Talavera, and the paternalism evident in Henry Cadogan of the 1/71st or Sidney Beckwith of the 1/95th; each in his way nevertheless ensured that he led a first-class battalion.[66]

Whether a commander was liked as well as respected depended to some extent on whether he favored the carrot or the stick when it came to motivating his command, but might also have much to do with the manner in which he treated his men on a daily basis. The bantering manner of address adopted at times by some of the commanders who emerge from the memoirs as being particularly popular with the rank and file stands somewhat in contrast to the more reserved conduct of men like Inglis and Donellan. It is certainly hard to envisage either of these gentlemen addressing their men as "my young tinkers" as Francis Tidy of the 3/14th did at Waterloo.[67] Command also had to be tailored to the men being commanded, and the methods that worked for Beckwith with well-motivated riflemen in the peninsula led to severe indiscipline when he attempted to apply them to the ragtag force making up the 1814 expedition to Hampton Roads.[68]

Some commanders won respect through their willingness to defend their commands against higher authority and take the side of their men when the circumstances dictated it. Both Sidney Beckwith and Hamlet Wade, respectively commanding the first and second battalions of the 95th, attempted to intervene when the conflict between Robert Craufurd's high personal standards and the different ethos of the rifles led to clashes over punishments. Craufurd being Craufurd, these interventions did not always have much effect, but the attempt was noted and appreciated.[69] In a similar case, Alexander Wallace attempted to protect his battalion from what was perceived by its members as the prejudice of their divisional commander Thomas Picton, although since the 1/88th was undoubtedly a turbulent unit, these interventions do seem somewhat less valid. Justified or not, they did Wallace's standing in the battalion no harm at all.[70]

If the above gives only a general concept of the ideals of good leadership, it is rather easier to identify those occasions where such leadership was lacking. In particular, a fine line existed between the "frequent, but not severe" disciplinary practices espoused by Inglis and overuse of the lash to make up for a deficiency in actual leadership.[71] Whilst the most serious crimes, including those carrying the death penalty, were restricted to General Courts Martial, most offenses were sufficiently minor to be tried at a regimental level, where the range of punishments, though less severe, was still extensive. Provision was also made for on the spot "drum head" courts-martial to try offenses rapidly under particular circumstances, but these were covered by limitations of usage.[72] However, whilst it was in the commanding officer's interests to maintain discipline, it was also in his interest to keep every available man fit for active service, and a soldier absent in hospital after a heavy flogging was one man less to carry a musket in the line. Thus, one finds that in well-run units many sentences were either commuted completely, or else carried out only in part. Repeated instances of corporal punishment being carried out in full with no remission generally indicate that something was wrong with the unit in question. Thus, William Lawrence of the 1/40th was sentenced to 400 lashes in 1809 having absented himself for twenty-four hours, but had received only 175 of these before Colonel James Kemmis ordered "the sulky rascal [cut] down," and the punishment stopped. There is a tendency in modern accounts to accept at face value Lawrence's suggestion that this was a minor infraction and a first offense, and the punishment still too harsh. Yet considering that Lawrence had absconded whilst on active service, he was lucky not to have found himself charged with the capital crime of desertion.

Unsurprisingly, considering his personal experience, Lawrence was one of several from the rank and file—Donaldson, as we have seen, was another—to use his memoirs to speak out against the practice of flogging. However, such methods of punishment were an inescapable part of life in early nineteenth-century Britain, and it is important not to let modern sensibilities obscure our understanding of disciplinary methods that, when enacted fairly, were no better or worse than what would have been encountered in British civilian life or in most other European armies of the era. The French, it is true, had abolished flogging, but since a Frenchman would have been shot for offenses that would have seen a Briton flogged, the value of this more enlightened approach is questionable when viewed from the ranks.[73] Even Lawrence was prepared to concede that hefty punishment of a first offense "prevented me from committing any greater crimes which might have gained me other severer punishments and at last brought me to my ruin."[74] Since he ended his service as a sergeant, Lawrence may well have had a point, and his case certainly counteracts the assertion that a flogging invariably turned a man bad for good. Nor was Lawrence the only man from the ranks to accept the necessity of the lash. Benjamin Harris—who was, however, never flogged himself—went so far as to profess it a good thing and the only means by which Craufurd's column was kept in order on the road to Vigo in the winter of 1808.[75]

The retreat to Vigo represented an exceptional circumstance, however, and commanders who resorted as a matter of course to the drumhead courts-martial used by Craufurd on that occasion were wont to find themselves incurring the displeasure of higher authority. A good example of this can be found in the case of the detachment of the 2/95th serving at Cadiz, which makes it abundantly clear that, for all the enlightened training of John Moore, Coote Manningham, and William Stewart, leadership in the rifles was not always benevolent or paternal. Having inspected the detachment during the summer of 1810, Brigadier General Daniel Hoghton noted,

> This detachment . . . has been commanded for the period under review by Major [George] Wilkins. . . . The proceedings of the Courts Martial are regular, with the exception of a drum head Court Martial held under circumstances which by no means called for such an extraordinary measure. I regret also to observe that within a period of four months 22 Courts Martial have been held in this Detachment, by which 4,700 lashes have been sentenced and 2,500 inflicted. I feel it the more necessary to enforce this remark, as I am persuaded that in some instances the necessity of corporal punishment might have been obviated by a proper discrimination in the Commanding Officer.[76]

Although Wilkins's brutality seemingly failed to dent the efficiency of the men under his command, it cannot have been conducive to retaining their respect or state of morale, particularly in a regiment with a larger-than-average proportion of genuine volunteers.

To place the case in context, in the same period Hugh Gough of the 2/87th, also at Cadiz, ordered some 12,250 lashes, of which only 2,720 were actually inflicted—this in a unit with more than three times the manpower.[77] Taking the rank-and-file strengths at the time of the respective inspections, this gives an average of twenty-five lashes per man ordered in Wilkins's detachment, as opposed to nineteen lashes per man in the 2/87th, but in sentences actually carried out this translates as thirteen lashes per man in the rifles but only four per man in the 2/87th. Thankfully, Hoghton's report seems to have been acted upon, since a penciled note in the report's margin records that "Maj. Wilkins has since returned to England." Wilkins's departure may well have been encouraged by his superiors, but it did not end his career, and he subsequently served with the battalion in Spain and Flanders, including another spell in command.[78]

Just as authority condemned the overuse of punishment, so too was credit bestowed on those officers who were able to run an efficient unit without recourse to the lash. An earlier report on the 2/87th, when the battalion was in garrison on Guernsey prior to embarking for the peninsula, not only bestows general praise upon Gough's predecessor as commander, Lt. Colonel Charles William Doyle, but goes on to specifically opine that "[P]erhaps the best criterion of this officer's conduct is that in a battalion of more than 1,000 men in a quarter where spirits are abundant and cheap there has been only <u>one</u> punishment since the last inspection and no punishment in the antecedent six months."[79] Evidently, the same humanitarian approach had been continued by Gough, whose attention to the comfort and well-being of his men was noted in Hoghton's 1810 report.[80]

Several senior officers—Henry Clinton in particular, as a divisional commander in the peninsula—used their position to limit the use of corporal punishment in their commands, and in March 1812 York himself caused a circular to be issued making it clear that he disapproved of excessive corporal punishments and drawing a direct comparison between a good commander and a lack of need for the lash. Issued above the signature of Sir Harry Calvert, who as adjutant general had the ultimate responsibility for military discipline, the letter was sent to all officers commanding regiments and read as follows:

Sir,

The Commander-in-Chief judges it expedient to transmit to you, with the enclosed documents, a few observations on the salutary effects with which it is reasonable to hope that an occasional recurrence to the powers with which you are hereby vested will be attended, amongst which the most obvious advantage is that of limiting the operation of regimental courts-martial strictly to the purposes for which they are designed by the Legislature, viz., for enquiring into such disputes and criminal matters as may come before them, and for inflicting corporal or other punishments for small offences, and, in order to prevent the possibility of any misunderstanding on this important point, it is His Royal Highness's command, that on no pretence whatever shall the award of a regimental court-martial hereafter exceed 300 lashes.

The Commander-in-Chief has commanded me to take this opportunity of stating, that there is no point on which His Royal Highness is more decided in his opinion, than that when officers are earnest and zealous in the discharge of their duty, and competent to their respective stations, a frequent recurrence to punishment will not be necessary.

The Commander-in-Chief is confident the officers of the army are universally actuated by a spirit of justice, and impressed with those sentiments of kindness and regard towards their men which they have on so many occasions proved themselves to deserve; but His Royal Highness has reason to apprehend, that in many instances sufficient attention has not been paid to the prevention of crime. The timely interference of the officer, his personal intercourse and acquaintance with his men, (which are sure to be repaid by the soldiers' confidence and attachment) and, above all, his personal example, are the only efficacious means of preventing military offences; and the Commander-in-Chief has no hesitation in declaring, that the maintenance of strict discipline, without severity of punishment, and the support and encouragement of an ardent military spirit in a corps, without licentiousness, are the criterions by which His Royal Highness will be very much guided in forming his opinion of the talents, abilities, and merit of the officers to whom the command of the different regiments and corps of the army are confided.[81]

Whilst some of the above is perhaps a little optimistic, and needs in part to be seen in the context of the contemporary political campaigns against flogging, it does make clear the prevailing trend in thinking at the highest levels, particularly so far as the link between effective leadership and the need for corporal punishment is concerned.[82] Of more practical import was the restriction of the number of lashes that could be awarded by a Regimental Court Martial, although, as we shall see, the spirit of this restriction remained open to legalistic challenge.

In practical terms, heavy punishment alone was not necessarily enough to ruin the efficiency and effectiveness of a unit, particularly if the penalty was seen as being fair. Inglis's tenure in command of the 1/57th saw the revival of the old regimental nickname of "Steelbacks," a title originally dating back to the eighteenth century when a previous commander had adopted a similar approach to discipline.[83] Whether this title was bestowed in grudging admiration or adopted out of perverse pride, it implies something of a manly immunity to the ordeal of physical punishment, and a general toughness amongst those to whom it referred. Conversely, arbitrary punishments served to destroy the element of trust that was essential for effective command, and all the more so if such measures were adopted by an officer already lacking respect through other deficiencies. One such officer was Richard Archdall of the 1/40th, who gained the dubious distinction of being the only commander to be formally relieved whilst on active service as a result of having awarded illegal punishments. Yet, even in this extreme example, the circumstances were by no means straightforward.

Having served with it for the bulk of the Peninsular War, Archdall assumed command of the 1/40th as a major in April 1812, replacing Lt. Colonel Charles Harcourt who had been wounded at Badajoz, and was himself brevetted lt. colonel in August the same year having distinguished himself at Salamanca in the interim.[84] Whilst the evidence shows Archdall as particularly harsh toward the men under his command, this in itself did not necessarily make him a bad commanding officer, and would alone have been insufficient to bring about his removal. But Archdall's harshness was arbitrary and was accompanied by a bending of the rules to enable punishments to be inflicted in excess of the three-hundred-lash limit instituted by York. What was more, so far as unit efficiency was concerned, Archdall did not restrict his ill treatment to the rank and file but also took a bullying approach in his dealings with his officers. In that his court-martial was due to charges preferred by some of the latter, this attitude may well have represented his undoing since when he was tried in February 1813 only half of the charges related directly to his abuse of the punishment system. The full catalogue of charges stood thus:

> 1st. For violent and oppressive conduct unbecoming an Officer and contrary to the rules of the service.
>
> 2d. For having, on the march from Villa de Ciervo [*sic*] to Cedovem [*sic*], and at the latter village, inflicted corporal punishment on several non-commissioned Officers and private men of the 40th Regiment, without any trial, sufficient Officers being present to have formed a Court Martial.
>
> 3d. For acting contrary to the spirit, and avoiding the intentions of the late act of Parliament, limiting the sentences of Regimental Courts Martial to 300 lashes, by inflicting, at one time, the sentences of two distinct Regimental Courts Martial, held for different offences.
>
> 4th. For using intemperate and improper language to Officers of the regiment, the same being in breach of good discipline, and unbecoming the character of an Officer and a Gentleman.
>
> 5th. For divulging the contents of a confidential letter, addressed by Captain Heyland, to the late Major Colquhoun, and falsely asserting that he had opened the same in the presence of Captain Phillips and Lieutenant J. Cook, being conduct unworthy the character of an Officer and a Gentleman.
>
> 6th. For having released soldiers of the regiment, sentenced by Courts Martial to receive corporal punishment, and having permitted those soldiers to do duty in the presence of the enemy, and at other times, the punishment still impending, and afterwards put in execution, the same being highly prejudicial to good discipline, and contrary to the orders of the army.[85]

Archdall was convicted in full on the second, fourth, and sixth charges, whilst the court found with relation to the first that although he was guilty of "violent conduct" this was neither "oppressive" nor "unbecoming." Since the actions that had brought about this charge were also deemed not to be "contrary to the rules of the service," it may be inferred that they did not relate to the illegal punishments covered by charges two, three, and six.[86]

Whilst the fifth charge, although indicative of the poor relations between Archdall and his officers, was thrown out as being patently frivolous, the third caused the court some difficulty, and exposed a clear, if legalistic, distinction between York's hopes of reducing corporal punishment and the actual letter of military law: "In regard to the third charge, the Court is of opinion the prisoner is Not Guilty, as no such Act of Parliament exists in the knowledge of the members of this Court, but is of opinion, that the prisoner is Guilty of acting contrary to the spirit of the circular confidential letter from His Royal Highness the Commander-in-Chief, to Officers commanding Regiments, bearing date, Horse Guards, 25th March, 1812." In other words, Archdall was guilty so far as the spirit of the charge was concerned, but could not be convicted by the law as the court understood it. Nevertheless, the convictions on the remaining charges, and the partial guilty verdict on charges one and three, were enough to ensure that Archdall was "sentenced to be dismissed from His Majesty's Service."[87]

This judgment, one might assume, should have been the end of it, but it was clear that Archdall had in fact managed to win considerable sympathy and support for his case, with the president of the court-martial, Major General Andrew Hay, composing a plea for clemency.

> I have the honour to acquaint your Lordship that I am requested by the members of the Court Martial, of which I am President, to recommend to your Excellency's favourable consideration the prisoner, Lieutenant Colonel Archdall, as after having performed the painful part of its duty, in consequence of his having transgressed against the Articles of War, yet it most humbly begs leave to submit, that there is a conviction on their minds, that he has been actuated by what appeared to him a most zealous discharge of his duty.
>
> The Court also begs leave to submit to your Excellency's consideration, the testimonials of Lieutenant General Cole, and Major General Anson, as to the prisoner's character, as an Officer and a Gentleman, as also that of Staff-surgeon Boatflower

to the same effect, and that of Assistant-surgeon Cartan, which latter evidence has also borne testimony to his humane attention towards the sick of the 40th Regiment.

The Court cannot close their proceedings, without feeling it their duty to remark upon, and to express their regret that the fourth charge should have been brought forward by the prosecutor, for their investigation, as it appears from the evidence of Lieutenants Butler and Lunn, that Lieutenant Colonel Archdall had made an explanation to their satisfaction; the Court cannot but remark also upon the frivolous part relative to Lieutenant Richardson's great coat, as adduced in evidence in that charge.[88]

Like Quentin of the 10th Hussars, Archdall may have benefited from an element of sympathy engendered by the fact that his juniors had spoken out against him. The timing of the events in question should also be borne in mind, since the march during which the offenses forming the second charge occurred had been in January 1813, in the aftermath of Wellington's infamous Burgos retreat circular.[89] Indeed, Archdall in his defense specifically cited Wellington's orders concerning the need to rapidly punish offenders on occasions when even a drumhead court-martial was not possible. He was, however, able to establish that such exonerating circumstances had been in force on only one of the instances for which he was tried.[90] Whatever his views on the necessity of appropriate recourse to the lash, Wellington most certainly did not advocate bending the rules to make illegal punishments, but one can readily believe that his words were fresh both in Archdall's mind when his battalion marched out of Villar de Ciervo, and doubtless too in the minds of the officers forming the court that tried him the following month. Archdall certainly appears to modern eyes as a deeply flawed character, but it is evident that at least some of his contemporaries judged him less harshly.

The terms in which Hay couched his appeal suggest that, to the officers of the court, Archdall had done the wrong thing but for the right reason. Whilst the regulations clearly required that Archdall be dismissed, nowhere in the court's findings is there an out-and-out statement that he was unfit to command a battalion.

However, the clemency plea, although successful, only ensured that Archdall was permitted to sell his majority and retire; the 1/40th, having suffered less than a year under his command, would subsequently go on and serve with effectiveness and distinction at Toulouse and Waterloo. Archdall, meanwhile, having escaped the complete disgrace inherent in being dismissed from the service, uniquely elected to begin his military career again from scratch, and purchased a cornet's commission in the 3rd Dragoons. He subsequently served in India with the 11th Light Dragoons and 17th Foot before finally going on half pay as a lieutenant of the 17th in 1822, dying seven years later.[91]

Assuming the man in question was removed before he could thoroughly destroy the morale and internal cohesion of his command, a single bad commanding officer need not necessarily result in a unit becoming militarily ineffective. However, some units were unfortunate enough to be commanded by a string of such officers, under which circumstances the results could be rather more serious. A prime case of this kind comes from the unfortunate career of the 2nd Foot during the period, wherein the regiment's single battalion suffered under a succession of poor commanders. The problem seems to have begun with the first of these officers, Lt. Colonel William Iremonger, who took command of the 2nd in March 1808 having been promoted from the 88th.[92] When the 2nd, then stationed on Guernsey, was inspected shortly afterward, Lt. General Doyle found a weak and inexperienced unit whose interior arrangements were described only as "tolerable."[93] This in itself was no great indictment of Iremonger, who had only recently taken command, but it soon became clear that he was not of a character suited to turning the situation around, and future events would demonstrate that his leadership style was both harsh on the men and unpopular with his fellow officers. At this stage, Iremonger was the regiment's junior lt. colonel, and when the 2nd went on active service in 1808 it was the senior man, the Hon. James Ramsay, who commanded. However, Ramsay left after Vimeiro, and during the Corunna campaign, with Iremonger in command, the battalion largely fell apart, losing 205 rank and file or 31 percent of its starting strength. These losses were far higher than the army average, and, since the unit was never heavily engaged, can only be attributed to straggling, and in turn therefore to poor internal management.[94]

Scarcely had the battalion begun to recover when it was redeployed as part of the Walcheren expedition. Fevers took a heavy toll, and, with no second battalion, this meant that replacement manpower had to be found from outside the regiment, either by transfers from the Militia or by direct recruiting. Men were found, but the sources from which they came meant that when the battalion was inspected in autumn 1810, as it prepared for a return to the peninsula, 355 of its 793 men had one year's service or less.[95] Furthermore, the problem was not so much obtaining recruits as keeping them; in addition to finding fault with the battalion's drill and internal economy, an exasperated Lt. General John Sherbrooke, newly returned from commanding the First Division under Wellington, wrote, "Twenty Two Recruits have joined out of which Two only remain with the Reg. the desertions in the period having exceeded the number enlisted."[96] Sherbrooke also noted that the adjutant had already complained to Horse Guards concerning Iremonger's conduct, but since the latter's actions had been within regulations there was no way in which he could have been removed. There are continued hints that the 2nd was a flogging regiment, although back in 1808 Doyle too could note only that "there seems much of the old system remaining in this Regt." and that punishments were "rather frequent"; there was nothing being done illegally.[97] Even so, the evidence from 1810 of men voting with their feet is a damning indictment of how low morale in the battalion had fallen.

At the time of Sherbrooke's November 1810 inspection, Ramsay was back in command but soon left for good, and Iremonger was again commanding the battalion when it sailed for Portugal, where it conspicuously failed to distinguish itself. Events reached a nadir when the 2nd, and its commanding officer, were implicated in the failure to prevent the escape of the French garrison of Almeida and were widely criticized as a result.[98] By this stage, Iremonger had already made arrangements to sell his commission and retire; indeed, a buyer had already been found by the time of the Almeida fiasco, although news of this had not yet reached the peninsula.[99] Unfortunately for the 2nd, the new commander was an officer equally unfit for the role, as Major General Henry Clinton made plain when he inspected the battalion a year later:

> This regiment has, during the last 7 or 8 months been commanded by Lt. Colonel Lord Robert Manners. From the present state of the Regiment his Lordship does not appear to have been very well qualified for the command of it, as he is however endowed with a very sufficient capacity, I hope by his further exertions the Regiment will recover, what, if it has of last year been in tolerable order, it has undoubtedly lost. The commanding officer has been accorded that assistance from the field officers which field officers ought to afford, and which should be the least dispensed with in this case as Lord Robert Manners succeeded to the command of it without having previously served in any corps of infantry.[100]

Unfortunately, the aid of his officers was likely to have availed Lord Robert little, since Clinton went on to state of the second-in-command Major John Kingsbury that "the length of his service constitutes his principal if not his only professional merit."[101] The junior officers also incurred his displeasure, albeit in part because of the faults of their seniors: "After such a report of the commanding and field officers it is not to be expected that the necessary attention has been paid to the junior officers as to qualify them for the useful discharge of their duty. [This] makes one fear that it will be some time before they acquire those correct military habits necessary to constitute good regimental officers."[102] As a result of these failings, the rank and file, though "a very serviceable body of men in . . . age and size," were "far from being well-drilled."[103]

This, then, was the official record so far as Clinton's opinion of the battalion was concerned. However, the general also wrote directly to his erring subordinate, emphasizing his displeasure and highlighting further problems. The letter makes clear that this was not the first time that Manners had been taken to task, and that the problems causing Clinton's concern had evidently been going on for some time.

Having been called upon to make a Confidential Report to the Commander of the Forces of the Brigade of which the Regiment under your command forms a part I am much concerned to say that the Report which I have felt it my duty to make of your regiment, is of a very unfavourable nature; I have so frequently spoken to your Lordship upon the subject of the serious irregularities which I have remarked in the conduct of the Regt. that even if I had not made the Inspection of it yesterday morning, you would not have been ignorant of the points of discipline to which your attention is most immediately called. For these points I must refer you to the several orders which I have found it necessary to issue particularly those of the 24th of February the 13th and 23rd of March and the 21st of April.

In addition to the irregularities which you use to these orders I have now to remark the [illegible] inattention of the greater part of the officers who commanded Companies yesterday morning to the state of these Companies, several were totally unable to account for the men altho' furnished with states signed by themselves, without consulting with the Sergeants and a great degree of negligence evident from the State of their Accoutrements and Ammunition. It is owing to [illegible] negligence in Commanding Officers that Accoutrements are suffered to become and remain unserviceable and that ammunition is damaged and lost by which not only a heavy expense is occasioned to the public but great difficulty is found in supplying a sufficient supply of ammunition.

There then follows a lengthy diatribe about the commander's responsibility to see to it that subordinate officers, both field and company, are fit for their duties before Clinton moves on to a second issue which is not explicitly addressed in his official report.

There is only one other subject upon which I think it necessary to say a word, and that is respecting the mode of punishment which appears generally to have been used in your Regt. Considering all things, I do not think the number of Court Martials great, but I observe that Corporal Punishment, when any has been judged necessary, has uniformly been resorted to; upon any account this ought if possible to be avoided, it should be inflicted only in very aggravated cases, and when an Instance of great Severity is loudly called for, more as an example by which to deter others from the Commission of Crimes, than as an ordinary mode of Punishment. The contrary system once established, which is unfortunately too much the case in our Service, it is more difficult to abstain from such instances of severity, but this most desirable object, be assured of it, from perseverance, and by a judicious dispensation of the power with which a Commanding Officer is by our Martial Law invested, is by degrees to be attained, and as nothing will contribute more to the well-being of yours and every other regiment than such a system of discipline, I earnestly recommend it to your serious and constant attention.[104]

Although Clinton was opposed to corporal punishment generally, as well as its use in the 2nd in particular, this is nevertheless pretty strong stuff and can lead only to the conclusion that, as Iremonger seems to have done before him, Manners was turning to the lash to make up for a lack of leadership.

Manners left the battalion in August 1812, having obtained a lt. colonelcy in a cavalry regiment; doubtless this had been his true goal all along. He subsequently transferred again, back into his old regiment the 10th Hussars with which he was wounded at Waterloo. Influential friends—he was brother to the Duke of Rutland and part of the Prince Regent's social circle—doubtless did much to ensure his steady rise through the ranks, which would eventually take him to major general, but it was nevertheless the misfortune of the 2nd Foot that its one and only battalion temporarily became a pawn in the promotions game that helped get him there.[105] Manners's replacement in the 2nd was another ex-cavalryman, Godfrey Mundy from the 3rd Dragoons, who transferred because his health no longer permitted him to serve in the mounted arm.[106] However, it transpired that his health was also not up to the rigors of infantry service either, and Mundy remained on leave in Britain. This left the 2nd under the unfortunate Kingsbury, whose effectiveness was further constrained by severe wounds sustained at Salamanca, the effects of which would eventually bring about his early death in August 1813.[107] Taking all this into account, it is not surprising that by the end of 1812 the 2nd was so shrunken as to be counted amongst those battalions that were reduced to four active companies and combined into part of a provisional battalion; nor is it surprising to see that the battalion thereby created was placed under the command of the staff of the 2/53rd, the unit contributing the other four companies. That this was more than just an arbitrary decision is made clear by the fact that Wellington enforced it notwithstanding the protests of Lt. Colonel George Bingham of the 2/53rd, who had wished to return home due to wounds sustained at Salamanca but instead found himself retained in the peninsula to command the combined battalion.[108]

Two inadequate commanders in succession can therefore be seen to have contributed to the failure to turn the 2nd into an effective battalion during the period 1808 to 1813, since neither Iremonger nor Manners seems to have had the ability or the inclination to lead the unit by example. Ramsay, in whom Doyle had identified "zeal and attention,"[109] commanded only at irregular intervals and made little mark on the battalion. The one common character throughout the story is Kingsbury, who remains somewhat elusive so far as the official record is concerned. It may well be that he, as the long-serving second-in-command, bears more responsibility than anyone for the poor state of the battalion, particularly with a commander like Manners who seems to have been so disinterested in his duties, but we have no way of knowing. In any case, even if Kingsbury were in any way at fault, the ultimate responsibility for the unit still remained with its commanding officer. In all fairness, the situation of the 2nd had not been aided by the circumstances of the regiment having only one battalion, and of its having become sickly through hard service during 1808 and 1809, which meant that effective strength rapidly deteriorated on active service. All this would have been bad enough, had the regiment not also managed to lose so many of its recruits during its home service; whilst the exact circumstances of the desertions remarked upon by Sherbrooke are unclear, they are a damning indictment of the internal economy of the battalion. Finally, once the battalion was again fit, Iremonger's incompetence, and Manners's inexperience, contributed to a decidedly undistinguished battlefield career—with a further negative impact on morale—and helped ensure the eventual disappearance of the battalion as an independent entity within the peninsular army.

The case of Lord Robert Manners attaining command of the 2nd with no prior experience as an infantry officer also highlights a significant flaw of the regimental system so far as command and leadership is concerned, in that regimental seniority alone was responsible for who succeeded to command a particular unit on service. Judging by his performance at Waterloo, Manners was at least a reasonably competent cavalry officer, yet in command of an infantry battalion he was a disaster. With regard to the selection of men for higher posts, York could use his knowledge of the British Army's lt. colonels, obtained through inspection reports, in order to select appropriate candidates. At the unit level, however, he had no choice, short of removing an unsuitable incumbent from the service, but to accept those battalion and regimental commanders elevated by the system to positions of authority. Whilst Manners's case is a little different and he himself should take an element of responsibility through having purchased into an arm of service with which he was professionally unacquainted, there were other occasions where the dictates of seniority placed officers in positions for which their character, inclination, and prior experience simply did not fit them. Perhaps the worst of these men was Lt. Colonel Sir Nathaniel Peacocke, who succeeded to command of the 1/71st after Henry Cadogan's death at Vitoria, and who was one of three peninsular battalion commanders removed during the winter of 1813–14. Of the three, Peacocke was the only one cashiered, and in light of his offense—desertion of his command during the Battle of Saint Pierre, he being subsequently found some distance in the rear affecting to be encouraging Portuguese ammunition carriers—his removal seems entirely fitting.[110] Yet the most shocking aspect of this situation is not that Peacocke was so flawed a character but that a man of his obvious limitations could obtain command of a frontline infantry battalion at all.

Indeed, the system of officer appointments was so rigid that just as there were no checks to prevent the likes of Peacocke being able to obtain a position of authority, so too did the lack of flexibility in the system prevent such an individual being shifted to a position where such talents as he might possess could be put to use. Even Wellington, whilst full of anger over Peacocke's conduct at Saint Pierre, recommended only that he, and Lt. Colonel Duncan Macdonald of the 1/57th, "be removed from the command of their Regiments to some other situation in which their want of fitness will not be so detrimental to the service," it being York who instead ordered Peacocke's cashiering. The third officer to go was Lt. Colonel William Bunbury of the 1/3rd, who preempted matters by resigning his commission before he could be stripped of it.[111] Ironically, many regiments unofficially took steps to ensure that more junior officers who could not cope with the stress of combat were placed in positions where they could still serve usefully—the baggage guard, for example—but there was no like provision for the more senior grades, unlike the contemporary French army where they could more easily be shuffled into depot commands.[112]

Whilst Peacocke had not been long enough in the peninsula for his failings to become apparent, the same cannot be said of Duncan Macdonald, who had already been the subject of adverse reports. Several months prior to Macdonald's removal, Brigadier General John Byng had concluded his summary of the 1/57th by stating, "From the foregoing report, it must be clear that I am not satisfied with the attention of the Commanding Officer. It is a delicate and unpleasant duty to report anything to the prejudice of so old an Officer as Colonel Macdonald but I must be neglectful of what is demanded from me, if I do otherwise. He means well, but he is easy and wants sufficient energy to command such a battalion."[113] Byng's reference to Macdonald being well-meaning related to his unwillingness to flog soldiers even when such punishment was merited, a tendency that may be the reason that Byng noted "some irregularities in the proceedings" of Regimental Courts Martial held under Macdonald's command. Whilst Macdonald obviously had his men's interests at heart, the stark contrast between his methods and those of his predecessor, William Inglis, led to the perception of him as "easy," and, since his removal was ultimately due to looting by his battalion during the advance into France, his men evidently took advantage of his good nature. Whilst the methods of Macdonald's leadership came from the highest motives, it was clear that neither Byng nor Wellington felt him suited to the command of a battalion on active service under their command. However, although unofficial pressure could be applied, and was in this case, there was nothing that could be done within regulations to forcibly remove Macdonald or any other officer found to be unsuited to the role in which they were serving.

An obvious parallel can also be drawn between Macdonald and the case of Mainwaring of the 51st, who also incurred Wellington's displeasure and left the peninsula after perceived misconduct in action, having lost his head under pressure at Fuentes de Oñoro where he ordered his battalion's Colours to be burned rather than risk their loss.[114] Yet both Mainwaring and Macdonald, unlike Peacocke, were popular commanders and effective trainers and leaders of men, and ideal candidates for a training or administrative post. Tragically in the case of Macdonald, whose care for his men and abhorrence of corporal punishment were both to his credit but made him an unfortunate successor to the command of the "Steelback" 1/57th, the circumstances of his removal, with its implication that his failings were on a par with those of the notorious Peacocke, led him to take his own life.[115]

But if a bad commanding officer had the potential to affect the performance of a unit adversely, so too could an underperforming unit be brought back to order through the provision of competent leadership. To complete the story of the 1/71st, having gone from the heights of Cadogan's leadership to the depths of Peacocke's, the battalion next passed to Lt. Colonel George Napier, who was able to get it back into order though only through resorting, against his inclination, to heavy corporal punishment.[116] Similar treatment was needed to restore the 2/73rd in the aftermath of Major Kelly's unfortunate stint in command. By the time the battalion was inspected again in the autumn of 1814, Lt. Colonel Harris was back in command, and the consequences of his return had already become apparent. The privates were now "well drilled, attentive and well behaved," but this had been achieved at a price, as the inspecting officer, Major General Kenneth Mackenzie, "had occasion to remark on the punishments of this battalion and . . . issued an order on the subject."[117] Although the rest accorded by six months of peacetime service must have helped, Harris had quite literally whipped his battalion back into shape, with a series of heavy sentences being carried out in full with no remittance. The cases of the 1/71st and 2/73rd both emphasize that there were occasions when leadership had to take a heavy-handed form, to ensure that men who had been allowed to grow slack received a reminder of just where the ultimate authority lay. However, the circumstances in each of these instances were unusual in that a poor commander had allowed a good unit to decline and, as Napier noted in the case of the 1/71st, "Had I been the cleverest commanding officer in the army it would have been impossible for me in so short a period, and under the circumstances of the time, to have made the progress I did had it not been for the former system established by Pack and Cadogan, which I found was still in the regiment, but, from the conduct of [Peacocke], had been totally neglected and almost forgotten; and I soon saw that I had only to be firm, and strict in enforcing my orders but with calmness and temper, and that I should in a short time bring all back to its former splendour."[118] Such methods, therefore, could only ever be effective if applied fairly and when used to restore, rather than enforce, discipline.

Although the sharp shock of a heavy hand might well be needed when a battalion had grown slack and undisciplined, in other instances the simple provision of new and effective leadership was sufficient to restore a unit to effectiveness. In this manner the 2/89th, after an unfortunate career in Europe, subsequently went on to distinguish itself in North America. Many of the initial problems with this unit stemmed from outside circumstances, for it bore the brunt of the disastrous Fuengirola expedition, where it lost forty dead and over two hundred prisoners.[119] However, there are also suggestions that things were by no means in a good order before the battalion went on that service, in part because of the extreme youth of many of its rank and file.[120] Thereafter, back in Gibraltar, things were made worse by the dispatch of a number of officers to the first battalion in India, which not only removed the remaining company-level officers but also cost the services of Lt. Colonel William Sewell, who had already begun to restore some order to it; his replacement was Major Miller Clifford, newly "promoted from the 28th Regiment."[121] Whilst Major General Sir Montague Burgoyne felt that the battalion was unfit for any duty at this time, his superior, Lt. General Colin Campbell, begged "leave to differ in opinion as to the unfittedness of the battalion for Service. For their numbers they are extremely good, and at least three hundred [of 478 total] are fit for any Service whatever. The men generally are stout and young, and the Boys will improve with care and attention."[122]

Clifford, a veteran notwithstanding his lack of seniority, at least had some good material to work with, but was still handicapped by lack of officers.[123] When Burgoyne next inspected the 2/89th, he again felt obliged to report that it was in a bad way, albeit this time rather more circumspectly in light of Campbell's correction of his previous findings. Nevertheless, he still affirmed that Clifford himself was "a steady zealous officer [who] has paid great attention to his Battn."[124] In the spring of 1812, the battalion finally received a new permanent commanding officer in the shape of Lt. Colonel Joseph Morrison, and was transferred away from Gibraltar, eventually, after a spell in Britain, being redeployed to Canada. Morrison, a commander both liked and respected by all ranks, completed the work begun by Clifford by restoring the battalion to a high standard of both training and morale, as evidenced in particular by its impressive performance at the Battle of Crysler's Farm, where Morrison, commanding the British forces, and Clifford, commanding the 2/89th in his absence, both distinguished themselves.[125] By 1814, when the battalion was distinguished again at Lundy's Lane, its reputation was such that it was described by its surgeon as composed of "wild tremendous Irishmen"[126]—a far cry from the leaderless collection of youths that had dismayed Burgoyne three years previously. Whilst an influx of recruits had obviously helped restore the battalion, it is also evident that good leadership had played a part, not only from Sewell, Clifford, and Morrison, but also from the company officers once sufficient of these were provided. By 1814, the latter had an average of twelve years' service for the captains and six for the lieutenants, and were clearly well set up to use that experience to good effect.[127]

As concepts of regimental tradition and identity grew and changed throughout this period, the exploits and characters of unit commanders became intertwined with that sense of identity, and further fostered its growth. Yet although, for the infantry, these commanders were operating at the battalion level, there does not seem to have been any great development of discrete battalion identities; rather, all exploits were absorbed into a regimental whole. Expressions of sorrow at transferring from one battalion to another are encountered in the body of memoir literature, but these generally represent officers, such as Grattan or Sherer, being unhappy at leaving an active unit with a chance for distinction for a battalion serving at home or in the colonies. Conversely, from the rank-and-file perspective, John Cooper was certainly not pleased that his status as a lance corporal was insufficient to prevent his transfer to the 1/7th in 1811, rather than going home with his friends in the rump of the second battalion, but this would seem to have been due to "the mortification . . . of seeing my old comrades march merrily from the camp for old England."[128] Clearly, and self-evidently, the desirability of active service had much to do with what one stood to gain or lose through participation in it.

In short, the regimental system during the period from 1808 to 1814 was responsible for a strong, and growing, sense of identity that, when fostered and developed through good leadership, created an extremely solid and durable esprit de corps. That this contributed a great deal toward the effective performance of the British Army cannot be denied, but due to the organizational deficiencies of the system it became necessary to rely on the moral over the physical to a great—at times a too great—degree. The most telling of these deficiencies was that the regiment remained the focus for manpower organization for both officers and rank and file, and, if in the former case the opportunity existed to transfer or purchase into a different unit, this did not prevent the inexorable rise, through seniority, of some individuals who were not fit for command. For the system to work, roles that required leadership needed to be filled by men capable of leading, and yet the same system was unable to ensure that this was the case. Organizationally, a proliferation of small regiments, each responsible for its own recruiting, rendered the effective distribution of man-power to meet global commitments increasingly problematic, and also limited the effectiveness of some of these units on active service. Of course, much of the problem was due to the established system addressing Britain's usual policy of small expeditions of short duration and limited objectives, rather than the prolonged and global conflict in which the nation and its army found themselves by 1808. Accordingly, whilst the regimental system had a largely positive impact in creating the British Army's sense of identity, it was also necessary, as we shall see in chapters three and four, to institute increasingly creative policies of unit rotation and manpower management in order to maintain the force as a functioning whole.

CHAPTER 3

The Regimental System in Practice

By 1808, the 20th Regiment of Foot was one of the most experienced units of the British Army, yet, when Major General Coote Manningham inspected the regiment's single battalion that May, he found deep cause for concern. There was certainly no problem with regard to leadership. Lt. Colonel Robert Ross, the future victor of Bladensburg, had been commanding since 1803 with good effect, and Manningham spoke highly of the seasoned officers and NCOs. However, many of the men in the ranks of the 20th had joined in 1799 as part of the great augmentation through which the regular army had been rebuilt by volunteers from the Militia after years of attrition on colonial expeditions. Back then, there had been men enough to form two strong battalions, but after Holland, Egypt, and Maida, only a few hundred were still fit for service, and the second battalion was long gone. In his 1808 report, Manningham therefore went on to suggest that, "[I]f H.R.H. the Commander in Chief would allow them to get 300, or 400, men from one of the militia Regiments, that have still men to give, they would undoubtedly become one of the finest Corps in the British service—but from not possessing the advantage of a second Battalion, they have no means beyond the common mode of recruiting, in which they have hitherto been very unsuccessful, of getting men."[1] Sadly for the 20th, there was no opportunity for this advice to be acted upon before the unit was again sent on active service, and it was not until after returning from Corunna that the battalion received any augmentation in its strength. Thereafter, the rigors of Walcheren served to render the battalion unfit until 1812 when it rejoined the peninsular army a little under

700 strong. With no further replacements available, rank-and-file strength had dropped to 561 at Toulouse, and, of these, only 288 were still listed as effective. Another campaign would surely have finished the 20th as a battalion fit for service, and the unit was in no state to accompany its old commanding officer to North America, nor to serve in the Waterloo campaign.[2]

The case of the 20th makes it abundantly clear that although there were many positive attributes to the regimental system, in terms of its fostering identity and esprit de corps, there were also obvious drawbacks so far as manpower management was concerned. With no opportunity to redistribute resources at an army level, some units languished understrength whilst others had more men than they required. The formalization of the regimental system, in particular with regard to the duties expected of first and second battalions, can now be seen as placing unnecessary limitations on the effective employment of manpower. This view has frequently been espoused by historians, to the extent that the impression that one gets from Oman and those who follow him is that organizational practices were inflexibly set in stone.[3] In reality, however, no hard and fast ruling was ever laid down, and the system was always manipulated to a certain degree in order to maximize its efficiency.

In fact, it is readily apparent that both Horse Guards and local theater commanders knew exactly how to work with the system, making use of its strengths and finding ways around its weaknesses even if this meant that the organizational ideal based on a two-battalion infantry regiment could not be followed. New organizational practices were developed and became close to doctrinal norms, but these methods evolved over time and represented a desire to work with the regimental system rather than being inherent in the system as such. Such methods, however, could only achieve so much, and the alternative solution was to implement measures that retained the spirit of the concept behind the formal regimental system but dispensed with those elements that could not be practically applied. In particular, this led to rotation of battalions between active, semi-active, and inactive commands in order to prolong the amount of time that they were able to remain effective. As opposed to working with the system, innovations of this sort can be seen as a means of making the system work.

Working with the System

In the earliest years of the Napoleonic Wars, the theoretical system of battalion deployments largely worked, with overseas duty being shouldered either by the first battalions of multi-battalion regiments, or else by the single battalion of those regiments that did not yet possess a second. Very few second battalions were therefore required to serve overseas; there were only eight instances between 1803 and 1807, and only four such units actually saw action.[4] In three of these early deployments, those of the 2/18th, 2/56th, and 2/62nd, the second battalion was sent to the station where the first battalion was already serving. This, then, was the two-battalion regimental system working largely as envisaged. However, from 1808 it became necessary to send second battalions overseas in growing numbers. Initially, this measure was resorted to in order to fill out the last batch of reinforcements for the peninsula, but whilst these deployments may well have been intended as a one-off measure to achieve a rapid resolution of affairs in that theater, it subsequently became necessary for second battalions to continue to serve in numbers as a result of so many first battalions being rendered temporarily hors de combat as a result of their experiences at Corunna and Walcheren.

In all, of the sixty-six regiments of the line that achieved the organizational ideal of two battalions, forty-six of them—just over two-thirds—deployed both battalions overseas during this period. In addition, the 4/1st, 3/14th, 3/27th, 3/56th, and 3/95th all also went on active service as the junior battalion of their respective regiments. Although many junior battalions went to garrison postings to render larger or more effective units available for campaign service, a significant number were sent to active theaters of war. Generally, deployments by second and third battalions were of short duration due to the difficulties inherent in a single regimental depot having to maintain two or more battalions overseas. There were occasions when junior battalions chalked up long periods of active service, as with the 3/27th and 2/83rd in the peninsula, but it was more typical for such a unit to last for a campaign or two and then be either shifted into a garrison role or else sent home. Alternatively, if more than one battalion of a regiment were serving in the same theater, the junior unit could draft its effective manpower into the senior, and go home to be built back up to strength. In an extreme example, the 2/52nd was deployed on no less than four separate occasions—the peninsula in 1808, Walcheren in 1809, the peninsula again in 1811, and Holland in 1813—totaling thirty-three months of active service in all. The battalion returned home intact in the first two instances, and on subsequent occasions as a cadre having given over its effectives to the 1/52nd.[5]

Irrespective of whether they represented the first, second, or only battalion of their regiment, all battalions would in any case leave a cadre behind to accommodate any officers and men unfit or unable to accompany the unit, and if all the active battalions of a regiment were overseas, then provision would also be made for the formation of a recruiting company or companies to take over the depot role. Thus, the 1/43rd embarked for the peninsula in 1808 leaving behind at Colchester two recruiting companies each composed of two sergeants, one drummer, and ten rank and file, and an additional depot detachment comprising six sergeants, one drummer, twenty effective rank and file, and ninety-four assorted sick. The depot detachment was in this instance commanded by a captain, assisted by two ensigns.[6] On this occasion, the sick could not be drafted into the second battalion, as would ordinarily have been the case, since the regiment was strong enough to deploy both of its battalions and, unusually, the second battalion had in fact gone overseas before the first.

More typically, when a second battalion was called upon to serve, the first would already be overseas, and on these occasions the junior unit would be forced to leave in Britain not only its own ineffectives but any that might previously have been drafted into it from the first battalion as well. Inasmuch as most junior battalions were on lower establishment strengths in the first instance—although this might be raised once the unit was slated for active service—the result was that they would generally be committed to service as relatively weak units. For example, of the seven second battalions sent out to the peninsula in 1809 only the 2/34th, with 996 rank and file, was comparable in strength to a typical first battalion; the remaining six units had an average strength of 622, with the strongest having only 664 men and the weakest 535.[7] For a typical single-battalion regiment, having only one set of ineffectives to leave behind, the situation was somewhat more tolerable, but, as we have already seen with the 20th, many of these small regiments experienced considerable difficulties in recruiting, and few were able to keep up to strength over prolonged foreign service.

Whilst the bulk of regiments either had to deploy their second battalion, or were never able to raise one, there were some twenty regiments that did function exactly as per the theoretical system, with a second battalion that never left the home station and served only as a depot-cum-feeder unit for the active first battalion. The bulk of these regiments are found amongst those whose first battalions formed the original disposable force of 1808 and went out to the peninsula with Wellesley or Moore. For the most part, the first battalions then either remained in the peninsula through to Toulouse, or, as was the case for most, went through Corunna and Walcheren before rejoining Wellington between 1810 and 1813. Hard service by the senior battalions meant that in the bulk of these cases all available manpower was directed there, maintaining the first battalions at a good strength but firmly confining the second battalion to the home station. Even amongst those regiments that missed Corunna and Walcheren, other misadventures created manpower problems in the first battalions, as with the heavy exposure to sickness suffered by the 1/40th early in its peninsular career, or the massive Albuera casualties of the 1/3rd and 1/57th. Therefore, far from representing the triumph of the system as designed, these twenty regiments are in fact somewhat atypical, and the fact that their second battalions were never actively deployed is indicative more of the poor relative effectiveness of their first battalions than of a desire to deploy units in accordance with any particular theoretical scheme.

Indeed, although working in this way ensured that the first battalions involved could remain in the field until 1814, and still be in a fit state for further service in North America and/or at

Waterloo, two obvious drawbacks ensued. Firstly, this situation was only achieved by permanently tying the second battalions to a home depot role, thus removing them from the pool of battalions potentially available for overseas service. Secondly, the poor state of a significant proportion of the manpower serving in the first battalions meant that these units had a very high proportion of noneffective manpower, so that although they remained numerically strong in absolute terms their potential battlefield strength was far lower. The experiences of the 40th Foot exemplify this point. In August 1808, the 1/40th, having returned somewhat ignominiously from the failed South American expedition, was part of the force assembled at Cork under Wellesley with a respectable strength of 926 men.[8] After Roliça and Vimeiro, October 1808 found the 1/40th still with 908 rank and file on strength, but of these no less than 241 were sick. This sickness rate amounted to 26.5 percent of the battalion's strength, as opposed to an average of only 10.4 percent in the force as a whole; no battalion had a greater actual number of sick, and only the battle-scarred 29th and raw 2/43rd were worse off in relative terms.[9] Lawrence was only exaggerating slightly when he asserted that the whole battalion "fell ill and was obliged to be returned unfit for service, which state of things lasted about two months."[10]

There was certainly no hope of the 1/40th being made fit in time to go into Spain with Moore, but it was ultimately detached, between February and May 1809, to help garrison Seville.[11] This respite allowed the battalion to recover its effective strength by the time it rejoined the main army for the Talavera campaign, but the 1/40th was nevertheless distinguished throughout its peninsular service for being an unusually sickly unit, and this largely accounts for the constant need for manpower from the 2/40th, which was received in the form of two major transfers in March 1810 and July 1811 as well as numerous smaller drafts. It may also be noted, in the context of the 2/40th's role as a source for manpower preventing it from going overseas, that the junior battalion had one of the worst desertion records to be seen amongst units stationed in the British Isles.[12] This naturally limited the availability of manpower to be drafted into the first battalion, and shows how problems with internal economy within one battalion could have an impact on the regiment as a whole even when encountered in those elements not on active service.

When the state of the 1/40th over the period in question is presented graphically, as in figure 1, the high levels of sickness, and resulting low effective strength in percentile terms, make it clear why all available manpower was needed in the first battalion and could not be spared to build up the second. However, simply packing more and more men into the 1/40th did not represent an ideal solution, and sickness in the battalion increased in proportion to its total manpower. Even the huge reinforcement draft of July 1811, which took total strength well over establishment to nearly 1,500 rank and file, only briefly raised the effective strength of the first battalion. Hard service through the winter of 1812, and heavy losses at Badajoz, soon had the figure for effectives back down even lower than it had been, with less than one-third of the battalion fit for duty as of April 25, 1812, and the situation remained poor throughout the rest of the year. Matters eventually improved during the first half of 1813, during the lull between the retreat from Burgos and the march to Vitoria, but the huge discrepancy between total and effective manpower continued throughout the rest of the Peninsular War, and was only truly eradicated when the 1/40th was redeployed, minus its ineffectives, to North America and then to Flanders. Whilst the 1/40th is an extreme case, and a little unusual in that it remained overseas and active throughout the period, getting the fever into its ranks via the marshes of the Tagus and Guadiana rather than the Walcheren swamps, similar trends may be identified in those units that owed their sickly condition to service with Chatham's ill-fated Grand Expedition.

Along with senior battalions of multi-battalion regiments, single-battalion regiments formed the other, secondary, group of units that the theoretical regimental system prioritized for active service. For the most part, regiments that remained at a single battalion establishment did so because they spent their war in the colonies and returned to Britain only toward the end of the period if at all, thereby never being in a position to be augmented by a second battalion. Nevertheless, some single-battalion regiments were employed more extensively in Europe, with eleven going to the peninsula between 1808 and 1814. Many, like the

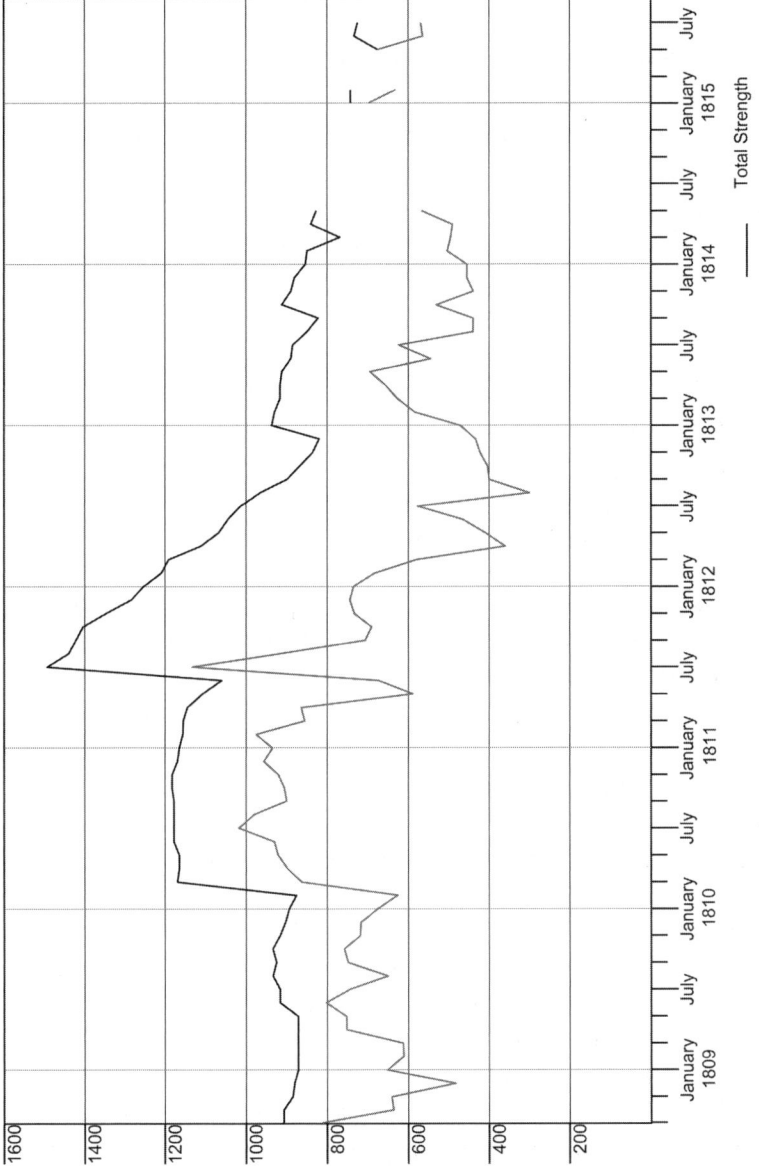

Figure 1. Campaign Manpower in the 1/40th. Data from Monthly Returns, TNA, WO17/2464–2465, 2467–2476, 1218, 1760, and Battalion Returns, TNA, WO17/150. The two blank periods represent firstly the passage from Bordeaux to the Mississippi, and secondly the return passage back to Europe.

20th, served two distinct stints in Spain and Portugal, and seven also took part in the Walcheren expedition. The problem with such deployments was in keeping the battalions up to a good effective strength, and, since it was in most cases lack of successful recruiting that kept these regiments down to one battalion in the first place, the difficulties involved were considerable. The 2nd, whose woeful story has already been told in the previous chapter, was cut down to four active companies as part of Wellington's project of provisional battalions, and two more single battalions, the 51st and 68th, were also considered for the same treatment. The remainder were mostly sent home as understrength, or else escaped this fate only by virtue of having come out to the theater in 1812–14 and thus not being on active service long enough to become completely worn down.

For the 2nd, poor leadership led to limited effectiveness, but this logic can hardly be applied to the 20th, with which we began this chapter. Nor—unless one includes the consequences of an ill-judged decision on the battlefield—can any failing of leadership be blamed for the difficulties encountered by the 29th, which, unlike the 2nd and 20th, remained in Portugal in 1808 and thus became part of the re-formed peninsular field army under Wellesley. However, unlike the 1/40th, it was not sickness but battle casualties that had rendered the 29th unfit to march into Spain with Moore. At Roliça, Lt. Colonel the Hon. Gerard Lake had unwisely led the battalion in an unsupported attack that sought to access the French position via a concealed gully. The stratagem was successful in enabling the 29th to penetrate the French first line, but left the unit confined and vulnerable to counterattack; assailed from all sides, the battalion was badly broken, with Lake amongst the dead.[13] Largely as a result of the 190 casualties incurred in this action, the 29th was, within a month of coming ashore, down from its embarkation strength of 863 men to only 592 effective rank and file, with a further 148 sick or detached.[14] Although substantial drafts of manpower were received from home in the summer of 1809, and again during the spring of 1811, hard service over the same period meant that the general trend of the battalion's strength was consistently downward, even as the experience and toughness of the survivors grew.

Moyle Sherer, then newly arrived in the peninsula, saw the surviving men of the 29th in the aftermath of the Talavera campaign; they were the first veteran battalion that he had encountered, and the sight impressed him greatly: "Nothing could possibly be worse than their clothing; it has become necessary to patch it; and as red cloth could not be procured, grey, white, and even brown had been used: yet, under this striking disadvantage, they could not be viewed by a solider without admiration. The perfect order and cleanliness of their arms and appointments, their steadiness on parade, their erect carriage, and their form and free marching, exceeded any thing of the kind that I had ever seen. No corps of any nation, which I have since had an opportunity of seeing, has come nearer to my idea of what a corps of infantry should be, than the old twenty-ninth."[15] Wellington too saw the value of this veteran battalion, but realized that it could not be kept in the field much longer unless something be done to augment its ranks. Indeed, Wellington had already written to Castlereagh, then still secretary of state for war and the colonies, in order to draw the matter to his attention: "I wish very much that some measures could be adopted to get some recruits for the 29th Regt. It is the best regiment in this army, has an admirable internal system, and excellent non-commissioned officers; but for the want of a second battalion, and somebody to attend to its recruiting, it is much reduced by numbers, by losses in the actions of Rolica and Vimeiro, in the expedition to the north of Portugal and at Talavera."[16] As well as giving an interesting view of the criteria by which Wellington judged his battalions, this passage makes it apparent, even as early as 1809, that he was already finding fault with the constraints of the regimental system insofar as it interfered with his hopes of keeping every veteran solider as long as possible in the peninsula.

For the 29th, however, little could be done. February 1811 saw an augmentation of one hundred additional rank and file from the regimental depot, but this was of little avail after Albuera, where the battalion formed part of Colborne's unfortunate brigade. The aftermath of that disaster found the 29th with a mere 144 effective rank and file, and by September it could still only put 193 effectives into the field, forming part of the Provisional Battalion (see chapter 4) with elements of the equally cut-up 2/31st and 2/66th.[17] The situation was not sustainable, and in October 1811 the 29th left the peninsula after just over three years of service. The General Order that announced the battalion's impending departure, along with that of the 85th and 97th, again stressed the battalion's good service record but also made two points of wider significance. The first was to make it clear that Wellington had sent these units home only under duress and as the result of a direct order, something that would prefigure the tussles over numerically reduced battalions over the next two years. The second, which again reinforces the want felt as a result of all three regiments lacking a second battalion, was that "the only chance of recruiting is to send them to England,"[18] because a depot detachment simply did not possess the manpower to recruit on the scale required.

Figure 2 demonstrates just how bad the situation had become, but the story does not end there, for there was time enough for the battalion to be rebuilt for further service with its ranks filled up with fresh recruits. The 29th was, like the 20th after Walcheren, ultimately built back up to a respectable strength, and in March 1813 went out to Gibraltar 713 strong. Thereafter, the battalion served successively in garrison at Gibraltar and Cadiz before joining Sherbrooke's force in the relative backwater of the Maritimes and Maine. On the face of it, this restriction to secondary theaters seems odd for a unit possessing such a good prior record, but the influx of outside manpower during the eighteen months spent at home had vastly diluted the quality of the battalion, rendering what had once been Wellington's "best regiment" unfit for anything other than garrison service. When Major General Cooke inspected the battalion shortly after its arrival at Cadiz, he found that "[The privates] cannot throughout be called a good body of men, and the Regt. is not at present fit for service in the field. They are very young and several of a low standard, particularly a bad description of recruits from the Irish Militia." However, Cooke went on to state that the rank and file were "well-drilled and appear attentive and well-behaved, and as sober as any other soldiers I have known at this station."[19] What Cooke's assessment indicates, therefore, is that the 29th had been posted to Cadiz because by this stage of the war it was an inactive garrison, where the battalion

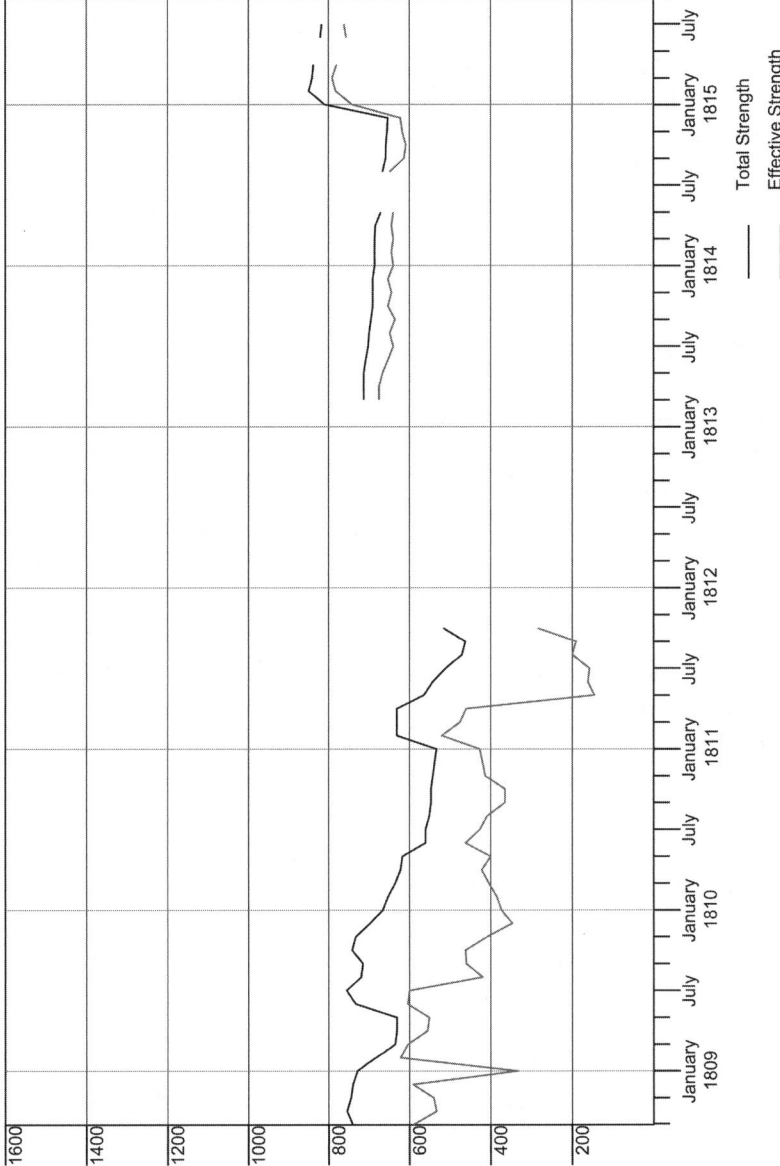

Figure 2. Campaign Manpower in the 29th. Data from Monthly Returns, TNA, WO17/2464–2465, 2467–2468, 1799, 1488, 2242–2243, 1760. The large gap in the center of the graph represents the time that the battalion was back in Britain, and the two smaller gaps toward the end relate to its passage to and from North America.

could complete its rejuvenation whilst still doing something useful in a secondary station. Only in 1815, after two years in second-line duties, was the 29th again considered fit to join a major field army; as it was, however, the battalion reached the Netherlands after Waterloo, and never got the chance to prove itself in action.

Because units like the 29th lacked any second battalion to serve in the depot role, not only did they have to take what they could get in terms of reinforcements but they also lacked the organizational infrastructure to process new manpower before it reached an active battalion. This in turn meant that a worn-down single-battalion unit was unavailable for active service whilst it absorbed and trained its replacements, needing instead to spend this time either at home, as with the 20th, or in an inactive garrison, as with the 29th. However, whilst these two units typify the experience of single-battalion regiments serving in Europe, the majority spent much of their war in the colonies. Of these twenty-one battalions, five spent the entire war in India, whilst the unfortunate 46th left there in 1813 only to be posted even further from home, to New South Wales. The bulk of the remainder were in similarly long-term postings to the West Indies and Canada, but were then drawn into more active service when 1812 brought renewed hostilities to the North American continent.[20] Although few saw service as long or as arduous as their peninsular equivalents, their experiences reflect, on a smaller scale, those already exemplified above. After the Niagara campaign of 1814, for example, it became necessary to send the two longest-serving single-battalion regiments—the 100th and 104th—away from the front and into garrison due to the level to which they had shrunk.[21] The latter unit suffered a steady decline from 908 rank and file in March 1813 to 572 when hostilities ceased, with effective strength in the field generally well below 400. Although this does not account for two companies left in garrison throughout, nor for the fact that the flank companies were also frequently detached, it is nevertheless a similarly low ratio of effective manpower to that found in the battalion's peninsular equivalents.[22]

Having considered the typical circumstances of those battalions that were intended to go on active service, and the problems they encountered, it is also necessary to look at those that were not originally envisaged as suitable for this sort of role—the second battalions and the handful of third and fourth battalions. Whilst some junior battalions did serve in North America, the bulk of these deployments were within Europe, and can be broadly divided into two groups. The first group comprises the battalions sent out to Portugal during 1809 and early 1810, which, along with some single-battalion regiments deployed at the same time, represented the only effective units then available. Although a few other second battalions did find their way into active theaters, having been moved from various garrison commands, the second major group comprises those battalions sent on service as part of the seemingly desperate attempts to deploy all available battalions, second, third or otherwise, as manpower systems began break down during the last twelve months prior to the peace of 1814. More relevant here, whilst we are still concerned with the workings of the regimental system rather than its temporary circumvention, are the experiences of the first group.

For the most part, the second battalions deployed to the peninsula in the early months of the war were those from regiments whose first battalions were deployed to the colonies, and whose regimental establishments were accordingly set at a high level. As such, the battalions in question had a not dissimilar experience to the single-battalion regiments deployed at the same time and in similar strength. Only the 2/83rd was able to serve throughout the entire Peninsular War, the battalion's manpower situation being detailed in figure 3, which shows a rapid decline in strength followed by an eventual stabilization with an effective strength of around 350. Whilst the maintenance of the battalion as an effective unit is a worthy achievement, there were certain factors that helped the 2/83rd escape the fate of its counterparts. Firstly, the 83rd Foot was, in 1809, on the highest establishment of any two-battalion infantry regiment in the service, 2,461 men all told, and therefore had a greater allowance of manpower at its disposal. Secondly, the 2/83rd benefited from a lengthy spell of garrison duty in Lisbon after the Talavera campaign, during which its effective strength grew by over 150 men.[23] Without these advantages, it must be assumed that the 2/83rd would have suffered the same fate as its sister battalions, and, at best, finished the war as one half of a provisional battalion. That said, its experiences are not too dissimilar to those of the 3/27th, which, as the junior battalion of a three-battalion regiment, served effectively not only throughout the Peninsular War but also in North America thereafter. Yet, with an established strength of 3,448 men in 1809, and a strong regional identity and accordingly strong recruiting potential, the 27th Foot was in a good position to sustain all three of its battalions. This was made easier by the fact that, whilst all three were overseas, the 1/27th and 2/27th were part of the Sicilian garrison and did not see active service until sent to eastern Spain in 1812.[24]

Part of the reason that so many of the initial crop of second battalions sent to Portugal in 1809 served for so long was that the first battalions of their respective regiments were not available to relieve them. Only in the case of the 2/5th, 2/7th, 2/28th, 2/39th, and 2/48th did the senior battalion also join Wellington and thus facilitate the eventual replacement of the junior battalion by the senior. Ideally, it had always been preferred to have both battalions of a regiment serving in the same theater, so that the second could be drafted into the first once it became impossible to sustain two separate units, but haphazard early wartime deployments meant that this rarely happened, and it took some time to get things unraveled. As late as 1814, York was planning to send the 1/83rd, which had been at the Cape of Good Hope since 1806, directly to join Wellington in order to combine it with the shrunken 2/83rd so that the latter could finally come home and recruit, and was also looking forward to bringing all three battalions of the 27th together in the same theater for the same reason.[25]

Ideally, the solution was either to deploy both battalions together, or else send the second battalion to join the first. Thus, even before the War of 1812, both battalions of the 8th Foot were serving in British North America, whilst one of the earliest reinforcements for Canada brought the second battalion of the 41st Foot out to join its senior counterpart. Although the 1/8th and 2/8th served to the end of hostilities as separate units, the presence of both battalions of the 41st facilitated their combination after the first battalion's heavy losses at the Thames on October 5, 1813.[26] From a total strength of only 178 in December 1813, the 1/41st could therefore be brought up to a far more respectable

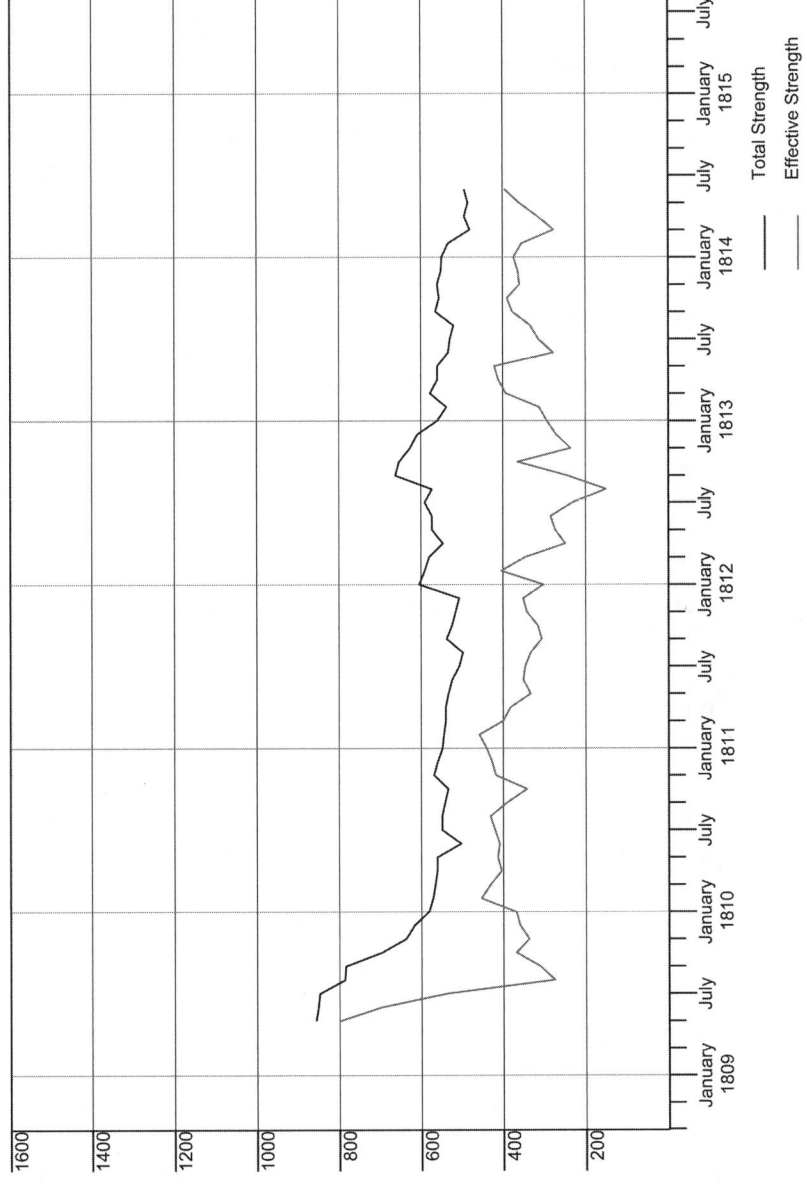

Figure 3. Campaign Manpower in the 2/83rd. Data from Monthly Returns, TNA, WO17/2464–2465, 2467–2476. The entire span of service relates to the Peninsular War.

561 rank and file the following month, having absorbed the 379 men of the second battalion in their entirety.[27]

Lengthy side-by-side service of both battalions of a regiment was generally confined to colonial or garrison stations. More typical of the European experience are the cases either of a second battalion serving in a theater being joined by, and rapidly absorbed into, its first battalion upon the latter arriving on station or, conversely, of a second battalion being sent to serve alongside the regiment's first for a few months before giving over its effective strength and taking a cadre home to recruit. The latter circumstance, in effect, represents the system working normally; even the strictest interpretation of the regimental system allowed a strong two-battalion regiment to deploy both battalions, the second returning home upon the combined strength falling below 1,600. Wellington tended to hang onto his second battalions rather longer if he could get away with it, but the concept remains the same and is encountered repeatedly throughout the peninsular and Waterloo campaigns.

The history of the 88th provides perhaps the best example of this practice, since it clearly shows how both battalions were affected. The 2/88th arrived in Portugal in July 1809, four months after the 1/88th, which was itself a creditable achievement since the former had been extensively drawn upon to rebuild the latter upon its return from South America. However, as a result of its rapid augmentation, the 2/88th was not fully effective and was accordingly sent away to do garrison duty, first at Gibraltar and then at Cadiz, before returning to Lisbon in August 1810.[28] From there the battalion was posted to join the 1/88th in the Third Division with effect as of September 4.[29] This transfer reflected a growing practice of keeping battalions of the same regiment together in the same division, which was already the case for the 48th and which was also applied, a little later, to the 7th. The logic was clearly to facilitate the eventual merger of the battalions if required, although it was not then felt necessary, as later, to place them in the same brigade.[30] Whilst the 7th and 48th were both badly cut up at Albuera and had their battalions amalgamated thereafter, the 88th soldiered on with two battalions in the field until the summer of 1811. By this stage sickness, exacerbated by the long march south to participate in the second siege of Badajoz, left the young and still relatively inexperienced 2/88th with only 274 effectives as opposed to 362 sick or detached. The 1/88th, with 746 rank and file on its strength of whom 585 were in the field, was still an effective unit but could clearly benefit from augmentation and so the inevitable merger took place.[31]

The General Order of July 10, 1811, explains the mechanics of how such an amalgamation was administered:

3. All the Private men belonging to the 2d Batt. 88th regt. in Portugal & Spain, are to be drafted into the 1st batt., and are to be distributed into companies of the 1st batt., with the exception of boys deemed at present unfit for service in the field.

4. All men missing from the 2d batt. Are to continue on the strength of the 2d batt.

5. The transfer is to be made as soon as convenient, and the officers commanding companies in the 2d batt. are to draw pay for their men up to the 24th inst., and are to account in the usual manner with the officers commanding companies in the 2d batt. to whose companies their men will be transferred under this order.

6. When the transfer shall be completed, the officers, non-commissioned officers, and staff of the 2d batt. are to proceed to Lisbon preparatory for their embarkation for England.[32]

Although Wellington concluded his order with the hope that he would see the battalion back, "in renewed strength," the need to keep the 1/88th supplied with reinforcements meant that such a return was never a realistic possibility. In apparent recognition of this, the 2/88th was amalgamated with the regiment's depot in June 1812, thereafter being used purely to process drafts to go to Spain. Only in October 1813 was a serious attempt made to recruit the second battalion up to strength in its own right, with the rump of the 2/88th being posted to Ireland to facilitate the incorporation of Militia drafts. Strength was rapidly boosted from eighty-eight men to in excess of eight hundred, but the war was over before there was any hope of turning this mass of manpower into an effective unit, and the battalion was disbanded in 1816 without seeing any further foreign service.[33]

Whilst sending the 2/88th home took a battalion out of Wellington's order of battle, it also facilitated the continued service of the 1/88th, as can be seen from the graph in figure 4. Immediately obvious is the huge increase in total manpower as a result of the drafting off of the 2/88th in July 1811, although the proportional increase in effective manpower is by no means as great. This influx of younger, less acclimatized, men from the 2/88th kept the sickness ratio in the 1/88th notably higher than previously, and this remained the case for the rest of the war. Also obvious is the draft of men joining in November 1814, sent out by the second battalion to bring the 1/88th up to strength for its new deployment to Canada. This was the same draft that took the 2/88th down to its nadir of eighty-eight men, indicating that, despite the active employment of many other second battalions at this time, the dictates of the system were still being followed in some cases and the role of the 2/88th as a feeder unit maintained.

After the successful integration of the two battalions of the 7th, 48th, and 88th, the practice of keeping battalions of the same regiment together was implemented on a more widespread level, with the army's brigading being manipulated to facilitate this. Beginning with the 2/52nd in March 1811, newly arrived battalions were placed not only in the same division but in the same brigade as their regimental counterparts, and steps were also taken to bring together the two battalions of the 7th.[34] The 48th and 88th remained as they were, with both battalions serving in different brigades of the same division, but this was evidently acceptable whereas the dispersal of the 7th, which had previously had one battalion in the Sixth Division and one in the Fourth, was not. As the war continued and regiments that had their second battalion in the peninsula sent their first out to join it, the two battalions were brigaded together for a time, and then, once the new unit was acclimatized and reasonably fit, the second battalions were drafted into the first as was done with the 88th. This took place with the 1/5th, 1/38th, and 1/42nd, all recovered from Walcheren, and also applied when the 1/28th and 1/39th joined

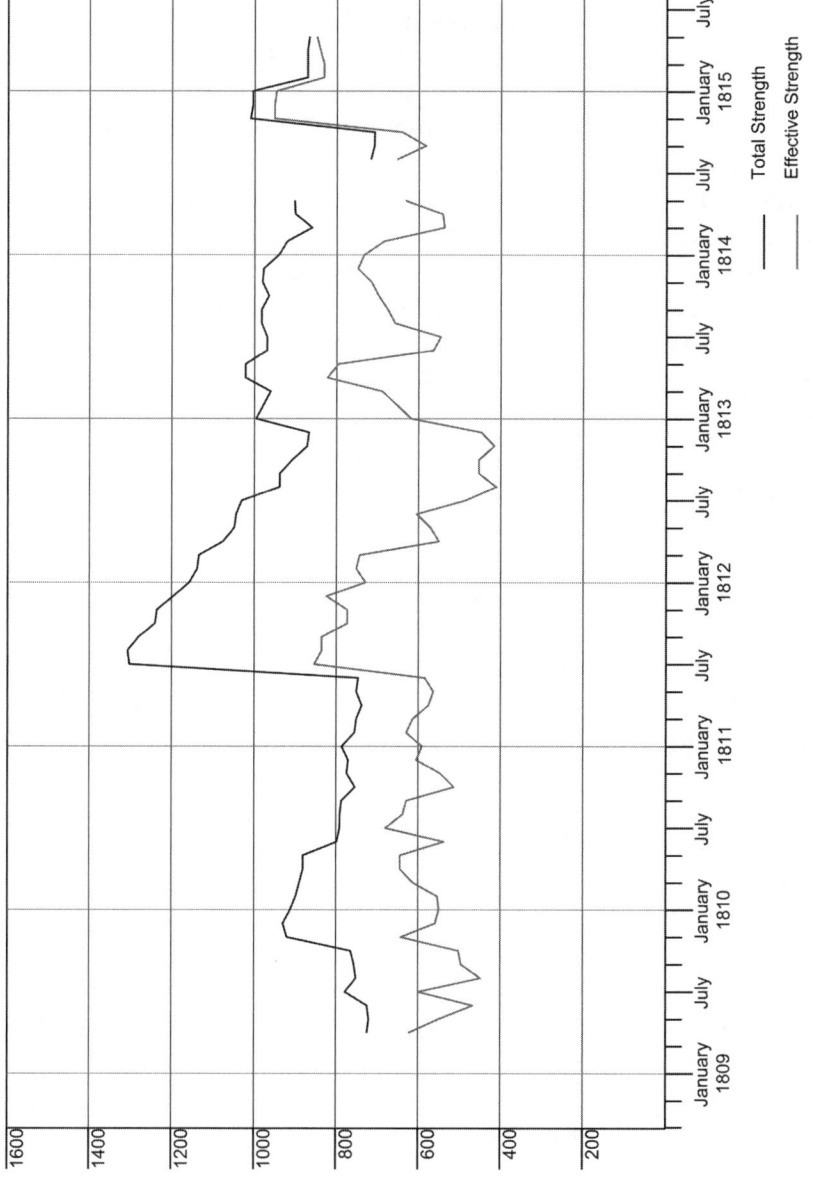

Figure 4. Campaign Manpower in the 1/88th. Data from Monthly Returns, TNA, WO17/2464–2465, 2467–2476, 1516–1519. The gaps toward the end of the run of data represent passage to and from North America.

from the Mediterranean, although in these two later cases the weakness of the relevant second battalions after Albuera meant that the period when both were in the field was brief.[35] In all these cases, once the second battalions had drafted off their effective manpower they returned home and, as with the 2/88th, saw out the remainder of the war in a purely depot role.

The utility of this system, in enabling the effective exchange of manpower, meant that it was also applied to the forces serving in Flanders in 1815. Even before Waterloo, the 1/52nd had joined the same brigade in which the badly understrength 2/52nd was already serving, rapidly absorbing the effective manpower of the junior unit and leaving it as a cadre to be sent home.[36] Reorganization post-Waterloo also placed multiple battalions of the 1st and 27th in the same brigades, although the unusual size of these regiments paired the third and fourth battalions of the 1st and the first and third battalions of the 27th. In either case, however, the option still existed to draft the junior battalion into the senior if the campaign had gone on long enough, and unit strengths fallen low enough, to warrant it.[37] All this showed that it was possible, to an extent, to work with the system so far as the infantry was concerned. At the same time, such measures did not eliminate all the system's flaws, and other steps were required to make the best of things. Before moving on to look at these, however, it is necessary to consider the organization of the cavalry, where rather more flexibility was built into the system from the outset.

Unlike the infantry, Britain's cavalry organization made a greater number of subregimental units available for deployment. The cavalry squadron was not the direct equivalent of the infantry battalion, being far smaller and lacking anything in the way of a staff. Furthermore, squadrons were not permanent entities, but were put together by pairing the mounted equivalent of the infantry company: the cavalry troop. For most of this period, the British cavalry was established with ten troops per regiment, identified by letter designations, although this number was cut back to eight after the peace of 1814. Until the reduction, this provided for five squadrons, of which one was retained at home in the depot role, leaving four available for service, although there were exceptions to this structure and practice. The 1st Dragoon Guards, as the senior line cavalry regiment, always maintained an additional squadron starting the war with twelve troops and six squadrons, and dropping to ten and five upon the 1814 reduction; alone of the heavies, it thereby mustered four squadrons at Waterloo. The four hussar regiments were also increased to twelve troops in December 1813, but were cut back to eight along with the rest of the light cavalry after the peace. Regiments serving in India also had different establishments, and the organization of the KGL cavalry was different again, with the regiments generally being stronger than their British counterparts.[38]

In contrast to the dispersal of infantry battalions, most cavalry regiments deployed overseas with all their active squadrons serving together. Only on a few occasions did circumstances require a regiment to retain more than a single depot squadron at home. The 20th Light Dragoons began the period with two squadrons in Sicily and the rest at home, with two of the home squadrons then going to Portugal before finally being reunited with the rest of the regiment after Oporto.[39] Also, only one of the hussar regiments— the 7th—was able to take advantage of the 1813 increment and deploy with five active squadrons, the other three regiments keeping four in the field and two at home.[40] Although squadrons had no permanent existence and thus no permanent designation, they were named in accordance with the seniority of their commander, which in turn dictated their position when the regiment was drawn up in line. Thus, the senior active squadron would be denominated as the Right Squadron, the next as the Left, and the junior as the Centre; for regiments with more than three active squadrons the designations Centre-Right and Centre-Left were also used.

This structure, however, relates only to the cavalry of the line, since the three regiments of the Household Cavalry were on a different organization again, and represent the most significant exception to the norm. These units had a lower establishment than the line, with the two regiments of Life Guards particularly weak with only six troops. Only toward the end of the period, when they were being readied for active service, were the household units brought up to a strength more akin to the line regiments, gaining an extra two troops apiece to give the Life Guards eight and the Royal Horse Guards ten. The Royal Horse Guards lost its additional troops after the peace, but the 1st and 2nd Life Guards both kept theirs.[41] These regiments were twice called upon to form a Household Brigade for active service—for the peninsula in 1812, and for Flanders in 1815—but on both occasions did so with each regiment taking only two squadrons apiece. The need for the household regiments to maintain their ceremonial and body-guard functions around the royal family meant that there was a legitimate reason for this weaker-than-normal deployment, but it served to place the Household Brigade, particularly in the penin-sula, as something of an organizational anomaly. It is nevertheless going rather far to say, as Oman does, that the 1812 deployment represented the equivalent only of "a large composite regiment,"[42] for their squadron strengths were sufficiently strong as to render all three household regiments viable in their own right, with at least as much effective manpower as many veteran line regiments. What was more, the presence of so great a proportion of each household regiment at home meant that replacement manpower was readily available, keeping the active element strong through-out its campaign service. Since, in addition, they were worked in the peninsula as a brigade of three regiments whilst most of the remaining cavalry brigades were down to two regiments each by the end of 1813, and those regiments down to three squadrons, the number of squadrons in a brigade generally equalized out at six either way. It was only in 1815, with the organization of the cavalry into larger brigades of nine or ten squadrons apiece, that the 1st Dragoon Guards had to be attached in order to raise the Household Brigade to the same standard.[43]

What was notable about the cavalry was the level of organiza-tional flexibility allowed by having so many subregimental units deployed together at once. For all its organizational imperma-nence, the squadron remained the basic tactical unit, and so a regiment that was numerically understrength could redistribute its effective manpower to create three workable squadrons out of four weak ones, sending home the cadres of the two superfluous troops. If the regiment in question was later able to raise addi-tional manpower, a whole squadron could potentially be rede-ployed as a reinforcement; indeed, within a two-squadron depot it was possible to prioritize one squadron for a potential return to service whilst the other remained in a pure training role.[44] In 1814 it was envisaged that the 10th and 15th Hussars and all three household regiments would each send an additional squadron out to the peninsula, although in this case the squadrons were ones that had been left at home at the time of the initial deployment.[45] This level of flexibility was furthered by the fact that a squadron could easily be commanded by its senior troop commander, so the availability or otherwise of field officers did not influence the process.

Due to the shrunken size of his mounted regiments, Wellington authorized a general reduction to three active squadrons for peninsular cavalry regiments as of October 1811.[46] Indeed, he would ideally have gone further and worked his four most reduced regiments at the outset of the 1813 campaign on a two-squadron establishment, rather than sending them home as he was ultimately forced to do.[47] All the above evidence tends toward the view that Wellington's priority was one of ensuring that his brigades contained a sufficient number of effective squadrons, irrespective, within reason, of how many regiments these came from. The case of the Household Brigade clearly demonstrates that two-squadron regiments were a viable and effective option on campaign even if this did potentially involve forming brigades comprising squadrons from four or more regiments.

The mechanics of restructuring a cavalry regiment on campaign are best understood through William Tomkinson's memoir of his service in the 16th Light Dragoons, one of the longest-serving peninsular cavalry regiments. The 16th went to Portugal in 1809 with eight troops organized into four squadrons; with 671 rank and file, this worked out to 168 per squadron.[48] C and H Troops were left at home to form a depot squadron. In October 1811, as a result of the order to reorganize into three active squadrons, A and I Troops were sent home, raising average squadron strength from 130 to 173, although this is a little deceptive as the effective strength was still only 339 for the regiment, or 113 as an average for each squadron.[49] The lack of demanding cavalry service during the last eighteen months of the war allowed the regiment's ratio of effective manpower to improve, and by April 1814 the average effective squadron strength had actually risen to 151. However, peacetime reductions meant that although the 16th could still field three squadrons for the Waterloo campaign, these squadrons averaged only 115 effectives.[50] The strength of the regiment is tracked in figure 5, which also shows how the number of horses was generally maintained at the same level—or ideally higher—as that of effective manpower. This practice was part of a deliberate policy of not sending drafts of cavalrymen to the front unless they could be mounted, and of controlling the regiments to which remounts were assigned, and there were accordingly only two short periods when the regiment could not mount all its fit men.[51]

The squadron echelon was not just important as a battlefield unit, but also had an importance so far as logistics were concerned. This can particularly be seen in the need to spread a cavalry regiment around billets in several villages in order not to exhaust supplies of forage.[52] Thus, squadrons could frequently be called upon to serve in a semi-independent role, on and off the battlefield. This imperative, and the changing composition of the officers present with a cavalry regiment, meant that reorganizations were occasionally necessary to ensure that the most senior troop commanders were commanding squadrons. By all the standards of military precedence, it would not be the done thing for the senior captain to command a troop as part of a squadron led by a field officer, at the same time as a more junior captain might be commanding a squadron.

As with most organizational changes, such alterations to the internal organization of cavalry regiments were made, where possible, outside of the campaigning season. Thus, on January 16, 1812, D and F Troops of the 16th Light Dragoons exchanged places due to the arrival of a new captain for the latter.[53] The newly joined Captain George Murray was second in regimental seniority amongst the five captains then present with the regiment, notwithstanding that one of them, the noted intelligence officer the Hon. Edward Somers Cocks, had brevet rank as a major. Murray's and Cocks's troops originally formed the Left Squadron of the regiment, but since Cocks was third in seniority amongst the regiment's captains, he was still entitled to command a squadron, which he could not do if his troop remained where it was. Thus,

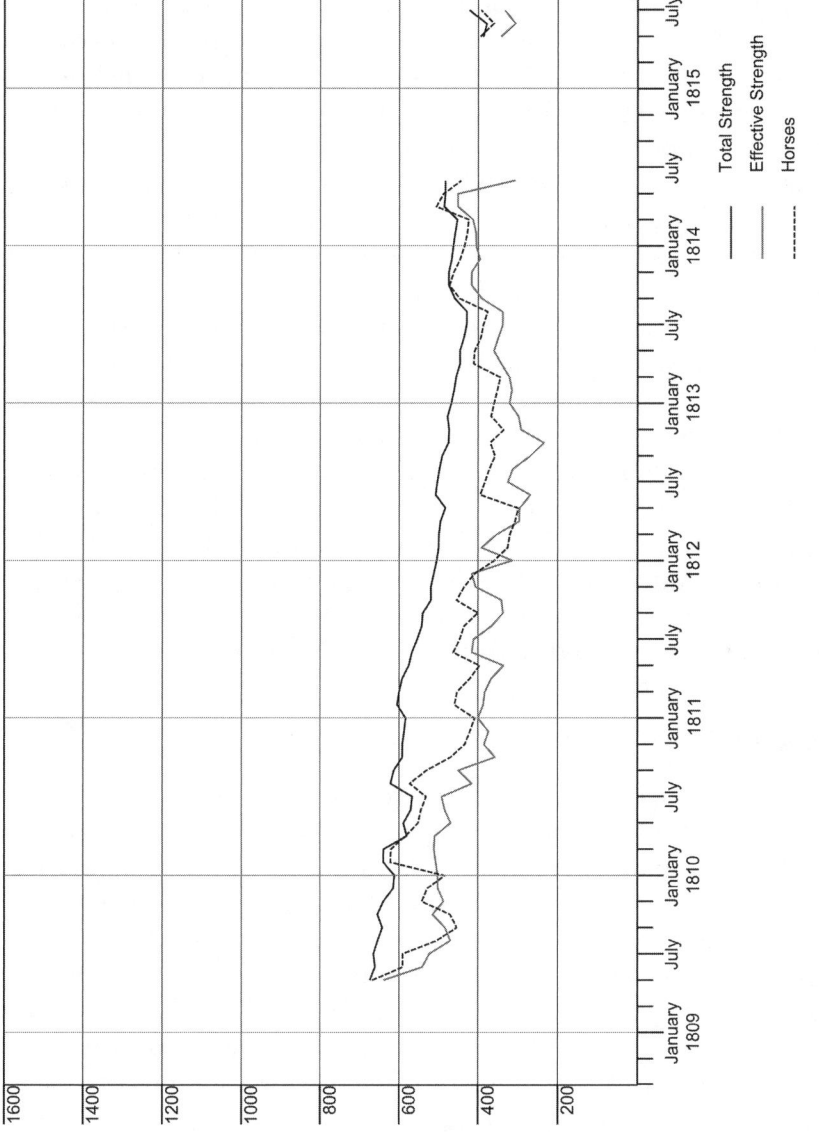

Figure 5. Campaign Manpower in the 16th Light Dragoons. Data from Monthly Returns, TNA, WO17/2464–2465, 2467–2476, 1760. The gap in data is the period during which the regiment was in Britain during the peace.

the regiment's six troops had to be shuffled such that both Murray's and Cocks's were paired with one commanded by a junior officer, allowing the senior man to command the resulting squadron. By shuffling the troops in this manner, Cocks became senior captain in the Centre Squadron and Murray likewise senior captain in the Left Squadron, whilst the Right Squadron remained unchanged in composition under James Hay of K Troop, who had been a captain longer than any of them. Since the important thing here was regimental seniority, Cocks's brevet rank had no bearing on matters although Tomkinson's notes confuse matters by consistently listing him as Major Cocks.

Because cavalry regiments contained more squadrons than infantry regiments did battalions, the greater amount of flexibility available as a result enabled a far more effective and efficient utilization of available manpower. In that the active elements of cavalry regiments almost always served together, and that it was possible to shuffle the component troops around within the squadron structure, a cavalry regiment could make more effective use of manpower than an infantry regiment. Even if the latter possessed three or four battalions—which few regiments did—there was still no guarantee that the subunits would all be in the same place to facilitate a similarly rapid redistribution of resources.

Making the System Work

With only a finite amount of manpower available, in a finite number of units, it was an inevitable and accepted part of the regimental system that these units would be moved around in order to maximize their effectiveness. As far back as the 1750s, a global policy of unit rotation had been successfully developed and implemented in order to prevent regiments potentially spending decades in one garrison.[54] Naturally, a large-scale war meant that peacetime policies had to be shelved, causing some units to again serve lengthy stints in India or in some of the remoter colonial outposts. Equally, the demands of war also forced commanders to implement a more localized form of unit rotation, within the confines of a single theater of operations. This was most pronounced in the peninsula, but it is also evident in North America and, to a lesser extent, during the Hundred Days; in other theaters, numbers simply did not permit this sort of luxury. Conversely, in the peninsula, the numbers involved meant that a more complex system could be developed. Not only were units rotated between front line or reserve formations within the main command, but a degree of fusion between intra-theater rotation and the wider, global, system was also possible. In its eventual form, this fusion enabled the movement of forces between Wellington's field army on the one hand, and the garrisons of Lisbon, Gibraltar, and Cadiz on the other, in order to make the best use of each and every unit.

That commanders within an individual theater would distribute the units under their command to make best use of what was available is both logical and obvious. Nevertheless, the level of sophistication ultimately attained, particularly in the peninsula, was considerable and gives a clear indication of how far commanders were able to make the regimental system work for them. By accepting that each unit had to be treated on its merits, and that it was simply not possible to assume that all infantry battalions or cavalry regiments were equal—in strength, training, or fitness—it was then possible to ensure that each individual unit was used to its full potential. If a unit was not fit to serve in the field, it could be posted to a second line duty, but by rotating units through such postings, rather than keeping them in the field until they deteriorated completely, it was also possible to prolong the amount of time a particular unit was able to remain in theater and thus squeeze the maximum level of efficiency out of the system.

For Wellington in the peninsula, the practicalities of this system were simplified by the fact that his command encompassed both the field army and various garrison forces. The border fortresses of Almeida and Elvas were largely left to Portuguese troops, but a British presence was continually maintained at Lisbon, which for the bulk of the conflict served as the main supply port and depot for the army. From June 14, 1809, the post of commandant at Lisbon was filled by Colonel, later Major General, Warren Peacocke, a guardsman who had briefly commanded a brigade in the Fourth Division but whose field service was cut short by illness. Peacocke rapidly became a key figure in the organization of the British forces, as it fell to him to report to Wellington on the condition of any reinforcements as they arrived, both new units and drafts for existing ones, and it is clear that Wellington trusted Peacocke's judgment and took his reports extremely seriously. In 1811, for example, he wrote to the secretary of state for war and the colonies as a direct result of Peacocke's comments: "I have the honour to transmit a letter from Colonel Peacocke with its enclosures reporting the imperfect state in which some Detachments have been sent from England and I shall be much obliged to your Lordship, if you will give such directions as may prevent the recurrence of such irregularities."[55] In most instances, reinforcements spent a short period of acclimatization at Lisbon and were then, when deemed ready, integrated into the main army.[56] In some cases, as with Craufurd's Light Brigade, which landed on July 3, 1809, and was with the field army by July 29, the demands of service meant that units had to go straight to the front, but, equally, units in poor condition frequently stayed longer. This was the case with the 77th, which had served longer in the peninsula than most single-battalion regiments but had spent much of its time confined to garrison duties at Lisbon as being unfit for service, having only two brief spells of active employment with the field army.[57]

Service under Peacocke at Lisbon could serve as a filter for incoming units, but, as with the 77th, units could also be sent back to Lisbon from the field army in order to recover after having fallen into poor condition. With reference to such units, Peacocke was particularly tasked with attending to their "habits of obedience to orders, subordination, regularity, and interior economy . . . as well as to their parade discipline and drill."[58] The 2/83rd benefited from this treatment, as we have already seen; the 3/27th was also detached for a time in similar fashion, as, for shorter periods, were the 1/26th, 2/58th and 2/88th.[59] Reference to the graph in figure 3 showing manpower strength in the 2/83rd makes the beneficial nature of a spell detached to Lisbon apparent. When sent to Lisbon, the battalion had only 274 effectives out of 789 rank and file and was in poor order, in part from having lost its commanding officer at Talavera.[60] Upon preparing to rejoin the field army a year later, the proportion of sick had dropped to less than 10 percent and the battalion was sufficiently restored in fitness and internal economy to serve actively throughout the remainder of the war.[61]

Whilst the Lisbon garrison remained important throughout the war, the shift to supply bases on the Biscay coast in late 1813 diminished its significance as many units now disembarked at ports closer to the fighting. At the same time, increasing strategic demands and shrinking manpower reserves meant that more poor-quality units were being sent on service, creating the need to replicate the functions of Peacocke's command without tying a significant body of manpower to a specific physical location. Wellington's solution was to create a new infantry brigade in July 1813, outside the established divisional structure, which would operate with the field army but which also served as an explicitly designated "nursery" for newly arrived regiments. As with his selection of Peacocke for the Lisbon post, Wellington appointed an experienced commander with substantial organizational experience. His choice was Major General Lord Aylmer, who had served throughout the Peninsular War as assistant or deputy assistant adjutant general, and who was additionally possessed of extensive campaign experience dating back to the early 1790s. Aylmer's Brigade was intended to compose the 1/37th, 76th, and 2/84th; however, the 1/37th was delayed, having to come from Gibraltar rather than Britain, and the 85th eventually became the brigade's third battalion. Later, the 2/84th was transferred to the Fifth Division, but the 1/37th did eventually join, as did the 2/62nd and 77th.[62] Because his new command was initially based around the Biscayan ports via which reinforcements were joining the army, Aylmer was also to take on Peacocke's old role of checking the state of drafts coming out to reinforce regiments already in the theater, and to "see that they are marched off without loss of time to their several divisions."[63] Although this assessment was an important task in itself, Aylmer's primary responsibility remained that of advancing the readiness of the battalions under his direct command.

Although Wellington's initial memorandum implied that the brigade would ultimately become part of the First Division, this was never formally implemented although both formations served as part of the army's Left Wing, under the overall command of Lt.

General Sir John Hope. However, retaining Aylmer outside the established chain of command made sense so far as the brigade's nursery role was concerned since it made it available for a series of second-line tasks that would hopefully enable its component units to gain experience and ultimately become fit for more demanding duty. This is made explicit in Wellington's instructions to Hope, upon the 2/62nd joining the army: "The 62d may be encamped and act as you please; but as they are very young, and but just arrived, I should think it best to keep them as a kind of reserve for some time, before they are put into one of the divisions. It was to nurse the newly arrived troops that I formed Lord Aylmer's brigade; and I think . . . it will be as well to treat the 62d in the same way."[64] The war ended before this plan could be carried through for the 2/62nd, but the 2/84th benefited from a spell of service under Aylmer to facilitate its subsequent integration into the line as part of the Fifth Division. Only the cessation of hostilities prevented the extension of the practice, which Wellington intended should henceforth apply to all new battalions joining his command.[65] The retention of the 85th with the brigade, when it was by all accounts an effective unit, may have owed more to the fact that as a light infantry corps the battalion was a useful addition to Hope's command, which was otherwise short of such troops, and, indeed, the 85th seems to have operated apart from the rest of the brigade for much of the time.[66]

Although Aylmer's Brigade subsumed much of the function previously exercised by Peacocke's command at Lisbon, the role of the outlying garrisons did not disappear entirely, and it is possible to see the existence of a nursery field brigade as an intermediate step between these forces and the main army. Indeed, a similar practice was observed when the forces in Flanders were built up during 1815, and two infantry brigades were created outside of the established divisional structure. Of these, Major General Mackenzie's Seventh Brigade was assigned to a permanent garrison posting at Antwerp, in a role somewhat akin to Peacocke's Lisbon command; further battalions were also independently assigned to rear-area garrison duties. In parallel to this, Colonel Sir Charles Greville's Twelfth Brigade, created after Waterloo, functioned in something like the nursery role that Aylmer's Brigade had filled in

1813–14, with newly arrived battalions being temporarily assigned to it before being forwarded to an active division of the line. In this case, though, once there were no more reinforcements for it to process, the brigade assumed a regular role as part of the Second Division.[67] Thus, by the end of the Napoleonic Wars a three-tier system had evolved, at least in the larger field armies, which not only allowed newly arrived units be fed into the active army via nursery formations but which also allowed poorly performing units to be taken out of the line for a spell in garrison before rejoining the line, either directly or via another nursery spell.

Yet further flexibility, over and above that obtained through the creation of posts like Peacocke's and Aylmer's, could be obtained by assigning veteran and/or garrison battalions to the major theaters, providing a fourth organizational echelon, albeit with only a one-way link to the rest of the system. Whilst these units could not provide manpower for the active elements, they could usefully absorb any individuals who were unfit to remain with their regiments, but who could still usefully remain overseas undertaking light duties. In this fashion, the 13th Royal Veteran Battalion was formed at Lisbon to facilitate the continued in-theater employment of those men unable to continue serving with active units. The battalion was embodied in April 1813 with an initial rank-and-file strength of 696, and was up to 957 by the end of the war.[68] Renumbered as the 7th Royal Veteran Battalion after the 1814 reductions, it was assigned to Flanders in 1815 along with the 1st Foreign Veteran Battalion, the latter unit having been formed in 1813 as an expansion of the KGL Garrison Company that had spent much of the Peninsular War doing duty at Lisbon.[69] With these two battalions, Wellington's 1815 command accordingly contained the facility for unfit manpower—British and foreign—to be usefully absorbed into rear-echelon formations and thereby still provide useful service within the theater.

To a lesser extent, but for the same reasons, this practice went on in other theaters, and the expansion of British forces in North America during the War of 1812 led to the institution of similar measures. The 10th Royal Veteran Battalion served throughout the War of 1812 much as the 13th did in the peninsula, doing garrison duty in Canada and providing a home for men unfit to

serve with active units there. However, a detachment was also created as part of the Nova Scotia command, entirely out of unfit
men transferred from the 64th and 98th already on that station.[70]
This detachment initially numbered only 24 men, but eventually
reached a peak strength of 197 rank and file in January 1815.[71]
Evidently, there were insufficient unfit men in so minor a command to create a whole new veteran unit, so making them part of
the existing 10th Royal Veteran Battalion represented an administrative shortcut to allow men to be usefully retained on the station
even though they were no longer fit to serve in an active battalion.
Even in distant New South Wales, a single company of veterans was
formed, officially attached to the 1/73rd but largely composed of
men left behind as unfit when the 102nd left the station. In this
case, it was felt that the men posted to the Veteran Company would
be capable of doing duty in the Australian climate, but that their
health would collapse if they were sent somewhere colder—since
the 102nd eventually ended up in Nova Scotia, this was doubtless
a wise decision. The New South Wales Veteran Company wore the
uniform of the 73rd, but with the blue facings of the Royal Veteran battalions, and, like the veteran detachment created on Nova
Scotia, represented a scaled-down version of what was being done
in the larger commands.[72]

At the same time as the organization of manpower within the
major theaters of war was being carried out with increasing levels of sophistication, these theaters were also being integrated far
more closely into the established global policy of unit rotation.
As the major focal points of the conflict shifted, the small permanent garrisons scattered around the globe were increasingly
employed as satellites of the major active commands, leading to
a far greater interlinkage of rotation policies. Even when a garrison was nominally independent of a larger adjacent command,
as with Gibraltar in relation to the peninsular or Nova Scotia in
relation to Canada, practical concerns overrode the restrictions
laid down by command jurisdictions. For Wellington in the peninsula, the situation was eased in this regard by the existence of
the Cadiz garrison as a semi-independent annex to his command,
and although the exact status of this posting took some time to resolve it was ultimately very closely linked into the peninsular unit

rotation system.[73] Gibraltar, on the other hand, was a station in its own right, with a commander in chief answering to Horse Guards rather than to Wellington. Although this independent command within the Iberian Peninsula occasionally led to tensions over conflicting views regarding unit assignments,[74] Gibraltar too ended up being successfully integrated into the local rotation system.

At the outset of the Peninsular War, Gibraltar was strongly garrisoned—understandably so, since Spain had so recently been an enemy. By late 1809, however, the pick of the original garrison had been sent to join the peninsular field army, and was replaced by weaker, less effective, units. September 1808 saw the core of the Gibraltar garrison composed of three strong first battalions, whose 2,514 effective rank and file made up over half the total force; a year later, these had been replaced by three second battalions, all numerically far weaker, and a second veteran unit to join the one that had been there from the outset.[75] Some of these replacements had come out from home, but others came from the peninsula command, with the 2/9th swapping places with the 1/48th in May 1809, and the 2/30th and 2/88th later coming down from Lisbon to release the 1/57th and 1/61st.[76] In similar fashion, the garrison on Madeira was also stripped of effective units to reinforce the troops being assembled in Portugal.[77] In the first instance, these transfers simply represented a redistribution of manpower to facilitate the mounting of a new campaign, but once it became apparent that the war in the peninsula represented a long-term commitment, battalions began to rotate duties on a more organized basis. This system was greatly aided by the Spanish finally allowing, in early 1810, a British contingent to share in the garrisoning of Cadiz.[78] Having a large garrison command within the peninsular command structure effectively created a three-tier system, in which units could be moved from sedentary garrison duties at Gibraltar, via semi-active service in the defense of Cadiz, to fully active duty with the main army under Wellington. Good cooperation between Wellington and the garrison commanders, and an effective working relationship with the Royal Navy, meant that troops could rapidly be shifted in times of emergency as in the autumn of 1810 when troops were rushed from Cadiz to Lisbon in response to Masséna's invasion.[79]

This system was never an exact one, even in comparison with the intra-theater arrangements already considered, and neither the forces at Gibraltar nor Cadiz existed purely as auxiliaries to Wellington's command. Cadiz was actively besieged by the French from February 1810 until August 1812, and it was therefore imperative that a core of effective units be maintained there at all times.[80] Additionally, units from both Cadiz and Gibraltar needed to be kept available for commitments elsewhere in the Mediterranean, as with the Barrosa campaign, the failed attack on Fuengirola, or the rather more successful defense of Tarifa. Nevertheless, examples do exist of units going successively from Lisbon, to Gibraltar, to Cadiz, and thence back to Lisbon. One such was the 2/30th, initially detached to Gibraltar as part of the reorganization of 1809. This battalion had not performed well during its brief spell of prior active service, and it is clear from the comments of Brigadier General Hoghton, who inspected the unit at Gibraltar, that Wellesley had had good reason not to retain it with the field army. "Sufficient attention," Hoghton wrote, "has not been paid to the drill of the regiment, and the field movements were by no means accurate," although the men were clean and steady.[81] The battalion eventually returned to the main army in October 1810, following ten months at Gibraltar and a further three at Cadiz. This return may have been rushed due to the need to for troops to defend Lisbon, but the resulting respite from active service did at least enable the battalion to put in a further two-and-a-half years' worth of duty before finally being recalled to Britain.[82]

More typical was the posting of a unit from the home station to Cadiz to enable it to become acclimatized before moving on to join Wellington. This was done with the 3/1st Foot Guards, 2/47th, 2/59th, and elements of the second and third battalions of the 95th Rifles and of the 2nd KGL Hussars.[83] Transient units were also shifted via Gibraltar on occasion, as with the 2/4th, 1/28th, and 1/82nd. Just as Wellington placed considerable reliance on the views of Peacocke when it came to reporting on the state of units passing through Lisbon, so too was much responsibility delegated to the commanders at Cadiz: successively Major General William Stewart, Lt. General Thomas Graham, Major General George Cooke, and Colonel the Hon. Edward Capel. In that these officers also had to contend with the need to manage their commands as part of an active force defending the city, a delicate balancing act was entailed. September 1810, for example, found Graham informing the secretary of state for war and the colonies, "I have the honour to send enclosed the present allotment of the troops, by which your lordship will see that all is left here [in the city of Cadiz] is 1 battn. 30th Regt. and two companies of the 95th. . . . I should not have reduced the garrison of this place so low but for the necessity I have to get on with the field works of Isla de Leon before the rainy season sets in."[84] In this instance, Graham's concerns went unheeded due to the threat posed by Masséna's advance on Lisbon. With Wellington pushed back into the Lines of Torres Vedras, Cadiz was stripped of yet more troops to reinforce the main army, an order arriving on September 23 "for the immediate embarkation of the 2nd battns. of the 30th & 44th Regts. for the purpose of joining the army under Ld. Wellington."[85]

Although it was also necessary to ensure that these movements were in accord with the desires of Horse Guards and the government, Wellington left it to the man on the spot to decide which units were fit for active service.[86] Thus, after the siege was finally lifted when the French withdrew from Andalusia in the aftermath of Salamanca, Wellington ordered Cooke to send a body of reinforcements to join the main army, to be marched overland as a column under Colonel John Skerrett. Whilst sending Cooke a detailed listing of those troops that he definitely wanted sending, Wellington nevertheless left his subordinate a certain freedom of judgment as to how best to dispose of his remaining forces: "In case the 2nd battn. 59th regt. should have arrived at Cadiz before you shall receive this order you will send either that battn. or the 2nd battn. 47th regt. with Colonel Skerrett's Detachment according to your judgement which of the two is best fit for service in the field. . . . Whenever the 2nd battn. 59th regt. shall arrive after you have carried these orders into execution you will send to Lisbon whichever of that battalion or the 2nd battn. 47th regt. is most fit for service, retaining at Cadiz one British battalion."[87] This instruction not only provides yet further evidence of Wellington's preparedness to place considerable trust on selected subordinates, at least regarding administrative decisions, but also emphasizes that battalions could not be judged simply in terms of strength alone and that distinctions in the quality of manpower and leadership had to be taken into account.

Self-evidently, the garrisons at Gibraltar and Cadiz did not exist purely as locations where units could recover themselves before joining the peninsular field army. Nevertheless, the value of such detached service was as pronounced, if not more so, than a similar detachment to Lisbon. Take the case of one of the longest-serving constituent members of the Cadiz force, the 2/87th. Although ultimately one of the most effective units in that command and distinguished at Barrosa and Tarifa, it was initially in a very poor state when first assigned there after hard service during the Oporto and Talavera campaigns.[88] Indeed, Wellington went so far as to say that the battalion should not be returned to the field army if Cadiz had to be evacuated, and should instead be sent to Gibraltar to continue its recuperation.[89] Happily, however, service at Cadiz soon enabled the unit to recover its levels of efficiency, as shown in figure 6. Although frequent the detachments during late 1811 and mid-1812 render the tracking of effective strength somewhat difficult, the general upward trend is still discernable. Whilst the numbers of sick in the ranks had already begun to drop after the battalion was detached from the field army in autumn 1809, the ratio continued to improve during the Cadiz posting, averaging just over 10 percent during the period as a whole. Although sickness levels rose again when the 2/87th returned to active service, this was in part because the battalion found itself faced with some hard marching in the closing months of 1812, and if the numbers of sick increased then they at least did so from a base rate that was a good two hundred men higher than when the battalion was detached two and half years previously.

Just as the case of the 2/87th demonstrates the utility of such detached garrison service in allowing a run-down unit to recover itself, so too does the wider body of statistical evidence confirm the value of service at Cadiz as part of an acclimatization process. The 2/47th and 2/59th, which both joined the field army from Cadiz in time for the 1813 campaign, subsequently went on to serve in good order for the remainder of the war; with 302 and

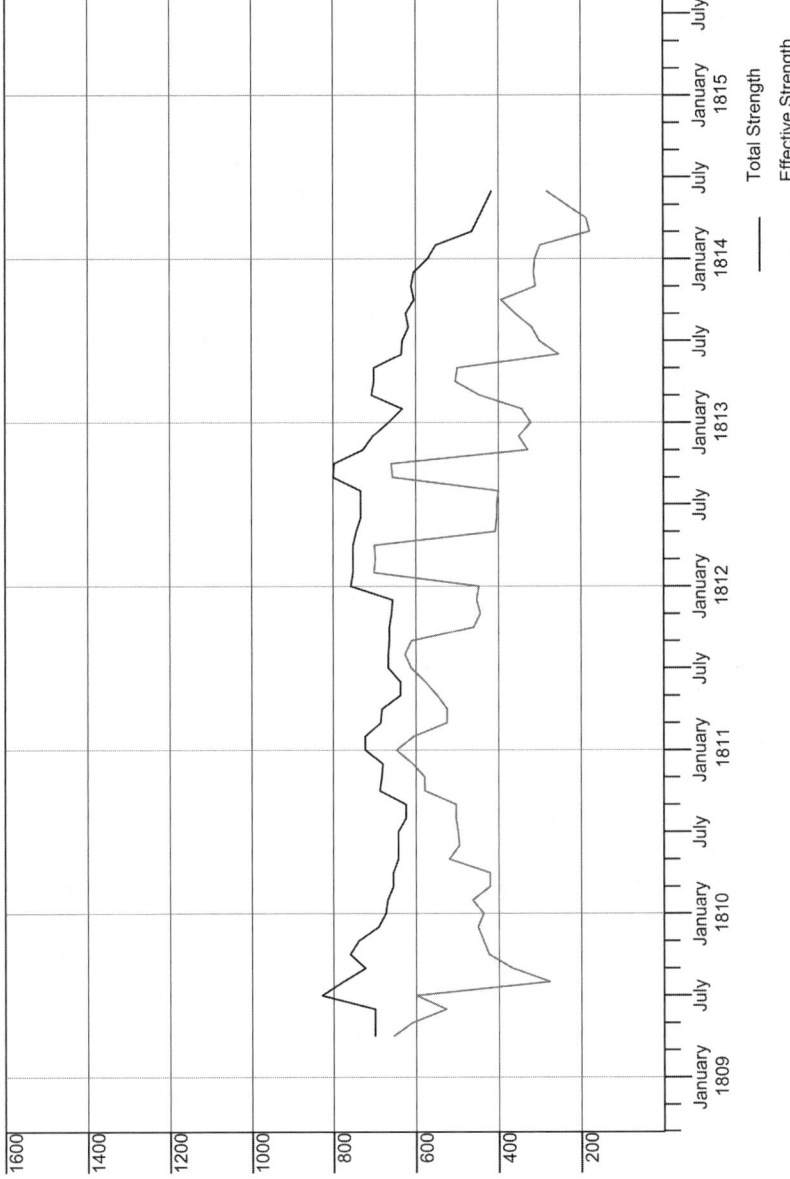

Figure 6. Campaign Manpower in the 2/87th. Data from Monthly Returns, TNA, WO17/2464, 2465, 1486, 1487, 2470–2476.

395 effectives respectively when hostilities ended in April 1814, both were still viable units even after a year's hard service.[90] However, a more direct comparison can be made with respect to the 1st Foot Guards, the third battalion of which joined Wellington's field army from Cadiz in October 1812, where it was reunited with the regiment's first battalion to reconstitute the First Guards Brigade. In early 1813, the brigade fell victim to a serious fever epidemic that removed it from active service for some eight months and that cost the lives of nearly eight hundred men. The whole outbreak created something of a mystery, with Captain the Hon. James Stanhope, deputy assistant quarter master general on the staff of the First Division, puzzling over the fact that "1,100 had come from England fine healthy young men from the militia and 1,000 from Cadiz where they had been gradually accustomed to the climate & who were steady soldiers; therefore it was impossible to account for it by any general rule, as the persons who suffered were so different."[91]

Insofar as the causes of the epidemic in the 1st Foot Guards are concerned, the mystery is as great today, but Stanhope had in fact hit upon, and yet discounted, an important point of distinction between the two battalions. He clearly expected the third battalion, having become acclimatized through service at Cadiz, to be better able to resist the epidemic, and was left at a loss when this seemed not to be the case. In fact, however, there *was* a considerable imbalance in the ratio of sicknesses and deaths between the two units, although with so many deaths in the brigade this was clearly not so obvious to Stanhope at the time. During the six-month period beginning in December 1812, when the illness was at its worst, 1/1st Foot Guards lost 534 men dead, as opposed to "only" 230 in 3/1st Foot Guards. Even allowing for the fact that the first battalion was 225 men stronger than the third at the outset, the imbalance is still clear, and is further highlighted by the ratios of sick within the two units. Taken as a monthly average across the six-month period, these ratios amounted to 184 men, or 26 percent of the total strength, sick in the third battalion, but no less than 410 men, or 40 percent, in the first.[92] The time spent at Cadiz by the third battalion, being the only aspect of their service in which the two units differed, clearly assisted in acclimatizing the battalion to peninsular service and ensuring that its members were better prepared for the rigors of such an epidemic than their comrades freshly arrived from Britain.

Following the eviction of French forces from southern Spain, the composition of the now much-reduced Cadiz garrison became far more static. The forces at Gibraltar were also left alone to a far greater extent than in the early years of the war, and the quality of troops in both stations seems to have tailed off as efficient manpower was reallocated to meet the growing demand for troops in the field.[93] Indeed, at Gibraltar it became necessary for the commander in chief there, Lt. General Sir Colin Campbell, to submit repeated reports to Horse Guards highlighting the increasingly poor state of the 4th and 7th Royal Veteran Battalions; this correspondence went on for several months before the units were eventually returned to Britain.[94] However, lack of active employment did permit the collection of understrength second battalions forming the remainder of the Gibraltar garrison to revert to the role assigned to them by the dictates of the regimental system. Thus, in the second half of 1811, the 2/11th at Gibraltar obtained a total of 124 new recruits from home, and was accordingly able to draft off 150 seasoned men to the regiment's first battalion with Wellington.[95] Similarly, when the 2/9th was finally recalled to Britain in April 1813 due to its shrunken state, 233 out of 431 rank and file, along with five sergeants and a drummer, were drafted off to the active first battalion with Wellington, leaving only a cadre to go home.[96] Although the battalions of the 9th and 11th were not themselves being rotated, the manpower within them was. The utilization of second battalions in this manner yet again demonstrates the amount of flexibility that could be worked into the system. The second battalions of both regiments were carrying out their theoretical role perfectly, but they were doing so in a location that made far more effective use of their manpower than if they had been retained at home.

It was quite possible, then, for the regimental system to be rendered far more flexible than a strict interpretation would imply. What was being done for the bulk of the period, both by the authorities at home and by commanders in the field, was to make use of its best features and find solutions to work around its shortcomings. After 1809, the assumption that second battalions would not serve actively overseas had to be dispensed with, yet at the same time the role of these units in respect to their senior counterparts was not forgotten, and this is evidenced in the deliberate policy of keeping first and second battalions together, or at least in close proximity, wherever possible in order to facilitate manpower exchange. Growing strategic commitments made it clear that the war could not be fought without making full use of all available units, leading to increasingly sophisticated measures to ensure that each unit was employed in a position best suited to its state of fitness and readiness, and to enable units to be moved around as these changed over time.

In short, with an element of judicious tweaking, the regimental system with which Britain entered the Napoleonic Wars served to meet the British Army's manpower needs for the greater part of the conflict. Even when strategic overstretch in 1809 and 1813–14 created pressures that meant that it did become necessary to deviate from the norms, care was still taken to ensure that as much of the established order was retained as was possible. How this was done forms the subject of the next chapter, but that it was done at all is, in itself, proof that the value of the system was recognized.

CHAPTER 4

The Limits of the System

On March 14, 1814, Field Marshal Frederick, Duke of York, commander in chief of the British Army, composed a brief memorandum for the attention of Lord Bathurst, secretary of state for war and the colonies.[1] In less than two hundred words, York seemingly conceded defeat in the struggle that had gone on for the previous eighteen months between himself, Bathurst, and Wellington regarding the best solution to Britain's growing military manpower problems. In its acceptance of elements of militiamen serving overseas alongside regular troops, and of forming detachment battalions from troops on the home station, York's proposals indicated a marked break from the tenets of the regimental system, to which he had previously clung. But rather than being a concession of defeat, York's proposals in fact represented an attempt to regain control of a situation that had become increasingly confused through lack of a single clear policy. Hamstrung by Wellington's attempts to retain every fit man in the peninsula, the British Army had struggled to meet the demand for troops for North America and northern Europe. Political pressures led to ill-judged and ineffective responses, which if anything served only to worsen the looming manpower crisis.

The events of 1813–14 represent the closest that the regimental system came to collapse during the Napoleonic Wars. However, the system had also been placed under extreme pressure during 1809, when so much was staked on the Walcheren expedition, and into 1810 when the bulk of its survivors were unfit for further service. Because the personalities involved were different, with Dundas as commander in chief, and first Castlereagh and then Liverpool filling the post of secretary of state, there was little or no direct continuation of policy between the responses to the 1809–10 manpower shortfall and that of 1813–14. Nevertheless, the experience gained, and the precedents set, can readily be identified as recurring in the policies espoused by York, Bathurst, and Wellington four years later.

1808–1812: Crisis Postponed

Dundas in 1809 found himself in a rather different position with regard to military manpower than York would in 1813. On the one hand, the British Army was less heavily committed, but, on the other, political initiatives led rapidly shifting strategic priorities that required maximum numbers of troops in the short term without considering what would happen if the war was not brought to a rapid conclusion. Unlike York, Dundas seems not to have involved himself in matters of strategy, and was willing to acquiesce, albeit sometimes under protest, with the schemes put forward by the Cabinet and to do his best to find the troops to support them. It is also telling that the secretary of state for war and the colonies during the first months of Dundas's tenure was Lord Castlereagh, whose involvement alongside York in the reforms of the previous few years meant that he was well placed to understand the nuances of the regimental system and how far it could be pushed. Until forced to leave his post in September 1809 through the intrigues of George Canning, Castlereagh was able to force his opinions through without undue debate, although he would be out of office before the nature of the resulting manpower shortfall became apparent.[2] Later, as foreign secretary, Castlereagh would become embroiled in the fringes of some of the same problems during 1813 and 1814.

For the campaigning season of 1809, the British Army had two manpower priorities. One was to rebuild the disposable force that had come back from Corunna and turn it into the nucleus of the Grand Expedition to Walcheren; the other was to build up Cradock's forces in Portugal, which would eventually take to the field under Wellesley in the spring. Although the latter was the secondary priority, it was the first area to see the adoption of ad hoc measures in the shape of the two battalions of detachments formed from the debris of the force that had taken part in the Corunna campaign.

Many of the regiments that had marched with Moore had left sizeable numbers of sick behind at Lisbon. These men had begun to return to fitness in the early months of 1809, by which time scores of stragglers who had been cut off or left behind on the retreat began to turn up at British outposts in the north of Portugal. Collectively, this amounted to a sizeable body of men, and since their units had returned to Britain there was no obvious organizational home for them in the peninsula. The forming of these men into temporary battalions was initially not so much to meet a manpower shortfall as a way to keep closer control over men who were isolated from their regiments and from all the support and sense of place thereby implied. Brigadier General Alan Cameron, commanding at Oporto in early 1809, found these detachments "scattered in all directions, without necessaries" and was particularly keen to prevent "some of them committing every possible excess that could render the name of a British soldier odious to the nation."[3] A formal battalion organization, however temporary, would expedite the issue of rations, clothing, and accoutrements, although with the limited resources that were available this was often done on only the most basic level. Only later, once the decision was taken to resume active operations, did it become obvious that these battalions could also serve as a means of eking out the meager force initially available in Portugal, and the battalions were accordingly retained until the end of the Talavera campaign.

Sufficient men were available from many regiments to form a provisional company within one of the two battalions of detachments, thus helping bring cohesion to the new units by ensuring that men remained largely with comrades and officers with whom they were at least loosely acquainted.[4] The First Battalion of Detachments contained companies from the 43rd and 52nd Light Infantry and 95th Rifles, adding to the combat utility of the unit since these men formed a useful addition to Wellesley's then extremely limited light infantry contingent. Practical considerations, and recognition of the importance of regimental "familial" ties in terms of motivation, clearly justified this form of organization, although the lack of a rifle officer to command the 95th's company meant that Lt. Thomas Munro of the 1/42nd Highlanders became its temporary—and somewhat conspicuous—commander.[5]

The successful conversion of stragglers and convalescents into combat manpower was exemplified during the campaigns of Oporto and Talavera, where both battalions performed creditably. The light infantry and riflemen serving in the First Battalion of Detachments won particular praise, being mentioned no less than three times in the Oporto victory dispatch.[6] Nevertheless, this positive view was not universally held, and Charles Stewart, now back in the peninsula as Wellesley's adjutant general, expressed a contrary view at least so far as the performance of the battalions' day-to-day duties were concerned: "I am sure that they are the cause of great disorder—no esprit de corps for their interior economy among them, though they will fight. They are careless of all else, and their officers do not look to their temporary field officers and superiors under whom they are placed, as in an established regiment."[7] This assessment may be taken as being indicative only of a period of adjustment rather than as an out-and-out denunciation of the battalions, particularly since it dates from a time when the whole army was in something of a state of disorganization during the rapid marches that took place before and after the Second Battle of Oporto.

The two battalions of detachments were never intended to be a permanent feature, and after the Talavera campaign they were broken up and the men shipped home; the exception was of men from regiments that again had a battalion in the peninsula, in which case the men were drafted to those units.[8] Wellington's comment on the departure of the two battalions expresses both his gratitude for their services, and a note of admonition for the army as a whole:

The Commander of the Forces cannot avoid to express his regret upon losing the services of the two battalions of detachments, which are about to join their corps in England. He will not flatter them by saying that he has not had, upon several occasions to be dissatisfied with their conduct, in their quarters, in their camps, and on their marches; but they have uniformly sustained, in an exemplary manner, the character of the regiments to which they belong, and of British soldiers, in the field against the enemy; and he trusts that the few, of whose conduct he cannot but complain, even upon this occasion, will discontinue and forget their bad practices and habits, upon their return to their regiments; and that they will endeavour to become an example of orderly and regular conduct in their quarters, as they must ever be of gallantry and discipline in the field.[9]

The element of admonition does at least partially justify Stewart's distrust of the two battalions of detachments, but, considering that much worse than this was said by Wellington of battalions forming part of the regular line, his commendation of these two temporary units is worth more than just the formalities expressed in the opening lines.

What is more, comparative statistics indicate that the two battalions were surprisingly effective units, with a ratio of strategic consumption hardly short of the averages for the peninsular army as a whole, as demonstrated in table 2. The numbers of sick are considerably higher for the battalions of detachments, but this is scarcely to be wondered at with formations containing convalescents from the Lisbon hospitals. However, this is balanced out in terms of its effect on the total effective strength due to the very small number of men returned "On Command." Self-evidently, detaching men from a unit that existed primarily to absorb detachments would be a counterproductive process. Differing death rates can be accounted for largely by differing exposure to combat at Talavera, but there is no obvious reason why the Second Battalion of Detachments should have been so free of desertion in this period. In terms of the regiments from which it drew its manpower, the Second Battalion of Detachments was marginally more homogenous in its composition than the First, which may have contributed to creating stronger interpersonal bonds at a company level, but this is sheer speculation. Indicating levels of

Table 2.   The Battalions of Detachments

	First Bn. of Dets.	Second Bn. of Dets.	Average, Bns. of Dets.	Regular Infantry
Effectives as %	76.0	74.9	75.5	81.0
Sick as %	22.0	23.4	22.7	15.5
On Command as %	2.0	1.7	1.8	3.4
Deaths/1,000 Men	11.0	5.0	8.0	8.7
Desertions/1,000 Men	1.1	0.0	0.6	1.3

Source: Data from Monthly Returns February to August 1809, TNA, WO17/2464, averaged over the seven months.

effectiveness similar to the norm, the statistics reinforce the view that these battalions were a useful, if temporary, means of extracting good service from men who would have otherwise remained a drain on resources whilst they awaited shipment home.

However, the retention of these men in the peninsula did not sit well with Horse Guards. It was one thing for Cradock to form temporary battalions as a means of ensuring that stragglers and convalescents were clothed, fed, and paid, but it was quite another thing for Wellesley to retain them as combat troops. With his attentions firmly set on assembling the Walcheren expedition, Dundas wanted the men back in Britain where they could rejoin their parent units, and told Castlereagh so in fairly blunt terms:

> I have the honour to represent to your Lordship that in consequence of the great delay which has taken place in the return of the two Corps of Detachments now serving with the army in Portugal, they have suffered greatly for want of clothing, their discipline as must be the case with corps similarly constituted may naturally be supposed to have relaxed, and much inconvenience has been occasioned to the Regiments to which these men belong, as most of them now being under orders for Foreign Service require the presence of all their effective men.
>
> Under all these circumstances, I have to request your Lordship will be pleased to direct that immediate measures may be taken to order these Detachments home.[10]

That Dundas sought to clear these operational matters with his political master serves in itself to give some idea of the relationship between the two men. With Wellesley already committed to the Talavera campaign, it was by this stage too late for Dundas or Castlereagh to do much to expedite the return of the battalions, but this first occasion of a clash of opinions between Arthur Wellesley and his commander in chief over the constraints of the regimental system, prefiguring far more significant disagreements with York in the coming years.

The two battalions of detachments were not the only provisional units existing during this first, organizationally confused, phase of the peninsular army, and one other in particular would go on to have a quite lengthy existence. This was the provisional light infantry detachment formed from the men left behind by the two KGL Light Battalions from Moore's army, which was attached to the 1st KGL Line Battalion.[11] Even when the battalions of detachments were broken up, this KGL contingent remained until reabsorbed into its parent battalions when they returned in 1811. The detachment's formation and initial retention for the Oporto and Talavera campaigns likely had more to do with Wellesley's shortage of light troops than anything else, but its continued existence implies a tacit approval from higher authority. Also independent whilst it remained in the peninsula was a squadron-sized detachment of the 3rd KGL Hussars, which formed part of Cotton's Brigade during the Oporto campaign and went home thereafter.[12] Since there were enough light infantrymen and hussars to form viable independent units, it was not necessary in either case to incorporate KGL manpower with detachments from regular British regiments. In a similar fashion to the attachment of the KGL Light Infantry to the 1st KGL Line, the substantial body of men left in Portugal by the 1/9th were attached to that regiment's second battalion rather than being put into either battalion of detachments. Forming a useful addition to the strength of an otherwise weak unit, these men remained with the 2/9th until it left the theater after Oporto, at which point they were finally detached and shipped home to rejoin their proper battalion.[13]

Although Dundas was slow to reclaim manpower from the peninsula, he made up for this by thorough and detailed use of all available returns in order to comb the British Isles for troops that could go to Walcheren.[14] Dundas did not lose sight of other priorities, producing a "Return of the effective strength of the corps of infantry serving in Portugal" that showed these battalions' deficiencies in manpower and how these could be made up by men from home,[15] and proposing the establishment of extra recruiting companies for regiments serving in the colonies.[16] Nevertheless, Walcheren was clearly the priority, and by the end of June Dundas was able to produce details not only of the forces that were to go but also of those that would be left behind in Britain and Ireland.[17] His investigations also revealed the existence of 3,102 men at the Army Depot on the Isle of Wight, mostly awaiting transport to go out as drafts to regiments overseas. As trained soldiers, these men could be a valuable addition to the Grand Expedition, but the restrictions of the regimental system prevented their being assigned to it. However, with Castlereagh keen to achieve success in the Netherlands through the application of overwhelming force, Dundas elected to bend the rules and take at least a portion of these men for immediate service.

In order to do so, Dundas authorized the creation of the unit sometimes known as the Battalion of Detachments, but more commonly as the Corps of Embodied Detachments. The decision to create this unit was clearly taken after the initial proposals for Chatham's command, with the battalion not making its first official appearance until July when Castlereagh was informed of "a Corps of Eight Hundred Rank and File with a proper proportion of officers and non commissioned officers, having been embodied from the Detachments at the Army Depot, for the purpose of embarking with the troops under order for service."[18] The officer placed in command was Lt. Colonel the Hon. Basil Cochrane of the 36th Foot, and his officers were drawn from regiments either serving with Chatham's forces, or from those that had supplied drafts to the unit, with the exception of the paymaster, William Armstrong, from the 2nd KGL Dragoons. The men were organized into ten companies on the standard model, but these do not seem to have included flank companies.

By July 7, the Corps of Embodied Detachments had been assigned to Picton's Brigade of Mackenzie Fraser's Third Division, with which it served for the duration of the campaign on Walcheren Island itself.[19] On August 25, 1809, the battalion had a rank-and-file strength of exactly 800 men, of whom 711 were fit for service, 5 present but sick, and 84 sick in hospital, equating to a sickness rate of 11 percent as compared with 7 percent for the force as a whole.[20] Both this higher level of sickness, and the corps's status as a provisional formation, would seem to be collectively responsible for the fact that the Corps of Embodied Detachments did not remain with the smaller force retained on Walcheren after the departure of the main body. Returning to England, the corps went first to Porchester barracks, and then back to the Army Depot, where it was disbanded, although by then many of the men were absent in various hospitals.[21]

No records have been found detailing the composition of the Corps of Embodied Detachments prior to its embarkation for Walcheren, but the battalion completed standard monthly returns for October and November 1809, and filed with these are breakdowns by parent-unit of the officers and men present after the battalion's return to England. The data for October is reproduced in table 3, and, whilst it cannot tell us the origin of those men who had already become casualties, it does give a reasonable idea of the composition of the battalion. The 12th, 19th, 22nd, 33rd, 53rd, and 78th Regiments all had a battalion in India, and between them account for to a considerable majority of the Corps of Embodied Detachments: 515 out of 712 rank and file, and 637 out of 793 overall, or 72 percent and 80 percent respectively. Of the remaining contributor regiments, the 41st had a battalion on long-term service in North America and the 15th and 64th were both in the West Indies, leaving only a handful of men contributed by regiments serving at home or in Europe.[22]

Although there was a subsequent proposal to create something similar in 1814, never again was the Army Depot plundered for manpower in this fashion. The bulk of the men drawn upon were from drafts due to go out and join their regiments in the colonies, and for battalions in India, where the dispatch of a single yearly draft was the norm, this diversion of manpower badly dislocated the usual system. More than half the total manpower came from two regiments, the 33rd and 78th, with battalions on the

Table 3.   The Corps of Embodied Detachments

Regiment	Officers	NCOs	Musicians	Rank & File	Total
6th	3	4	2	0	9
8th	1	0	0	0	1
12th	0	0	0	42	42
15th	1	0	0	0	1
19th	2	0	0	61	63
22nd	2	0	0	55	57
29th	0	3	0	34	37
33rd	4	3	0	150	157
36th	3	1	2	0	6
41st	0	1	0	37	38
48th	0	1	0	13	14
50th	3	12	2	0	17
53rd	0	0	1	9	10
64th	1	0	0	0	1
78th	9	1	0	298	308
79th	3	13	3	0	19
88th	0	0	0	13	13
Total	32	39	10	712	793

Source: "Return showing the different Regiments which the officers, non-commissioned officers and rank and file belong who are serving with the Corps of Embodied Detachments," October 24,1809, TNA, WO17/234.

subcontinent; the redirection of the planned reinforcing drafts could only have had a detrimental effect on those battalions. The 78th had its first battalion in India and its second in the process of returning from active service in the Mediterranean, leaving the 1/78th wholly reliant on the home depot for its reinforcements; however, these were diverted in their entirety to the Corps of Embodied Detachments. Whereas 400 men had gone out to India in 1808, 1809 saw the 320 available diverted to Walcheren and no draft sailed east at all; accordingly, in 1810, the 2/78th had to be stripped of every fit man in order to keep the senior battalion up to strength, thus itself becoming unfit for service for the next three years.[23] Inasmuch as the 78th supplied some two-fifths of the total men assigned to the Corps of Embodied Detachments, not all contributing regiments would have suffered to the same degree. However, the example of this regiment does confirm that this policy of robbing of Peter to pay Paul could ultimately have only a deleterious effect on the British Army as a whole, far outweighing the initial gain to be had from a single 800-man battalion as reinforcement for an expedition numbering over 30,000. That Dundas was prepared to countenance such a step can only be understood in the context of the political pressure that the British Army was under to provide a maximum amount of manpower as part of an attempt—albeit inspired by the hope of a final settlement—to throw every man into the field with little thought for the future.[24]

The problem, of course, was that the allied campaigns of 1809—in the Netherlands, in the peninsula, and on the Danube—did not bring about the collapse of Napoleonic France. This in turn meant that the British Army was forced to plan again for a longer haul, with a manpower system that was badly dislocated both by the maximum effort of 1809 and by the illness that the Walcheren regiments brought back with them. Under these circumstances, even the far more modest manpower demands of 1810 would require a continued flexible response—the more so when at least some politicians were beginning to question the value of the huge sums invested in the armed services.[25] In order to meet part of this demand—for troops to help hold Cadiz against the French—Dundas oversaw the creation of another major provisional formation. This time, however, there was a rather more measured response, and advantage was taken of a more flexible model to manpower distribution than was normally available.

This flexibility was possible since the troops in question came from the Foot Guards. As of 1804, the seven battalions had been formed into three semipermanent brigades. The First Guards Brigade, comprising the first and third battalions of the 1st Foot Guards, served with Moore and subsequently went to Walcheren, and the Second Guards Brigade, composed of the first battalions of the Coldstream and 3rd Foot Guards, had been part of the peninsular army since March 1809. Whilst these two brigades were intended for active service, the Third Guards Brigade was, by contrast, designed to fulfill domestic ceremonial obligations and to serve as a depot for the two active brigades, being composed of the second battalions of all three regiments.[26] Despite its depot status, the Third Guards Brigade did not entirely escape Dundas's 1809 manpower hunt and was called upon to send its flank companies to Walcheren, where they were combined with those of the First Guards Brigade to form a battalion apiece of light infantry and grenadiers.[27] Now, in response to the need for reinforcements for Cadiz, the brigade was called upon again—this time not for a few companies but to provide a whole composite field force for service at Cadiz. Accordingly, a total of eleven companies were made ready to go overseas. Although the 2/1st Foot Guards did go out as a battalion, it was an understrength one of only six companies, whilst the 2/Coldstream Guards and 2/3rd Foot Guards sent two and three companies respectively, which were combined to form a single Detachment Battalion.[28] Whilst this organization would have been an unusual measure for the times if adopted by regiments of the line, such sub-battalion organizations had precedent in the Foot Guards, as when a composite regiment was sent to fight the rebellion in North America and again when detachments were drawn off for the 1798 raid on Ostend.[29]

The Foot Guards contingent at Cadiz formed the First Brigade of Graham's little force, under the command of Brigadier General William Dilkes, whose inspection of his units that summer reveals an interesting choice in the men assigned to this service:

The men of [2/1st Foot Guards] look older than appears by the return, nearly 60 have served 18 years and upwards and above 200 have been transferred or exempt from service in the 1st Brigade as not being fit for active service, or on account of having large families, the return of wives and children is considerable, all left in England. . . . The Detachment Battalion is composed of 224 rank and file of the Coldstream Regt. formed into two, and 342 of the 3rd Guards formed into three companies; of the former 100 have been transferred from or exempted service in the 1st battn. of the Regiment as being unfit for active service,

or on account of having large families, of the last three companies only 48 exempted or transferred.[30]

The returns show far more men in their twenties and thirties than
typical for line units, and with longer service. Such an organization could be implemented thanks to the more flexible system his-
torically enjoyed by the Foot Guards, in which the company rather
than the battalion maintained greater significance and autonomy,
combined with the organization of the seven battalions into three
standing brigades to add a higher organizational echelon than
was available for the line. Taken together, this organization permitted a three-tier system to maximize effective use of the available manpower, with the best men going to the First and Second
Guards Brigades for active service, whilst the better men from the
Third Guards Brigade undertook less arduous duty in an overseas
garrison and the rump of that formation held the unfit, processed
recruits, and generally served the role of depot for the remainder.

The provisional Foot Guards units served for a year as part of
the Cadiz garrison, before being replaced on the station by a single
full-strength battalion, 3/1st Foot Guards, after taking fairly substantial losses at Barrosa. Upon its arrival, the 3/1st Foot Guards
absorbed the bulk of the effective rank and file of the 2/1st Foot
Guards, whilst the effectives of the other regiments went home.[31]
The fact that, at Barrosa, the two elements of the Detachment Battalion divided and fought on different parts of the field is often
taken by some historians to imply that the contingents of 2/Cold-
stream and 2/3rd Foot Guards were in fact independent units, but
this is hardly conclusive since the 2/67th also inadvertently broke
into wings during the confusion of Graham's deployment. Indeed,
Oman's explanation of how the two companies of 2/Coldstream
split off from the rest of the brigade in order to provide cover
for the British artillery actually confirms that, though ultimately
ending up on the left of the whole army, these companies had
initially formed the right, senior, wing of the Detachment Battalion.[32] In qualification of Oman's stance that the detachments were
independent units in his casualty figures, it should be noted that
they are likewise represented in the monthly returns for Cadiz;

however, this relates to the internal economy of the detachments, not their tactical use.[33]

The flexibility engendered in the Foot Guards by a greater reliance on the company echelon paid off in a similar fashion in the 95th Rifles. A full-strength rifle company was a valuable combat asset in its own right, leading to the regiment frequently having detached companies serving in a variety of locations. In October 1810, to take a rather extreme example, there were eight companies of the first battalion with the Light Division of Wellington's army, along with a single company from the second battalion; a further company, from the third battalion, was part of Erskine's Brigade of the First Division. Meanwhile, a further seven companies were with Graham at Cadiz—two from the second battalion and five from the third.[34] Rather than allow its companies to shrink in proportion to wastage suffered on active service, the 1/95th reduced the number of subunits present in the field, sending the company cadres back to the regimental depot. The 1/95th accordingly dropped from ten companies in the field to eight in early 1810, and then to six after further losses in the 1812 sieges. Conversely, the 2/95th and 3/95th deployed by odd wings and companies, and each was only able to concentrate a battalion's worth of manpower in one place relatively late in the war.[35]

Whilst the 95th shared similarities with the Foot Guards at a company level, the existence of semipermanent Guards Brigades mirrors the formation of standing two-battalion brigades of KGL infantry, which also facilitated operational flexibility. Although this system began to break down as the war went on—with the two line brigades in the peninsula being combined after the 1809 campaign, and the five KGL battalions remaining by 1814 eventually forming a single large brigade—the close ties created as a result were clearly demonstrated by the ease with which it was possible to draft effective manpower from the 7th Line Battalion into the 1st, 2nd, and 5th, after all four units had been reduced by wastage by 1811. Indeed, it is hard to escape the conclusion that the KGL brigade was treated as akin to a multi-battalion regiment of the type fielded by most continental armies. This situation is emphasized by the arrangements for the manpower transfer: "Major General Low will accordingly give orders that this transfer be made, taking care to transfer to each battalion such a number of effective men as will nearly equalise the strength of each, [and] direct that the soldiers transferred may be placed in battalions and companies with their countrymen and comrades."[36] The first point serves to emphasize the ease with which the greater number of battalions into which the men could be drafted simplified the process—a regular British battalion in a similar condition could, at best, give up its effectives to one other unit, assuming that another battalion of the same regiment was at hand. The second also emphasizes the existence of a common identity across the KGL battalions that could not exist within a similar British brigade, in which there would likely have been no previous connection, nor common experience and origin, to link the battalions prior to their brigading.

For the Foot Guards, Rifles, and KGL, being something apart from the norm allowed for more creative approaches to the management of manpower, but the extension of such practices to the regular infantry of the line was far rarer for the bulk of the period. Nevertheless, there are a few earlier examples that provide an element of precedent for what would follow. Thus, when the 2/23rd was preparing for service as part of the Walcheren expedition, only months after its return from Corunna, the available fit rank and file—403 of them—were concentrated into five companies for active service, and the remaining five left at home with only 172 rank and file between them, of whom 62 were sick.[37] A similar expedient was resorted to in the case of the 2/8th, which was originally intended to embark as a full-strength battalion of 544 men. This figure was then revised down to 400, in an unspecified number of companies, but this target too could not be met, and eventually only the flank companies were slated to join the expedition with an anticipated strength of 200 men. Exactly how many deployed in the event is unclear, since the detachment is not included in the monthly returns for the campaign.[38] It is unclear why such a weak unit was not made into part of the Corps of Embodied Detachments, unless the hope was still being entertained that the remaining companies would in time be made fit to join the flankers. The regiment's manpower shortage was down to its first battalion having just been brought up to strength for service in America by drafts from the second, leaving the junior unit with insufficient manpower to take the field in its own right.[39]

This practice of deploying only parts of battalions continued after Walcheren. In much the same way that Corunna had sapped the strength of the 2/23rd, so had Walcheren sapped that of the 85th, and this influenced its state of readiness when being prepared for deployment to the peninsula a year later. Although the battalion was apparently healthy again by late 1810, its training had suffered whilst its manpower had been on the sick list and so too had its recruiting.[40] Again, effective manpower was concentrated into five active companies, totaling 414 rank and file, leaving only 82 fit rank and file at home, with a further 27 sick.[41] In a sense, what was being done here was a division at battalion level that would normally have been applied on a regimental scale. The five active companies were in effect the "first battalion" whilst the five inactive ones took the role of the "second battalion." It may be inferred that the intention was to send out the remaining companies once they were back up to strength, but in practice the situation never arose. The detachments of the 2/8th and 2/23rd were soon returned to Britain once the Walcheren campaign became bogged down, whilst the 85th lasted less than a year in the peninsula before its numerical weakness and poor internal economy saw it ordered home. On the other hand, the 1/71st, having initially gone out to the peninsula with only six companies in September 1810, for much the same reasons as the 85th, did regain its missing companies as fit men became available; as we have already seen, the reunited battalion rendered good service for the remainder of the war.[42] By this time, though, a flood of manpower was again becoming available for active duty as the Walcheren men returned to fitness, and throughout 1811 and into 1812 the manpower shortfall diminished and the forces in the peninsula were steadily built up.

1813: Pushing the System to Its Limits

As units returned from Walcheren began to regain fitness from late 1810 onward, finding additional reinforcements to go overseas became less of a problem, at least for the time being. Furthermore,

Table 4.   Battalions Deployed to the Peninsula, 1810–1813

	1810	1811	1812	1813
Number of Battalions	10	9	6	4
Total Strength	7,334	5,868	5,083	2,646
Average Unit Strength	733	652	847	662

Source: Compiled from data, TNA, WO17/2465, 2467–2474, using total strengths upon arrival in Portugal. These figures do not account for units redeployed to the peninsula from elsewhere or to sub-battalion deployments such as the trickle of rifle companies.

thanks to the policies of unit rotation outlined in the previous chapter, more effective use could be made of troops already overseas thus extending the amount of service that could be obtained from them. However, with manpower demands steadily increasing, the redeployment of recovered Walcheren survivors was not, in itself, sufficient to maintain an indefinite supply of new units, particularly once 1812 saw the opening of a second front in America. Similarly, unit rotation could only do so much; once all the effective units were assigned to active theaters, further transfers could achieve nothing. These factors led to an inevitable reduction in both the quality and strength of reinforcements being sent to the main theater of war, as shown in table 4. The tailing off in both size and number of units toward the end of the period is clear, particularly after the peak of 1811 and 1812 once the ex-Walcheren units had joined. Reinforcements were in similarly short supply for North America, and the forces there were initially built up largely through moving units from elsewhere.[43]

Although the strategic priority remained the peninsula, both Horse Guards and the Cabinet now had other areas to consider—not only North America, but the Mediterranean and northern Europe as well. Unfortunately, it was at this point that Wellington's tendency to emphasize his own needs above all others began to become apparent. Previously, his frustration with the regimental system had been confined to barbed asides in General Orders; now, he began to oppose it more vigorously. In Wellington's view, it was far better to retain seasoned veteran troops in the peninsula; they were hardened to campaign service, could be relied on in action, and were less likely to fall ill. On the other hand, the end of the 1812 campaign saw many of the units containing these veterans so reduced in numbers as to be verging on ineffective. Wellington hoped to work around this problem by reducing the number of subunits, working weak cavalry regiments with only two squadrons, and cutting weak infantry battalions down to four-company detachments that could be paired to form the equivalent of a full-strength unit. Wellington's proposals for the cavalry were rejected out of hand, with four worn-down regiments going home but six new ones, later joined by a seventh, replacing them.[44] In a sense, this was a fight that Wellington could afford to lose; the new troops of the Household and Hussar Brigades gave him trouble, but were ultimately integrated into the field army without undue problems. The issue of the infantry was, to Wellington, rather more important, but here he did at least have precedent on his side because for two battalions he had already been using a version of the proposed organization for over eighteen months. Thus, he knew that what he was proposing would work, and he knew too that he had been able to implement it without anyone in London finding fault.

The initial concept of what would become the creation of provisional battalions stemmed from a typical piece of Wellington pragmatism in the aftermath of the Battle of Albuera. So badly had the Second Division suffered that the seven battalions of what had been Colborne's and Hoghton's Brigades were all far too weak to continue as independent units. In the immediate aftermath of the battle, all effective troops from these two brigades were formed into a single battalion, commanded, as every account feels required to remind us, by an émigré French captain with the unfortunate name of Cimetière.[45] The only exceptions were the two battalions of the 48th Foot, which were combined into a single unit, although in due course the 1/3rd and 1/57th were rebuilt by drafts from their second battalions in England and resumed a separate identity. This left three battalions—the 29th, 2/31st, and 2/66th—still being worked as a single unit but now on a rather more formal basis, with the 2/31st providing four companies and the other battalions three each.[46] Ultimately, the shrunken state of the 29th resulted in its being ordered home, but the expedient of working the other two half-battalions together was maintained, formally designated as the Provisional Battalion. The combined unit came under the lt. colonel commanding the 2/31st: first Guy l'Estrange and then Alexander Leith. Field and staff officers from both battalions remained in the field, with Lt. Colonel Charles Nicol of the 2/66th still with the battalion in early 1813, although his function was restricted to maintaining order and discipline in his own proper command.[47]

Even before the Provisional Battalion had been organized, Wellington was speculating on the possibilities of extending the concepts it entailed. When Liverpool wrote to him in the spring of 1811 discussing plans for offensive operations, Wellington made it clear that he would need to build up his forces if such operations were to be viable, and that this would mean retaining units in the peninsula that might otherwise have gone home. In order to do this, Wellington proposed that he "form into six companies the 2d battalions of the 24th, 31st, 38th, 42nd, 58th, and 66th, and to send home to recruit, or to form the recruits, the Officers and non-commissioned officers of the four companies drafted."[48] The proposal does not make it clear whether the resulting cut-down battalions would have been paired or not; possibly they would have formed combined units for tactical purposes, whilst retaining a distinct organizational identity. At this stage, Wellington still accepted that he would be unable to entirely escape the strictures of the system, and that he would likely lose some weak units. Nevertheless, he went on to try and obtain Liverpool's blessing for an expansion of his proposals, sugaring the pill with a suggestion that his schemes would also lead to financial economies: "According to this plan we should reduce in some degree our expense in this country. We should keep here Officers inured to the climate and accustomed to the service; at the same time that we should send to England Officers and non-commissioned officers to raise and train recruits. Indeed, it would be desirable if I were authorised from time to time to incorporate the ten companies of a regiment into eight or six companies, according to their numbers, and to send home to recruit, or to train recruits, the Officers and non-commissioned officers of the drafted companies."[49] Whilst blessing was not forthcoming—unsurprisingly, since it was in the gift of the commander in chief, not the secretary of state—it is clear that the concept behind the later extension of the provisional battalion concept had been developed for some time in advance of its implementation. In the short term, Wellington got his reinforcements for the campaign of 1812 and could thus shelve the scheme retaining only the original, still unnumbered, Provisional Battalion to test the validity of the concept.

Although the pairing of the 2/31st and 2/66th seems only to have been envisaged as a temporary expedient until the two units could be brought back to strength, drafts were not forthcoming and the paired organization was maintained. Nevertheless, the combined battalions formed an extremely effective unit. Colonel John Byng, in whose brigade they were placed, reported in spring 1812, "I have now to remark, that a proper understanding and good spirit pervades both Corps, they agree well together, [which] reflects much credit on their commanding officers . . . I consider it a very satisfactory battalion to have under my command, and think it in general good order and fit for any service."[50] The ease with which the two battalions were able to merge may well have had much to do with their having served alongside each other for eighteen months prior to their amalgamation, and was also evidenced by the good relations between their officers, with Lt. George l'Estrange of the 2/31st's "friend and chum and invariable shooting companion" being Lt. Stepney St. George of the 2/66th.[51]

Thus, in the aftermath of the Burgos retreat, Wellington had both a solid precedent to prove the effectiveness of the provisional battalion organization, and several other units that he believed would benefit from it. Some of these battalions had been noted as early as 1811 for potential reduction, and were now so reduced by wastage on campaign to be nearly as weak as the 2/31st and 2/66th had been after Albuera. Other second battalions that had been noted as weak in 1811 had since been absorbed into the first battalion of their respective regiments, but six units remained where there was neither a sister battalion to which the fit men could be drafted nor any chance of reinforcements from home. Accordingly, as of December 6, 1812, Wellington ordered these battalions to implement a similar organization to that of the

2/31st and 2/66th.[52] Incorporating the ideas that Wellington had proposed to Liverpool in 1811, Wellington's scheme incorporated steps to aid the rapid reconstruction of the component battalions and expedite their return to the field as complete units. Thus, the six battalions were required to draft their effectives into four companies of equal strength, sending the unfit men home with the officer and NCO cadres of the six remaining companies. The intention was that these companies would recruit up to strength and then return, allowing the battalions to resume their original independent identities. Meanwhile, the four companies remaining in the field were paired to form three new provisional battalions, numbered 2nd through 4th with the original unnumbered unit becoming the 1st. Rather than keep the headquarters of both component battalions in the field, the commanding officer and staff were transferred complete from one regiment and that of the other sent home: as an additional advantage, this enabled Wellington to retain whichever battalion commander he thought more useful. The only downside of the new system over the 1811 prototype was that, because the weak battalions were taken from across the army, the prior relationship that had aided the amalgamation of the 2/31st and 2/66th did not always exist. In fact, it was paralleled only for the 2/30th and 2/44th, which did not last long together as a provisional battalion but which did continue to maintain good relations with each other thereafter.[53]

The units involved were the 2nd and 2/53rd, forming the 2nd Provisional Battalion with the staff of the latter; the 2/24th and 2/58th, forming the 3rd Provisional Battalion with the staff of the former; and the 2/30th and 2/44th forming the 4th Provisional Battalion with the staff of the former.[54] Inasmuch as the monthly return for December 1812 caught the 2/58th in the process of dividing itself, we can see the mechanics of the new arrangement in action, with the returns showing 23 rank and file going home, along with 5 staff and 9 company officers, 18 sergeants, and all 16 musicians; 354 rank and file remained in the peninsula along with 9 officers and 20 sergeants. Only 2 men recorded as sick were in the detachment set to leave, but there were 147 with the four companies staying.[55] Although retaining so many sick seems contrary to the intention of keeping only fit men, this proportion was in line with most other battalions in the aftermath of the Burgos retreat, representing men who could reasonably be expected to recover in time for the next campaign.

Wellington might have thrown the rulebook of the regimental system out of the window by organizing the provisional battalions, but he was careful to maintain the individual identities of the units involved with companies within the new battalions each containing men from only one regiment, under their own officers. Interestingly, the record of courts-martial for the 2nd Provisional Battalion, which shows the company and battalion of origin of all those brought to trial, demonstrates that here the ordered eight-company structure was dispensed with in favor of a standard ten-company organization including flank companies. Coming from the senior unit, men from the 2nd formed the grenadiers and right-wing battalion companies, whilst the light company and left-wing battalion companies came from the other component battalion, the 2/53rd.[56] Unit seniority was also maintained at a higher level, since the seniority of the provisional battalions within their respective brigades—important so far as the combat deployment of the battalions was concerned—was taken from the identity of the senior element of the provisional battalion. Thus, the 2nd Provisional Battalion was the right-hand, or senior, unit in Anson's Brigade of the Fourth Division since it contained elements of the 2nd Regiment of Foot, irrespective of the fact that it also contained men from the 2/53rd, which would otherwise have been the brigade's most junior unit.[57] Wellington always took care, when mentioning the provisional battalions, to stress the identities of the component units by name, in preference to that of the single combined unit. Whilst this no doubt aided the continued fostering of regimental esprit de corps, there may also have been an element of discretion involved, with Wellington not wishing to draw attention to a scheme that he knew was unpopular at Horse Guards. He also wrote in this fashion to Bathurst after the Vitoria campaign, emphasizing the quality of the units composing the 2nd and 3rd Provisional Battalions, justifying the correctness of his decision to retain these men in the peninsula by telling the secretary of state "it is impossible for any troops to behave better."[58]

Table 5.   Peninsular Provisional Battalions

	1st Prov. Battalion	2nd Prov. Battalion	3rd Prov. Battalion	Average, Prov. Bns.	Regular Infantry
Effectives as %	75.5	70.4	63.0	69.9	66.1
Sick as %	18.9	26.0	32.3	25.4	28.8
On Command as %	5.5	3.5	3.5	4.6	5.0
Deaths/1,000 Men	9.9	12.4	14.6	12.2	18.8
Desertions/1,000 Men	0.6	0.4	1.2	0.7	1.5

Source: Data from Monthly Returns, TNA, WO17/2470–2476. The short-lived 4th Provisional Battalion has not been included in the averages.

By this stage, Wellington had already been forced to break up the 4th Provisional Battalion, and plans to create a 5th out of two weak single-battalion light infantry units, the 51st and 68th, were shelved so as not to provoke further complaints from Horse Guards. From Wellington's point of view, however, his decision was justified by the campaign performance of the three remaining provisional battalions. Looking at the period from the formation of the 2nd and 3rd Provisional Battalions in December 1812 through to the end of hostilities in April 1814, the results show an improvement over the average in a variety of areas, emphasizing Wellington's belief in the value of acclimatized veterans over troops fresh from Britain.

Whilst the overall averages for the provisional battalions, as shown in table 5, work out as consistently better than for the body of regular infantry, there are two partial qualifications. Firstly, the performance of the 3rd Provisional Battalion is worse than the average so far as the maintenance of a good effective strength is concerned. However, a closer examination of the monthly data for this battalion offers something of an explanation, since this unit began its life with a higher number of sick than did the other two provisional battalions.[59] Heavy casualties at Orthez also skew the totals.[60] Regrettably, unlike the 1st and 2nd, no inspection returns for the 3rd Provisional Battalion survive to corroborate these assumptions, but this combination of a lower effective strength to begin with, and higher combat casualties, seem between them to fully account for the discrepancy.

The second qualification relates to the way that this data should be interpreted, since a comparison with Wellington's infantry as a whole gives the deceptive impression that the provisional battalions were more effective than all other battalions. This was simply not the case, for the average figures given in the table include data from strong, experienced battalions on the one hand, and weak, sickly, units on the other. That said, direct comparison with other units does demonstrate that the provisional battalions were a good bargain for Wellington and certainly vastly to be preferred to the manner of battalion that would have replaced them if he had sent them home. For example, to take a sample of long-serving regular battalions then with Wellington, the 1/7th averaged an effective strength of 62 percent of its total manpower over the same period as the provisional battalions were in existence, whilst the 1/11th averaged 65 percent, the 1/45th averaged 64 percent, and the 1/71st averaged 68 percent. All these figures fall in the same region as those for the provisional battalions, although these units had not had the same opportunities to send their ineffectives home, and their figures are therefore skewed by the presence in the line battalions of a greater proportion of convalescents still on the unit strength in-theater.

Comparing the provisional battalions with the sort of battalions that would have likely replaced them is a rather more abstract exercise, since they were retained in the field and no direct replacements were sent out. However, we can compare figures for battalions that did join Wellington in the last eighteen months of the Peninsular War, on the assumption that they typify the units likely to have replaced the provisional battalions. Because these new battalions spent only the last few months of the war at the front, it is impossible to produce an equivalent monthly average for comparison. However, sample figures for this time demonstrate the limited effectiveness of these units, which were mostly those for which Aylmer's Brigade was formed as a "nursery." In April 1814, the 76th had just over 54 percent of its rank and file fit for duty, and the 2/62nd, only 52 percent—neither having seen much in the way of heavy fighting since their arrival at the front.

What was more, these represented drastic reductions over a short period of time—the effective strength of the 2/62nd dropped by four hundred rank and file between October 1813 and April 1814—whereas the ratio of effectives in the provisional battalions, battle casualties excepted, remained steady over the same period. Quite evidently, they were not troops in the same class as those in the provisional battalions, and this would seem, on the face of things, to justify Wellington's retention of his veterans.

The problem with this reasoning is that it addresses only the problems facing Wellington in the peninsula, rather than the global demands that York and Bathurst were dealing with. Looking at the three-sided struggle between these men, a feel for these wider issues soon becomes apparent that also opens up the question of just where the real power lay within Britain's military hierarchy. No single individual emerges from the controversy with an unblemished reputation, but it is impossible not to feel that York was rather more sinned against than sinning, in particular so far as the modern interpretation of events is concerned. From Oman onward, historians have worked from the assumption that Horse Guards was collectively too inflexible to react to the demands of nineteenth-century warfare, and that York's stance over the provisional battalions exemplified this.[61] Following this line of thinking, it then becomes logical that Wellington, in creating provisional battalions against the wishes of Horse Guards, was acting in a manner that was entirely right and proper, even though such analysis takes no account of Britain's military manpower needs across the rest of the globe. Even Rory Muir, addressing Britain's global strategy, states that in wanting to bring battalions back from the peninsula, York was correct only "in theory," and accuses the commander in chief of a lack of flexibility "given the state of affairs in 1813," quite as if there were no other competing demands for troops.[62]

In fact, when the reasons behind York's stance are taken into account, the reasonable nature of what he was proposing, and the extent to which he was willing to compromise, suggest a far more balanced and objective appraisal of the situation than do Wellington's arguments at the time or those of York's detractors since. York's response to Wellington is usually presented via the garbled and paraphrased rendering created by Oman, which is by no means a fair representation of the commander in chief's stance.[63] Having begun by acknowledging his receipt of Wellington's notification that three provisional battalions had been created against his wishes, York went on to concede that he would be

> extremely sorry to withhold my concurrence from an arrangement which your Lordship has already carried into effect, particularly where an immediate objection would tend to reduce your effective force in the field. But when I state the grounds upon which I consider this formation detrimental to the well being of the regiments included in it, I trust your Lordship will see, that while I have every disposition to support and agree to any arrangement of yours, I could not possibly concur in the expediency of the present measure consistently with the opinion I entertain of what is essential to secure the continued efficiency of the army in general; and that you will not be surprised that I should strongly urge the recall of the detachments forming these provisional battalions, when I have the means of proposing an adequate reinforcement for their relief.[64]

York then went on to explain what the knock-on effects of such a policy would be, noting,

> Experience has shown that in every instance wherein a battalion has been brought home a skeleton, composed of its officers, non-commissioned officers, and a certain foundation of the old and experienced soldiers, the greatest success has attended its re-formation for any service within a short period; and this is fully exemplified in the case of the 29th regiment, which is now complete in numbers and ready for foreign service; but if a corps reduced in numbers be broken up by the division of its establishment, it is quite impossible but that such an interruption must be occasioned in its interior economy and esprit de corps as a regiment as would effectually prevent that speedy completion of numbers and reorganization which I look to as the only means of providing for the demands for foreign service.[65]

There was certainly logic in York's stance so far as the big picture was concerned, but when he went on to argue against the provisional battalions as a concept in their own right, his logic was rather more dubious since it was grounded on the unfortunate experience of the 85th's first peninsular deployment when, as York reminded Wellington, "Five companies of this corps were sent to the Peninsula, and the other five were retained as skeletons in this country. A degree of irregularity, contention, and every species of indiscipline, ensued from this division of the officers and establishment, and which can only now be rectified by the adoption of a very strong measure against the whole corps."[66] But as York must have been aware, considering the need for his intervention into the matter, the problems in the 85th stemmed from poor leadership and internal squabbling amongst the regiment's officers. Although the battalion had ultimately been sent home as numerically ineffective, this was hardly a fair comparison with the weak battalions that Wellington sought to retain for the 1813 campaign since they, unlike the 85th, were being paired into provisional battalions. York's argument also took care to ignore the successful deployment of the 2/23rd and 1/71st without all ten of their companies, neither of which by any means represented a failure. York also ignored the successful precedent set by the pairing of the 2/31st and 2/66th, although his language earlier in the letter suggests that he may not have been fully aware of this organization until Wellington chose to place it on the same formal footing as the other three provisional battalions. In any case, if York was attempting to discredit the utility of the provisional battalions as a workable concept in the field, he was struggling for a convincing argument.

York concluded his reply by seeking to convince Wellington that a minor local advantage in the short term would not be worth the wider long-term problems: "I am aware that it has been, and is justly urged, that men who are seasoned to the climate and experienced in the field are more valuable than a greater number Fresh from England without these advantages; and I concur with your Lordship in the opinion you have expressed to such an effect: but I think it would be inexpedient and improvident, for the sake of a present, and comparatively trifling advantage, to sacrifice the only foundation upon which we can look to the eventual efficiency of the army."[67] Because of this need to focus on the bigger picture and longer term, York believed that Wellington would have to give up the bulk of the battalions that he had sought to retain, keeping only the 2/31st and that only until the 1/31st might be transferred from Sicily to replace it. He instead sought to demonstrate that the established regimental system, as it stood, could solve the perceived manpower shortfall in the peninsula: "I transmit herewith, for the purpose of more easy reference, the copy of the last return of your force, made up at the Adjutant-General's office, with the present means of reinforcing the different corps. And your Lordship will perceive that, according to the present effective strength of the regiments you have proposed to retain as provisional battalions, no great sacrifice will be made by their return to England; and you will also bear in mind, that if once my object of giving you 1st battalions only could be attained, you would never be required to part with any experienced soldiers."[68] York's accompanying figures show that although Wellington's infantry were some 12,609 men short of their full establishment strength, much of this shortfall could be made up by drafts taken from regimental depots and/or second battalions. With York still maintaining as a long-term hope the expectation that the field army in the peninsula be composed entirely of first battalions, Wellington's interference was counterproductive and, in York's eyes, served to delay, rather than advance, the moment that Wellington's forces could be brought to maximum efficiency.

Of equal significance, however, were York's points relating to the difficulties in rebuilding units, for it was in this regard that Wellington's scheme began to cause problems for the regimental system. York's choice of the 29th to exemplify this point was apt. As we have seen, that battalion had to be recalled from the peninsula in mid-1811 after having been reduced to a state where its numbers prevented it from functioning as an independent unit, but, as York pointed out to Wellington, by the end of 1812 it was again available for service. True, the 29th was not immediately fit to take the field, but by posting it to Cadiz it was in turn possible for the 2/59th to leave that station and join Wellington. York was trying to make Wellington see that by hanging on to weak but experienced battalions in the short term, he was hazarding the availability of reinforcements being available in the future. Furthermore, Wellington's unilateral decision to suspend the British Army's systems of manpower rotation had the potential to do harm well beyond the confines of the peninsula, since cutting off the stream of attenuated battalions going home to be rebuilt further limited the already-small pool of battalions available for deployment from the home station. Having made such a good case, it is hard to see why York waited so long before exerting his authority on Wellington over this issue. Although he achieved an early success in obtaining the return of the 2/30th and 2/44th, it was not until January 1814 that he gave a direct order to break up the remaining provisional battalions. Even then a caveat meant that Wellington need not do so until it could be achieved "without inconvenience to your present operations," with the additional proviso that men from these battalions could remain in the peninsula if they could be persuaded to volunteer into other regiments.[69] With this loophole, Wellington was able to retain the provisional battalions until the end of the war.

York's January 1814 order also permitted Wellington to retain the 2/53rd, since a draft had been assembled at the regiment's depot large enough to enable the battalion to again take the field as an independent entity. Tellingly, however, such a restructuring was not possible for any of the other five battalions that Wellington had held on to, although York's order did nevertheless permit his retention of the 2/58th, so that its rank and file could be drafted off into the 1/58th once the senior battalion could be reassigned from Catalonia.[70] Reinforcements were unavailable for any of the other units involved in the provisional battalion project, and, since the 2/31st and 2/66th had been awaiting such augmentation for two and a half years, they were unlikely to be forthcoming. Prioritization of the first battalions of the regiments concerned certainly did not help—although the 2nd, as a single battalion regiment, could hardly use this excuse—but the cadres sent home by Wellington were for the most part too weak to recruit effectively for the men left behind.

Conversely, the two battalions that York had recalled in January 1813 were both able to recruit successfully once reunited back on the home station. Indeed, when York ordered the breakup of the three remaining provisional battalions, the constituent units of the erstwhile 4th, although by no means at full strength, were certainly fit for further employment with the 2/44th already on active duty in the Netherlands and the 2/30th about to embark on the same service. With this in mind, it is hard to escape the conclusion that York's rejection of the provisional battalion concept was correct, and that the dislocation to the regimental system as a whole more than outweighed the small advantage gained for Wellington's field army.

Since York was never fully able to force the provisional battalions issue until it was too late to matter, the commander in chief had to find other ways of keeping the system functioning in the face of what was rapidly developing into a manpower crisis to match that of 1809. With new demands for troops for northern Europe, and an escalation of the fighting in the Mediterranean and North America, the British Army was being pulled in several directions, and York, whose correspondence throughout the period shows a marked concern with unit strengths, potential reinforcements, and drafts going overseas, was beset by conflicting demands from different quarters and obliged to make difficult choices in which his own military judgment was not necessarily the primary influence.[71] That said, the first demand for troops to be sent to northern Europe in July 1813 did not create too great a burden; indeed, sending second-rate battalions to the Baltic could be seen as a positive measure that would ultimately enhance the effectiveness of the units in question and make them available for more extensive deployment. Thus, of the battalions sent with Gibbs to Stralsund, the 33rd and 54th were single-battalion regiments recovering from long service in the Indies whilst the remainder were junior battalions of regiments whose first battalion—or first three, in the case of the 4/1st—were already on active service elsewhere. Whilst rather weak, the two single-battalion regiments were effective enough, with Gibbs characterizing the 54th as "a very good serviceable body of men with very few young boys,"[72] and the 2/73rd, which fought at Gohrde, was also a good unit. The three other battalions, however, were in rather worse condition, with inspections of the 4/1st and 2/91st identifying major flaws. In the former case, the rank and file were sound enough, but Gibbs found fault with both the officers and NCOs, many of whom were either worn out or newly appointed and raw. Unsurprisingly, the battalion's ability to maneuver was inadequate.[73] Conversely, the 2/91st could maneuver with some competence, but its rank and file were "very bad, with the exception of about two hundred stout able men" and the NCOs again poor.[74] These two battalions were initially retained in Germany when the other four redeployed to the Low Countries, effectively confirming that they were the worst of the six. Conversely, the other four units would later perform well in the field and so evidently made good use of the opportunity to put themselves into readiness.

Also sent to Germany was a detachment of the KGL, including the 3rd KGL Hussars and two horse artillery troops, to which was attached part of the 2nd RHA Rocket Troop. Like the 2/73rd, these troops served in the multi-national corps commanded by Lt. General Ludwig, Graf von Wallmoden-Gimborn, although the rocket-artillery was later detached to the Army of the North and fought with it at Leipzig.[75] The remaining KGL units, and the 2nd Rocket Troop, eventually joined Graham's forces in the Netherlands just as hostilities came to a close.[76] Along with these formed units, substantial drafts were also sent to Germany from the KGL depot at Bexhill, eventually amounting to 550 rank and file, plus officers and NCOs. However, most of the latter, and some of the former, were promptly reassigned to help train and lead the new levies of the reraised Hanoverian army. The remaining 400 or so men formed a provisional half-battalion of four companies. Commanded by Captain Phillip Holtzermann, this largely functioned as a headquarters guard but was in the thick of the action at Sehestedt on December 10 where Holtzermann was taken prisoner by the Danes; shortly thereafter, the unit was broken up.[77] Because these detachments had primarily a training role, it would be unreasonable to consider Holtzermann's half-battalion as a provisional unit in the same manner as those created by Wellington. Nevertheless, even the creation of this relatively small formation was not without its wider impact, since the redirection of drafts from Bexhill—the bulk of which came from the 1st, 2nd, and 5th Line and 1st and 2nd Light Battalions—diverted men who would otherwise have joined their parent units with Wellington. Each measure that made men available for one theater took men away from another, as would become increasingly apparent in the months to come.

For the moment, however, the established system was coping with the new demands. Nevertheless, the supply of battalions fit even for second-line duties was running out, and the lack of unit cadres coming back from the peninsula to be rebuilt began to be felt. Under these circumstances, York's first response was, on the face of it, entirely logical. If Wellington was retaining existing battalions in the peninsula, York would simply raise new ones. Although new infantry battalions had been created in large numbers during the early years of the Napoleonic Wars, the practice had tailed off, and between 1808 to 1812 only five new battalions were added to existing British regiments.[78] Now, in the twelve months leading up to the first abdication of Napoleon, York raised as many again. Thus, during the second half of 1813, York authorized a second battalion for the 37th Foot, and third battalions for the 14th and 56th. This was followed in early 1814 by the raising of second battalions for the 22nd and 86th. In all of these cases, the existing battalions of these regiments were overseas— mostly in the East Indies although the 2/14th and 1/37th were at Gibraltar—and manpower for the new battalion was provided by redirecting drafts that would otherwise have gone overseas and taking recruits from the regimental depots. There was certainly no intention of creating new second battalions wholesale, and the expedient was preferred of giving a third battalion to a regiment already having two, but with good prospects for recruiting, to that of giving a single-battalion regiment a second battalion unless the latter had the manpower to justify it.

York's 1814 proposals indicate how he identified regiments suitable for augmentation, and of how regimental establishments could be juggled to help create new battalions. Explaining his choices to Bathurst, York first noted that both the 22nd and 86th Regiments had enjoyed particular success both in ordinary recruiting and in attracting volunteers from the Militia, and then went on to detail the particular circumstances of the two units. The 22nd had 752 men on "Isle de France" and 105 on passage to join, giving 857 total, but there were an additional 373 rank and file at the regimental depot and a further 100 recruits on their way there. York therefore proposed that the establishment of what would become the 1/22nd be reduced to 1,000, removing any need to send further reinforcements to that battalion, and "that the numbers within this country, which appear by the statement in the margin to consist of 473 Rank and File, shall be formed into six companies in the first instance with two field officers, and to be augmented in the usual progression until they shall arrive at the Establishment of 1,000 Rank and File." In similar fashion, York noted that the 86th had 604 rank and file in India, 117 on passage and 194 "on march to embark" totaling 915, whilst the depot had 275 men with a further 51 more on the way to join. York on this occasion proposed four companies and one field officer for the new 2/86th, but again expected an increment to 1,000 in due course.[79]

What York was seeking to do was not to abandon the regimental system but to force it to work in his favor, by creating additional battalions at home in order to make effective use of existing manpower from the depots. This was partially reminiscent of Dundas's creation of the Corps of Embodied Detachments, in that York, by plundering regimental depots, was taking manpower that would otherwise have largely been destined for the colonies. Like Dundas in 1809, York was leaving the Indies to look after themselves for a year with the manpower already deployed there. Unlike Dundas, he was taking a far smaller risk in doing so since, the usual rumblings in India excepted, colonial hostilities were effectively over, and, again unlike Dundas, he was using the manpower made available within the existing system rather than in a provisional unit. York's scheme mirrored the raising of the two battalions embodied in 1808, the 2/73rd and 2/84th, both of which were created to meet the demands of the first manpower shortfall by converting the recruiting company of a regiment whose existing battalion was in India. But whereas the 1808 battalions had been used to bolster the home station, thus releasing more seasoned units to go overseas, those of the 1813–14 creation would be deployed almost immediately—ironically, near simultaneously with the first deployments of the 1808 battalions. Because of the demand for manpower in the field, time was too pressing for the new battalions to reach anything like full strength or to come together as cohesive units. Indeed, by the time York penned his proposals for the 1814 battalions, two of those from the 1813 augmentation were already on active service, and in light of this one cannot help but identify an element of naïveté in York's expectation that the 2/22nd and 2/86th would have time to recruit up to 1,000 men apiece. That these battalions went overseas so soon was due to the need to find troops for Graham in the Netherlands; this demand, along with those of the existing theaters of war, saw the first steps toward the temporary abandonment of many of the basic tenets of the regimental system.

1814: Breaking with the System

The decision to send a force to the Netherlands was made in November 1813, with the first wave of troops sailing during December. The struggles to assemble and maintain this force make it abundantly clear how limited Britain's remaining manpower reserves were at this stage in the war, highlighting in turn the damage done by the failure to agree on a manpower policy during 1813. Because York had failed to obtain the return of unit cadres from the peninsula, and was unable to create replacement units quickly enough from troops on the home station, the last five months of hostilities before Napoleon's abdication in April 1814 would see the employment of increasingly dubious and desperate measures to keep up with the manpower demands of what was now a three-front war.

In the case of the Netherlands campaign, however, matters were made worse by the political significance of this theater, which led to Bathurst taking a close interest in the composition of the force being sent there. On December 4, Bathurst sent a series of instructions to Graham, outlining the scope of his command and the objectives he was expected to pursue.[80] Included was a memorandum outlining the composition of his forces, decided upon two weeks previously, and by the time Graham received his orders steps were already under way to get the first two of Graham's four infantry brigades on their way to Holland. One of these brigades would be formed of the best four of Gibbs's six battalions from the

Baltic; another, which would form the vanguard of the expedition under the command of Major General George Cooke, was drawn from the second battalions of the Foot Guards—the same standing Third Guards Brigade that had provided men to serve under Graham at Cadiz three and a half years previously. Apparently drawing on the 1810 organization, Bathurst envisaged the Guards providing two battalions totaling 1,600 men. In this conception, 2/1st Foot Guards would go out as a strong battalion of 800 rank and file, but 2/Coldstream and 2/3rd Foot Guards would contribute only half that number, combined to form a single provisional battalion.[81] However, the numbers did not work out as Bathurst had hoped; 2/1st Foot Guards had insufficient men, but the other two battalions met and surpassed their smaller quotas. Accordingly, all three regiments were able on this occasion to put small but workable battalions, of six or seven companies each, into the field. All three subsequently had additional companies sent out to bring them up to establishment before Waterloo.[82]

Meanwhile, with Gibbs and his men on the way back from the Baltic and the Foot Guards rushing to complete their preparations, Bathurst was scouring the country in search of other units available for deployment, ultimately producing nine additional battalions to send to Holland, which, in his proposed organization would, between them, form two further brigades. Graham would wisely break up this organization as soon as he had the chance, mixing these battalions in with Gibbs's more seasoned troops from the Baltic, but could do nothing about the remarkably poor quality of many of the battalions sent. Indeed, the state of some of these battalions was so poor as to place a question mark over the level of awareness that the secretary of state for war and the colonies actually possessed with regard to the practicalities of mounting a campaign.

For a start, of the nine battalions one was a single-battalion regiment, five were second battalions, two were third battalions, and the last was a veteran battalion. Two of these battalions had been raised as part of York's creation of new units discussed above. Bathurst's November memorandum envisaged these battalions providing 3,950 men, but, although Bathurst had evidently made use of the appropriate monthly returns in compiling his memorandum, he was either unaware of, or chose to ignore, the distinctions between total and effective manpower. Admittedly, in all but one case—that of the 2/35th, where it fell short by forty-three men—Bathurst's figure was within the total strength of the unit in question, but by no means all of these men were effective due to sickness or absence. Furthermore, comparison with the figures for men considered by their units to be "fit for immediate service"—that is, also discounting boys and recruits at drill—produces an even wider discrepancy. The 2/37th was now left 176 men short of Bathurst's 500-man target, whilst the 2/52nd, down to provide 300, reported nobody fit for service as the bulk of its 219 effectives were boys.[83]

Irrespective of the palpable unfitness of the majority of these battalions, all nine were sent out in December 1813, mostly grossly understrength. Neither of York's two new battalions could yet muster a full complement of companies, with the 2/37th having six and the 3/56th, only embodied as a unit a month before sailing, five.[84] A manpower shortage in the five companies of the 3/95th Rifles then in Britain, serving as depot for the active companies in Spain, also meant that to meet Bathurst's quota of 250 riflemen it was necessary to take men from the depot of the regiment as a whole to form a provisional battalion with companies from all three battalions of the 95th—one each from the first and second and two from the third, under Brevet Lt. Colonel Alexander Cameron.[85]

With so many units falling short of Bathurst's targets, the nature of these discrepancies are made most readily apparent when presented in table 6. Whereas the regular battalions all provided as part of their monthly returns a figure detailing the number of men fit for immediate service, no such distinction exists in the return for the 1st Royal Veteran Battalion; no doubt this was because under normal circumstances the unit would never have been earmarked for service in an active capacity. Nevertheless, since this unit would, by virtue of its role, contain neither boys nor new recruits, it has been given the benefit of the doubt and the assumption has been made that the full effective strength would be available. When inspected, the 1st Royal Veterans were found to contain many "fine men [who] would be found fit for active service, & certainly many more, tho' not fit for very active service,

Table 6. Battalions Sent from Britain to the Netherlands, December 1813

	Strength per Memo	October Strengths			December Strengths	
		Total	Effective	Fit	Total	Effective
1st RVB	500	554	489	489	461	459
2/35th	600	557	487	463	461	453
2/37th	500	533	324	281	298	279
2/44th	500	663	394	333	422	406
2/52nd	300	412	219	0	197	191
55th	400	476	373	276	356	340
3/56th	400	418	310	247	280	262
2/69th	500	589	426	366	487	487
Rifle Bn.	250	625	386	216	305	287
Total	3,950	4,827	3,408	2,671	3,267	3,164

Source: October strengths from Battalion Returns, TNA, WO17/262–263, 267–268; December strengths from Monthly Return, TNA, WO17/1773. October data for the 3/56th refer to the regimental depot out of which that unit was formed; those for the Rifle Battalion relate to the home companies of all three battalions of the 95th. "Fit Strength" refers to men listed as fit for immediate service.

would be able to under go fatigues in our foreign garrisons."[86] From the evidence available, this would suggest that they were probably of more use than some of the regular battalions.

The number of fit men available fell short of Bathurst's expectations by nearly one-third, and the total deployed—which, by default, must therefore have included men deemed unfit by their units—still fell short by nearly a quarter. What was more, over and above the units' being understrength, the state of these battalions was frequently poor, with many lacking much of a cadre of old soldiers, and many having particularly young NCOs. In the case of the 52nd and 95th, the units in question were configured for the depot role, with the 2/52nd sending off all fit men to the first battalion as soon as they were ready for service.[87] Despite this, the light infantry units performed well on service, and the 2/35th and 2/69th also stood out as being better than the remaining line battalions with Lt. General Doyle finding the 2/35th composed of "a good body of men very fit for service, with a healthy and cleanly appearance," whilst those of the 2/69th were seen to be "in general fit for service, but very young" when inspected by Major General Hawker.[88] Even the better units, however, were numerically weak and were further reduced because Graham detached the unfit men to garrison duty for the sake of their health. Since his force lacked such necessities as camp kettles, blankets, and shoes, sending unseasoned and unfit men into the field in the middle of winter would have been, in Graham's opinion, tantamount to a death sentence.[89]

It soon became apparent that Bathurst's original scheme had failed to provide the manpower needed to meet Britain's aspirations for the Netherlands, leading to the dispatch of the 2/30th and 2/81st from Jersey and of the 2/21st and 2/78th from Leith.[90] These battalions were, if somewhat weak, in good order and possessed of good officers and NCOs, and the 2/30th had the benefit of a veteran commanding officer in the shape of Lt. Colonel Alexander Hamilton.[91] The last two battalions from the Baltic were also marched overland to join Graham, although neither was much of an addition in terms of quality. In the meantime, original battalions were left dispersed and shrunken. By March 1814, the 2/37th was down to 144 rank and file, of whom only 61 were actually effective, and several others were in nearly as bad a condition.[92] Accordingly, Graham resorted to Wellington's peninsular expedient and formed his two weakest units, the 2/21st and 2/37th, into a provisional battalion, in part because the 2/21st had lost its field officers at Bergen op Zoom.[93] What was more, the quality of replacement manpower was even worse with many drafts composed of boys completely unfit for active service; in the 33rd, a virtue was made of necessity and the bulk of the unfit youths reassigned as officers' servants.[94]

Sending Graham more battalions helped matters in the short term, but it further reduced the pool of units available in Britain and did nothing to inject extra manpower into the system. Recruiting had not noticeably tailed off, but nor had the supply of recruits increased in proportion to the growing need for active manpower. Accordingly, thought had to be given to greater employment of foreigners—not just in specifically foreign units but also to fill the ranks of the British line. Generally, foreign man-power was actively employed to a far greater degree in the last year of the war, but this was achieved largely by switching existing units of foreigners in British pay to active duty from garrisons. This shift in policy is reflected in the deployment of the Regiments de Meu-ron and de Watteville to North America, and in the large number of foreign units assigned to the forces on the East Coast of Spain. There was also an increment to those nominally British regiments that were, in fact, largely foreign: most obviously the 60th, which gained two extra battalions during the last twelve months of the war—a seventh raised on Guernsey largely from ex–prisoners of war, and an eighth at Cadiz through the re-designation of a bat-talion of foreign deserters originally formed in 1811. In October 1813, York proposed that since this body was 700 strong it should be taken into the line, and this was done the following January, although with only 590 rank and file, which suggests that York had again been rather optimistic.[95] The new 8/60th, being the older unit notwithstanding its junior status in the line, had a bet-ter class of deserter in its ranks; Graham, when forming the unit, had sought to ensure that no French were enlisted, preferring Poles, Germans, and Italians.[96] Conversely, the 7/60th contained a large number of Frenchmen and Italians who were the cause of considerable trouble. Around half of the surnames of those ap-pearing in the abstract of Regimental Courts Martial are French, and the nature of the trials suggest that little discrimination was made respecting the character of the men enlisted. One such, Philippe Le Brun, was broken from sergeant after an outbreak of violence and subsequently, as a private, received 250 lashes for "Biting Pr. Darsen [and] threatening the lives of his officers and comrades."[97] So bad was Le Brun's conduct that he was the only man to be singled out by name when Horse Guards instructed that the battalion's worst characters be discharged to the Foreign Depot, prior to the unit embarking for North America.[98] There were also difficulties with training, on account of the fact that "the greater part [of the men] have been drilled in the French manner & it is difficult to break them of habits that they have contracted in that service."[99] No doubt as a result of the turbulent mixture of men assigned to these battalions, care would seem to have been taken to place them under steady commanding officers, and the conduct of both Lt. Colonel Henry John of the 7/60th and Major John Fitzgerald of the 8/60th was favorably reported on, with John's absence on sick leave during late 1814 considered a particular blow for his battalion.[100]

The 8/60th never saw active service, going to Gibraltar when Cadiz was vacated, but the 7/60th served in Nova Scotia and Maine, remaining in Halifax on garrison duties until disbanded in 1817. But if the composition of these battalions was suspect, worse still was that of the final body of foreigners recruited from the captives and cast-offs of the Grande Armée. These were the five Companies of Independent Foreigners, established from July 1812 out of French prisoners and deserters. The measure was prompted as much by lack of suitable prisoner accommodation as anything else, but their indiscipline caused more problems than it solved, and they were hastily disbanded upon the coming of peace after achieving a characterization as "desperate banditti."[101] The employment of the First and Second Companies of Independent Foreigners in the expedition to Hampton Roads, during which at least two murders were committed along with several cases of rape, is a particular stain on the honor of the service during the period.[102] Whilst such conduct cannot be justified, it should however be seen primarily as an indication of how short of manpower Britain had grown that such manner of troops were employed in active service, even on a minor expedition in a secondary theater.

The year 1813 also saw a sizeable draft of foreign manpower into another British regiment, the 102nd Foot. However, as with the 60th, the 102nd's British status was largely nominal; the regiment had begun life as the New South Wales Corps, and still retained something of the character of a colonial regiment. Under these circumstances, the drafting in of German volunteers to bring the battalion up to strength for garrison duties on Bermuda makes perfect sense.[103] In May 1814, its German element numbered 189, or 26 percent of the strength, and it was noted that these Germans were "the best conducted men in the Battalion," something reflected in the fact that fourteen of them had gained NCO rank. By then, other battalions were benefiting in a similar fashion, albeit on a smaller scale and without the sanction of Horse Guards. The first instance of this practice came in the peninsula, where in 1812 Wellington authorized the recruitment of a limited number of Spaniards into understrength battalions. Subsequently, this practice was extended to all British battalions serving in the theater, which, assuming all reached Wellington's quota of 100 Spaniards per battalion, would have added 4,100 recruits. In practice nothing like that figure was achieved, since numbers were small in the first place and many of those who did enlist subsequently absconded; September 1812, for example, saw the 2/31st recruit three Spaniards but have seven desert.[104] The majority of those who did stay, estimated at no more than 300, served primarily in the Light Division; they were considered able soldiers, but they were far too few to make much difference to the grand scheme of things.[105] Gibbs's battalions in the Baltic also began to add foreigners to their strength, with an inspection of the 2/91st finding ten Germans—"good sized stout men"—amongst the battalion's rank and file in October 1813.[106] Nor was this practice restricted to the 2/91st, since a return for the troops joining Graham in February 1814 includes "100 Foreign Recruits for the 33rd Regiment."[107] The exact fate of this draft is unclear, for the total strength of the 33rd increased by only sixty-one men between February and March 1814, and the battalion was expecting fifty recruits to join from its depot as well as this draft of foreigners.[108] Nevertheless, when the 33rd was inspected in April 1814, it had seventy-eight foreigners in the ranks, including two sergeants, a corporal, and a musician, although sixty-nine of them were "not yet supplied with arms."[109]

At the same time as measures were being undertaken at the instigation of local commanders, Castlereagh was advocating a more significant increment to Graham's army by proposing to Lord Clancarty, ambassador to Holland, the large-scale enlistment of Dutchmen.[110] Since Castlereagh sought to add a hundred men to each of the fifteen British line battalions then either serving with Graham or earmarked to do so, this was something on a par with Wellington's planned enlistment of Spaniards, although the relative weakness of Graham's battalions would have ensured that a greater percentage of the total force were foreigners. Castlereagh recognized that adding too many new recruits would have an adverse effect and likely render the battalions ineffective, but did not specify how the language barrier was to be overcome. In any case, the Dutch took an unsurprisingly dim view of British attempts to recruit their nationals and nothing came of the project, which Graham considered provocative and unworkable from the start.[111] On the other hand, Graham was keen to recruit foreign deserters from the French service, informing Bathurst, "There are, however every day considerable numbers of deserters coming in at different posts. I should suppose there w'd be no objection to my enlisting the Germans &c & incorporating them in some of the very weak Battn's here—I shall detain those who are willing to serve with us at Willemstadt till I receive your L'dship's directions."[112] In the end, nothing came of this either, and the only battalions under Graham's command with substantial numbers of foreigners in the ranks by mid-1814 were the 33rd and 2/91st, both of which had already started recruiting Germans whilst in the Baltic.[113]

Nevertheless, the fact that such measures were even considered makes it clear that the bulk of the battalions deployed in 1813 were weak and unfit for service, to the extent of having to take pretty much any available manpower. In a particularly poor state were the two new battalions raised during late 1813, and Cooke's report on 2/37th makes a specific condemnation of deploying battalions as raw as these.[114] That the other new battalion in the field, the 3/56th, was less badly off would seem to be due largely to the exertions of its commander, Lt. Colonel John Browne.[115] In Britain, meanwhile, another of the new units, the 2/86th, was still in such a state of organizational flux that its sick had had to be left in the care of a Militia regiment in the same garrison due to a complete lack of medical staff, and was clearly in no fit state to go overseas.[116] York's attempt to force the system to suit his purposes had failed, and more drastic measures were therefore required if the manpower embodied into these new battalions, and that remaining in the home depots of other regiments, was to be employed effectively.

These steps, however, were to be taken in conjunction with another move that would represent an even greater shift away from the army's established methods of manpower management, and which would in turn indicate an increasing belief in the failure of those systems. This was the introduction of a bill into Parliament on November 4, 1813, to allow the Militia to volunteer for service in Europe. The policy was not well received by Horse Guards or by Wellington, and it indicated that, for the first time in thirteen years, the British Army's own established methods had failed to meet the country's needs for active military manpower. Bathurst initially hoped for some 20,000 militiamen to volunteer into the line, but explained to Wellington that although this remained the preferred option in the long term, it would take some time to implement.[117] If, as was hoped, existing Militia battalions could be deployed on active service as complete units—as they had been during the Irish Rebellion of 1798, although that had technically still been home service—then this would get men into the field more readily. Yet one can well understand that the scheme did not sit well with York. Sending the Militia overseas did not simply circumvent the regimental system; it circumvented the country's whole military organization.

In practice, Militia volunteering fell far short of the levels expected, and no one Militia battalion was deployed as such, with the only units going overseas being three provisional battalions formed into a brigade under Major General Sir Henry Bayly, ready for service in March 1814. Such a limited response stood in stark contrast to the favorable results achieved in 1800–1801 when the Fencible infantry was called upon to go overseas and sent units as far afield as Portugal and Egypt.[118] In conjunction with the reduction in the numbers of militiamen prepared to volunteer directly into the line, the difficulties inherent in finding sufficient manpower to form the Bailey's three battalions in turn raises questions as to how far the war-weariness and overstretch existing in the regular army by this time also extended back to the home front. Initially earmarked to join Wellington, the Brigade of Provisional Militia was briefly reassigned to the Netherlands before this order was also countermanded and its original assignment restored; the three battalions eventually reached Bordeaux on April 12, a day after the news of Napoleon's abdication.[119] Each provisional militia battalion was built around men from a single core unit, topped up with volunteers from others, but this arrangement introduced complications, with Colonel Sir Watkin Wynn obtaining command of the 3rd Battalion, largely drawn from the 2nd West Yorkshire Militia, because the 135 volunteers from his own Royal Denbigh Militia refused to serve under anyone else. Since George Canning described Wynn as the "worst man of business" he had ever met, either the colonel's military talents were considerably greater than his political ones or those 135 Denbighshire men came at a potentially high cost—the more so when it transpired that although armed as riflemen they were lacking instruction in the drills for such weapons.[120] Politicking of this degree—Henry Torrens claimed the three Militia battalions were "more troublesome than the whole army put together"[121]—and fears over the response of the rank and file ensured that the militiamen were retained as a distinct brigade. Since they mustered 2,450 rank and file when they prepared to re-embark in May 1814,[122] this represented an appreciable infusion of manpower, but meant that as a body their potential combat value was considered by Wellington to be fairly negligible.[123]

With the Militia schemes failing to have the desired effect by the spring of 1814, and demands for reinforcements from all theaters still coming in, Bathurst was forced to turn to York for new ways of getting men into the field. After the failure at Bergen op Zoom, the Netherlands once more became the focus of Bathurst's attentions. One result of this was the soon-to-be-rescinded decision to redirect the Brigade of Provisional Militia from Wellington to Graham, provoking a petulant outburst from the former.[124] The other result was that pressure was put on York to find more men. York's acceptance of the inevitable came on March 14, 1814, when he presented his compromise plan for the creation of detachment battalions at home.

> In reference to the conversation I have had with your Lordship respecting the necessity of sending a reinforcement to Holland, I have to acquaint you, that, exclusive of the Brigade of Provisional Militia, now embarked at Portsmouth, there is no alternative but to make up weak Battalions of the Line by Detachments from other Corps which, although injurious to

the service generally, I cannot hesitate to recommend to the Prince Regent under the present emergency.

The following is the force proposed to proceed on this service

3 Provisional Battalions – 2,400
Foot Guards – 3 Companies – 300

5th Foot – 4 Coys. – 400
63rd Foot – 3 Coys. – 300   } 800
39th Foot – 1 Coy. – 100

14th Foot – 5 Coys. – 500
86th Foot – 2 Coys. – 200   } 800
4th Foot – 1 Coy. – 100

22nd Foot – 5 Coys. – 500
9th Foot – 1 Coy. – 100   } 800
19th Foot – 2 Coys. – 200

Total 5,100[125]

Within York's 5,100-man total, the three companies of Foot Guards were drafts already under orders; the "Provisional Battalions" refer to the Militia then embarked at Portsmouth, not the new battalions whose composition York then details. York's initial draft has the 22nd providing four companies only, giving just seven hundred men for the last detachment battalion, but this figure has subsequently been scored out and the totals changed to those given above.

Not only did York submit the proposal under protest, stressing that it was "injurious to the service generally," but also, even in extremis, he did not abandon entirely the tenets of the regimental system. In no sense does the proposal imply a return to the hated eighteenth-century practice of permanently drafting men from one regiment to another, although logically one might expect the three battalions central to the scheme—the 2/5th, 3/14th, and 2/22nd—to have absorbed the smaller detachments. Clearly, so far as York was concerned, the proposed organization of these three battalions was a temporary step only, and the detached men would ultimately return to their units. The maintenance of distinct, regimentally based companies within these detachment battalions emphasizes the importance placed on maintaining unit identity, ironically mirroring the similar arrangements in Wellington's provisional battalions. In effect, York was seeking to bring in extra manpower as a way of advancing the readiness of the three "core" battalions, thereby retaining the intention to revert to normal methods as a long-term objective, and the central role of these units was emphasized by their each being called on to provide two of the three field officers assigned to each provisional formation.[126]

The scheme indicates York's acceptance that battalions of three hundred to five hundred men required some form of augmentation before being put into service. This in turn signals a realization, after the poor performance of the 2/37th and 3/56th, that his newly created battalions were not fit for service in their own right, hence the utilization of two of them—2/22nd and 3/14th—as core units in this scheme. Far from being fit on its own, the last of York's new battalions, the 2/86th, was called upon to contribute only two companies. The number of men actually available in each of the units concerned, including where appropriate those in the regimental depots, is shown in table 7.

As the figures indicate, York's proposal was by no means unreasonable. The 2,400 men required were easily available, and although a comparison between the numbers demanded by York and the effective strength of the units finds the 2/22nd badly deficient of its required total and the 2/63rd and 2/86th slightly so, other battalions still had men available over their quotas. Only when the required figures are compared with the numbers "fit for immediate service" does the proposal begin to fall down, with the total fit strength of all nine units amounting to barely two-thirds of the numbers required. Additionally, the figures for the 2/5th transpired to be out of date, and the battalion had fewer men available than York had initially believed.[127] To make the scheme work, men who were not thought fit do so by their units would therefore have had to go overseas, but by no means to the same degree as had been the case in the battalions sent to Holland the previous

Table 7.   York's Provisional Battalion Scheme

	Numbers Required	Total Strength	Effective Strength	Fit Strength
2/4th	100	363	163	37
2/5th	400	739	543	497
2/9th	100	228	134	88
3/14th & Depot	500	662	521	359
Depot/19th	200	383	262	119
2/22nd	500	419	348	227
2/39th	100	213	134	28
2/63rd	300	312	264	221
2/86th	200	301	178	95
Totals	2,400	3,620	2,547	1,671

Source: Unit strengths from Battalion Returns for February 1814, TNA, WO17/274, 275, 277, 279, 281.

December. There were some men who certainly could not have gone—the 2/39th's eight pioneers, for example, listed as "Unfit for active service but essential for depot purposes."[128] Equally, in a time of maximum effort, there is no reason why the sixty-nine men variously excepted by the 2/9th as being "Corporals, Taylors, Officers Servants, &c, &c." should not have been called upon—indeed, all three categories might be deemed essential for the effective functioning of the detachment.[129]

Eventually, by way of compromise, the strict quotas were abandoned and units unable to meet their targets instructed to send "all such Recruits as were equal to the fatigues of a Campaign" to Colchester, where the three "Detachment Battalions of the Line" were assembled under the superintendence of Major General Acland.[130] In order to partially address the manpower shortfalls, the 2/4th had its quota doubled to two hundred men, and the depot of the 29th Foot was ordered to supply the same number; when the latter proved impossible, the order was rescinded and the 2/15th, on Jersey, was called upon instead.[131] Despite such problems, and difficulties in arming and equipping some of the detachments, the three battalions were brought together surprisingly quickly. It was nevertheless recognized that the initial hoped-for strength was unlikely to be attained, and when Graham was informed that these units would be joining him their combined strength was set at a more achievable 2,000.[132] When peace intervened, the detachment battalions were essentially complete in their modified form, awaiting only the men from Jersey, but it was no doubt with some relief that York was able to authorize the breaking up of these units before they had the chance to see any active service, the orders being sent within days of the news of Napoleon's abdication.[133]

For a scheme produced at short notice and against the opinions of its author, much can be said in favor of York's scheme, both in conceptual and practical terms. As one would expect, it also shows that York, the professional, was far more conversant with the subtleties of the monthly returns than his political master. Unlike Bathurst, putting together his initial proposal for Graham's command, York's proposal indicates a far more realistic grasp of the significance of the information contained in these documents— tempered, perhaps, with a slight element of optimism as to how many men the 2/22nd might recruit. All in all, the scheme suggests a measured approach and a careful weighing of short-term gains against long-term problems. Most probably, York would not have proposed the scheme without pressure from Bathurst. Nevertheless, his responding to political demands in this way indicates a flexibility of mind and approach not traditionally associated with York's management of the British Army; Calvert's staff work, in ensuring that the scheme was so speedily enacted, also merits praise. Unlike Bathurst, seeking only to get numbers overseas, and unlike Wellington, seeing only the local picture, York's response to the problems of 1813–14 imply a mind well aware of the qualities required to enable a unit to function effectively, but also of the wider strategic imperatives that inevitably forced a compromise when political and military priorities clashed.

Because there were clashes of priorities, and because the overall shortage of manpower meant that not every demand could be met, the events of 1809–10 and, in particular, of 1813–14, force questions to be asked as to where the real power lay within this web of relationships. During Dundas's tenure as commander in chief, Horse Guards can be seen as essentially subordinate to political control, willing to readily acquiesce to measures that deviated from the established system in order to generate the manpower required for the government's schemes. Yet these measures were generally still built on existing regimental identities and made the most of the fact that administration of manpower was most effectively carried out at regimental level, something that is reinforced by the effective performance of the battalions of detachments. With the return of York as commander in chief and the appointment of Bathurst as secretary of state, the situation should have equalized, with less scope for a single dominant individual, but the increase in Wellington's powers as commander in the peninsula now complicated matters. Delays in communication meant that ostensibly temporary solutions could in practice continue as semipermanent for months if not years, with the commander in chief having little practical recourse unless he were to give a direct order overturning that of the man on the spot. Quite rightly, Wellington's primary focus was always set on winning the Peninsular War, but, equally rightly, that of Horse Guards was set on winning the Napoleonic Wars as a whole—a point that is often discounted by those who defend Wellington's stance.[134] In this context, the provisional battalions, although undoubtedly effective in the field, can only be seen as counterproductive so far as the struggle as a whole is concerned. Had all eight units involved in the scheme been returned to Britain during 1813, they could have been rebuilt as were the 2/30th and 2/44th, perhaps in time to provide an effective nucleus for the Netherlands expedition. For all that he pleaded shortage of manpower, Wellington in 1813 and 1814 frequently made major detachments from his main army, which rather negates his plea that he needed every experienced man he could muster and suggests that the outcome of the Peninsular War would not have been radically altered by the replacement of the provisional battalions with less seasoned units.

Toward the end of 1813, a clear strategic shift can be seen toward a favoring of political priorities over military concerns, and this in turn gave greater weight to Bathurst as secretary of state.[135] Whatever his many good qualities as a man and as a politician, there were undoubtedly occasions when he displayed an

unfortunate grasp of military realities, which in turn led to his pre-
senting or supporting rash and impractical schemes.[136] Bathurst's
responses to the manpower shortfall certainly tends to confirm
this assessment, and the volte-face inherent in his involvement
in the Netherlands expedition is remarkable. From meddling in
military matters to the degree of outlining Graham's brigade organization, Bathurst was soon reduced to effectively begging Horse
Guards to find him some effective replacements post haste. Under
these circumstances, with conflicting demands coming in from all
sides, it is hard to fault York's clinging to the established system
as the best means of fending off these pressures. It is, however,
equally to his credit that he was able to recognize by spring 1814
that this approach had been pushed as far as it could go, and
to produce a feasible and well-reasoned alternative. Even in the
desperate days of March 1814 the best elements of the regimental
system were largely maintained in York's Detachment Battalions
of the Line, along with precedents drawn from five years' experience of coping with that system's flaws. Far from indicating inflex-
ibility, the events of 1813 and 1814 in fact show a commander in
chief and staff fully capable not only of absorbing the lessons of
previous expedients but also of applying that experience to the
new crisis at hand.

CHAPTER 5

Beyond the Regiment

In May 1815, Lt. General Sir Lowry Cole received news that he had been offered a divisional command in the army then being assembled in Flanders to deal once and for all with the resurgent Napoleonic threat. Whilst his impending marriage meant that Cole would be delayed in taking up his new command, he was keen to ensure that he had first choice of the two vacant appointments, requesting of Wellington that he

> be appointed to the 6th division in preference to the 5th, as I cannot help feeling a very strong partiality for those regiments which composed the 4th in Spain; and I understand that General [John] Lambert's brigade, of which the 27th and 40th form a part, are already attached to the division, and that the brigade daily expected from America, among which are the 7th Fusiliers, are likewise to be attached to the 6th division; and if, without inconvenience to the service, the 23rd Fusiliers, which have been some time in Flanders, could be added to it, I should have four of my old regiments with me, a circumstance by which I should feel much gratified and obliged.[1]

Cole was in part motivated by a desire to build on established command relationships, noting that he was convinced that he was "more likely to meet your future approbation with officers who are well acquainted with me, and whose merits I can appreciate." However, his letter also indicates the existence of a close connection between the identity of a regiment and the identity of the division to which it was assigned, something that was in turn closely wrapped up in the function of the divisional commander as a duplication, on a larger scale, of the role played at a regimental level by the unit commander.

That such connections could exist by 1815 was a clear indication of how far the divisional system had become an accepted part of the British Army's organization and psyche, having become more and more widespread during the period covered by this work. Eighteen months previously, Graham had reorganized his forces in the Netherlands into two small divisions, telling Colonel Bunbury at Horse Guards that such an organization was "so much more convenient than any other," and doing so in conscious imitation of the system with which he had worked in the peninsula.[2] It would be wrong to imply that Wellington introduced the British Army to the divisional system, but it is certainly the case that his five years of peninsular command saw existing systems of organization developed to a far more sophisticated level than hitherto. By delegating both command and staff functions to an intermediate level of command, theater commanders were left with far more time to deal with their wider responsibilities. But there was more to it than this. Careful choice of divisional assignments, in order to position battalions most effectively within the army's order of battle, allowed for the effective functioning of the policies of manpower rotation outlined in chapter 3, but the concept could be extended still further by ensuring that the battalions with a brigade, or the brigades within a division, were composed of troops of varied levels of experience in order to homogenize the performance of the force as a whole. Alternatively, specialist troops could if necessary be concentrated together, and Wellington in the peninsula long maintained the Light Division in this way as something apart from his numbered divisions of the line. Naturally, if they were in existence for any length of time, such formations developed a sense of individual identity and esprit de corps as self-defined elites; so too, however, did the peninsular line divisions, with a form of rivalry developing that was very similar to that seen at the regimental level.

It is, therefore, all the more surprising to find that permanently organized divisions were a comparatively new innovation so far as the British Army was concerned. Until Wellington began to reorganize his forces during 1809 and 1810, field command arrangements above the brigade level had generally been short-lived and their autonomy limited. Nevertheless, many of the concepts that reached fruition during Wellington's tenure in the peninsula had their roots in earlier campaigns, and their development is therefore best understood by first considering the precedents from which they stemmed.

Higher Organization before 1809

At least as far back as the War of the Austrian Succession in the 1740s, the British Army had gone to war with its organization up to brigade level generally remaining as constant as the circumstances permitted. Brigade commanders might come and go, but the same units would frequently serve side by side for the duration of a campaign, and not infrequently for longer than that. At a higher organizational level, however, matters continued to be rather more fluid, with brigades assigned to ad hoc wings, lines, or columns depending on the requirements of the ongoing campaign or anticipated battle.[3] This worked for the relatively formalized nature of warfare of the early and mid-eighteenth century, but the more flexible tactics of the American and French Revolutionary Wars indicated the need for a more substantial subordinate organization that could shoulder some of the burden of command and staff duties.

French commanders and theorists as early as the 1760s had argued for the adoption of a divisional system in which a single senior commander would assume control on a more concrete basis of two to four brigades of infantry, along with supporting artillery and sometimes cavalry. The vastly expanded armies of the Revolutionary Wars saw this theory put into practice, and by the opening of the nineteenth century, France had not only adopted this organization as a matter of course but had taken the concept further by experimenting with the *corps d'armée* of two or more divisions.[4] Other nations followed suit, often in direct response to defeat by the French, but Britain lagged behind. Its own experiences in Germany and America during the second half of the eighteenth century had taught the effectiveness of such methods, at least in a primitive form, but in the new struggle against revolutionary and Napoleonic France, the opportunity to field an army large enough to warrant the implementation of a divisional system did not immediately arise.

The Duke of York did, it is true, eventually come to command substantial forces in Flanders during 1793–94, but this was a multinational force, and, although its British contingent contained enough troops to form several divisions, higher organization remained within a rigid eighteenth-century structure of Advance Guard, First and Second Lines, and Reserve. When York again commanded British forces in the field, during the Helder campaign of 1799, his troops were organized into three divisions, but this related only to the groupings in which they were shipped to the Netherlands, and once committed to battle this organization was dispensed with and the troops assigned to various ad hoc columns; in size and combat function, these formations equated to divisions, but their impermanent nature meant that staff functions remained at army level.[5] A similar organization—or, rather, lack thereof—can be seen in Lt. General Sir Ralph Abercrombie's army in Egypt in 1801. However, although this force largely fought as independent brigades of three or four battalions, it did include one larger formation, designated the Reserve, in which a subordinate general officer, Brigadier General Hildebrand Oakes, assisted the commander, John Moore, then still a major general. Moore's Reserve had no internal brigading, and this development cannot truly be seen as influential in the movement toward divisions; nevertheless, as a grouping of picked troops, the concept would recur in subsequent campaigns.[6] Lastly, in this survey of precedents from the era of the Revolutionary Wars, it is important to consider the organization of the forces in India during Arthur Wellesley's service on the subcontinent. Lt. General George Harris's army during the 1799 campaign against Mysore had a left and right wing of infantry, each having three component brigades, and a two-brigade cavalry command; Harris's army operated in conjunction with Lt. General James Stuart's divisional-sized Bombay Army and assorted smaller contingents, including the Hyderabadi Subsidiary Force overseen by Wellesley. Significant in light of subsequent practice was the mixing, by brigades, of regular British

Army units with those of the East India Company, which parallels the subsequent incorporation of Portuguese and Hanoverian troops into British divisions during the period of this study. During Wellesley's later Indian service, when he himself commanded in the field but with smaller forces, this practice of mixing troops was continued at a brigade level.[7]

During the early Napoleonic Wars, a steady development of a divisional organization is evident in the organization of those expeditions embarked for the continent, but the short duration of these expeditions limited the progress that could be made so far as doctrinal developments were concerned. The force sent to North Germany in 1805 and 1806 went out by divisions, but these were largely administrative and transportation groupings only, much as has been the case for the Helder expedition.[8] Brigade organization remained fixed during these campaigns, but higher organization was still fluid. Much the same can be said of the Copenhagen expedition under Lt. General the Earl of Cathcart two years later, which nominally had a Left Division, a Right Division, and a Reserve, the former two being divided into two and three brigades respectively but the latter having all four battalions directly under its commander, Wellesley, who was assisted by Colonel Richard Stewart as "acting Brigadier General." As had been the case when Abercrombie designated a Reserve in the Egyptian campaign, this formation contained the army's light troops. Troops from the KGL formed an additional division of their own, outside of the British structure. However, when Cathcart detached troops from his main force for field operations, the resulting organization was fragmentary and did not even preserve the original brigading, let along divisional assignments.[9]

The remaining campaigns prior to 1808 were so small as to render any large-scale organization unnecessary; the expeditions sent from Sicily to Calabria in 1806 and Alexandria in 1807 were made up of small brigades formed from troops already serving in the Mediterranean, whilst that to South Africa was assembled at home and shipped out with a brigade organization already implemented.[10] The same can be said of the successive expeditions fitted out for South America to reinforce the initial unauthorized foray.[11] In none of these cases did the forces warrant a divisional

organization; only Moore's Baltic expedition of 1808, which ultimately formed part of the first peninsular army, had a divisional
system, but the three such formations created were numerically
very weak.[12] In the majority of these organizational schemes, light
troops if present in any numbers, were generally concentrated
into a single command—usually under the designation of the Reserve—and a senior officer appointed as second-in-command of
the whole force. Both these precedents can be seen again in the
early organization of Britain's forces in the peninsula: the former
to good effect, the latter less so.

It was with these rather hazy precedents in mind that Horse
Guards was forced to consider how best to organize the substantial
field army that would take the field in 1808 upon the union of the
smaller contingents sent out under Wellesley, Spencer, and Moore.
From the outset, York envisaged a divisional organization for this
force, but circumstances meant that it was never implemented. Instead, the continual changes of commander from Wellesley, to Burrard, to Dalrymple, and finally, after Cintra, to Moore, meant that
each man—with the exception of Burrard whose tenure was temporary and brief—implemented the organization that he thought
best. None of these much resembled that initially planned by York,
but the various schemes do give a good overview of this transi-
tional stage in organizational doctrine, which provided a basis for
further refinement.

Wellesley's initial organization, after he and Spencer disembarked, created a traditional organization with Spencer as second-
in-command and the infantry distributed into six brigades, one of
which contained all the army's light infantry. This was just about
workable at Roliça, but a little unwieldy by the time of Vimeiro
where the number of brigades had risen to eight.[13] In the aftermath
of the fighting, the troops and commanders now arrived such that,
had he chosen to do so, Dalrymple could have implemented the
original organizational plan set out by York. This had envisaged a
final organization of four line divisions, each having two brigades,
and a three-brigade Reserve under Moore comprising the light
infantry and cavalry.[14] The KGL line infantry, organized as a small
division in Moore's corps, was now to become a large brigade, but
in effect it retained its old organization; until after Oporto, the

four battalions were intermittently treated as either a large brigade or a small division in successive organizations without, in reality, altering their initial organization.[15] However, for Dalrymple
as newly arrived theater commander, York's scheme was less than
ideal. Not least of its failings was its accordance of a junior command, the Fourth Division of seven battalions, to Wellesley who,
whilst indisputably the junior of the seven lt. generals present, was
nevertheless not only the recent victor of Vimeiro but also a member of the government.[16] Not wishing to implement a wholesale reorganization, or to risk snubbing the influential hero of the hour,
Dalrymple came up with own alternative organization; this can
now be found in a beautifully watercolored chart, detailed even
to the point of showing facing colors, filed with the monthly returns of Dalrymple's forces.[17] This organization split the army into
two unequal corps, but these, confusingly, were denominated by
Dalrymple as divisions. The larger, under Moore, comprised that
officer's original command from the Baltic, retaining its original
divisional organization, plus a new division formed by combining
Acland's and Anstruther's brigades, which had joined Wellesley's
force prior to Vimeiro. These two brigades were to come under
the command of Lt. General Mackenzie Fraser, who had previously commanded the First Division of Moore's Baltic corps; Fraser was in turn replaced by Lt. General Sir John Hope, previously
Moore's second-in-command. The smaller "division," having six
brigades and no intermediate organization, was Wellesley's command largely as it stood at Roliça, with Spencer as second-in-
command. Both Moore and Wellesley would also control cavalry
and artillery assets. Judging by the layout of the diagram outlining
the organization, the intention was that Moore's "division" would
form the army's first line and Wellesley's its second—if this formalized, Frederician deployment represented Dalrymple's conception of tactical organization, it is perhaps as well that he never had
chance to implement it in the field.

In any case, Dalrymple's initial backward-looking arrangement was short-lived, as a General Order of September 5, 1808,
created a new organization of six divisions: First through Fourth
and Reserve, each having two brigades of line infantry, and an Advance Corps similar to, but smaller than, the Reserve envisaged by

York. As in the previous scheme, Burrard remained as second-in-command with no direct troop command.[18] This reorganization stripped considerable authority from Moore and Wellesley, who went from being commanders of quasi corps to leading divisions of six and a half and six battalions respectively; both, the latter in particular, had incurred Dalrymple's displeasure.[19] Again, the full implementation of this organization was overtaken by events: first the departure of disgruntled senior officers, and then the recall of its instigator. With two abortive schemes to his name, one can readily understand why Oman castigated Dalrymple for "spending a great deal of time over the redistribution into brigades and divisions of his army."[20] That said, the same criticism could also be made of Moore, whose forces underwent a tortuous series of reorganizations before assuming, after being joined by the reinforcing corps sent out under Lt. General Sir David Baird, a final organization of four infantry divisions, two independent "flank brigades" of light infantry, and one division of cavalry. Five of Moore's light infantry battalions were posted to these flank brigades, and two more formed part of Major General Edward Paget's Reserve Division, along with three picked line battalions; only one light battalion—the 2/43rd—formed part of a line infantry division.[21] Mention of Baird also brings up two precedents set by that officer. The first of these was his decision to implement a temporary divisional organization as he moved to his rendezvous with Moore, lack of senior officers notwithstanding, although in numbers he had fewer troops than Wellesley had fielded at Roliça. The second was his preference, once his forces were united with the main body, for a role as senior divisional commander that left him with a direct troop command, rather than the nebulous role of second-in-command of the whole.[22] Baird's preference for a direct command was typical of this crusty fighting soldier, but in any case made far better use of his talents.

As can be seen from the various organizational schemes outlined above, by 1809 a fairly good idea existed within the British Army as to how a large force could be organized into multiple divisions. Whilst the system would ultimately be taken further by Wellington, it is instructive to look how far it had already evolved prior to the development of the peninsular divisional system, as exemplified by the organization of Chatham's army sent to the Scheldt in July 1809. The Walcheren expedition represents the single largest British contingent of troops dispatched as an entity during the entire Napoleonic Wars. As such, unlike the peninsular army, the Walcheren expeditionary force clearly demonstrates how Horse Guards then thought it best to organize a large army. Chatham's army had five numbered divisions of line infantry, each of two brigades; a "Reserve of the Army" under Hope with three brigades, one of which was composed of Foot Guards; and a Light Division under Lt. General the Earl of Rosslyn with two brigades of light infantry and one of cavalry. The similarity between this latter formation and the Advance Corps envisaged by York and Dalrymple is marked; Rosslyn's Light Division certainly has more in common with them than it does with that formed in the Peninsula army. There are, however, other areas where a greater commonality with later peninsular practice is evident, in particular the assignment of divisional artillery and the attachment of independent rifle companies to larger formations. The junior line division—Lt. General Lord Paget's Fifth—also contained a larger-than-normal concentration of light troops, much as would the junior Seventh Division in Wellington's peninsular organization.[23] However, Paget's command was soon broken up when the forces on Walcheren Island were redistributed into a new organization under Chatham's second-in-command, Lt. General Sir Eyre Coote.[24] This apparent lack of preference for a divisional organization is reinforced by the fact that the force left on Walcheren under Lt. General George Don after the departure of the main body reverted to a brigade-level organization, although with 15,566 men in six brigades there were troops enough for a more sophisticated command system.[25] However, Don's command was now a garrison force, whose main enemy was fever, and the imperatives for a combat divisional system were no longer present.

Thus, although many important precedents had been set, and ideas tested, by mid-1809 the British Army still had no accepted norms for divisional organization. On the other hand, organization of the component brigades had settled into a standard form that would be retained for the remainder of the conflict, even if there was still no clear policy on how these brigades should be grouped to form higher organizations. As had been the case back in the previous century, brigade composition had again begun to take on a semipermanent nature that even lasted through multiple campaigns; the 1/4th and 1/28th, for example, remained together through a variety of organizations from Moore's Baltic expedition right through until the final reorganization before the retreat to Corunna, and although separated for a time were then reunited for the Walcheren expedition. The Foot Guards and KGL naturally retained their standing brigade structures throughout, but it also became increasingly the case that other battalions having an obvious affinity to each other were brigaded together. In the case of light infantry, this made sense for operational reasons, as such troops were generally concentrated within the army's order of battle. However, both Moore in the Baltic and Wellesley in his original command as formed in Ireland kept their highlanders together to form distinct brigades, and York sought to retain this in his unenacted proposal for Dalrymple's order of battle.[26] These organizational methods would continue in later years, and would form a key way of reinforcing and manipulating regimental identity; Wellington would later extend them by brigading together his three fusilier battalions, reinforcing their self-identified elite status.

Irrespective of composition, however, infantry brigades by 1809 had settled quite firmly into an established structure, with—on average—three battalions serving side by side under a permanent commander no junior than a colonel and no more senior than a major general. On occasion, in smaller forces or when individual battalion strengths were very high, brigades of only two battalions continued to be seen; conversely, when unit strengths fell toward the end of the period, brigades of four or even five battalions are encountered. Cavalry brigades, in similar fashion, rarely contained more than three regiments. Irrespective of the size of his command however, the officer commanding a brigade had, to assist him, only a single aide-de-camp and his brigade-major; the latter officer, title notwithstanding, was usually, in fact, of captain's rank, seconded from regimental duty to act as a junior staff officer.[27] For a sustained campaign, it was therefore quite evident that something more substantial and permanent was required.

The Divisional System under Wellington

It should not be thought that Wellesley arrived at Lisbon in 1809 with an organizational blueprint tucked into his pocket. Rather, the first eighteen months of his peninsular command would see the steady evolution of a divisional system that did not acquire all the identifying features associated with it until mid-1810. The fact that Wellesley took his army into the Oporto and Talavera campaigns almost immediately after his arrival meant that any desire to make sweeping organizational changes had to be subordinated to the practicalities of organizing troops for active operations. Thus, the first months of Wellesley's command would see a steady progress toward an organizational ideal, with ideas being tested and either incorporated or discarded depending on how well they stood the test of service.

When Wellesley reassumed the peninsular command, he found an army with no formation above that of the brigade, although Sir John Cradock, his predecessor, had promulgated a new organizational scheme during March that placed his eight weak line brigades in pairs under the senior brigadiers.[28] However, it is unclear to what extent this scheme was implemented, and it certainly had little bearing on Wellesley's subsequent plans. One thing that had prevented Cradock from producing a more sophisticated command system was a lack of senior officers, which was also initially a problem for Wellesley; other than the second-in-command, Lt. General Sir John Sherbrooke, the peninsular army was initially lacking senior officers with experience of commanding larger formations. Lt. General William Payne soon arrived to command the cavalry, but the initial retention of half the available horsemen around Lisbon precluded the operation of the mounted troops as a division in any practical sense. However, when the newly promoted Lt. General Sir Edward Paget joined the force, it was possible to assign that officer and Sherbrooke command of an infantry wing apiece—these formations in effect representing two large divisions of four brigades each.[29] This arrangement was short lived, and for the final advance on Oporto, Major General Rowland Hill also stepped up to divisional command, as reported by Wellesley in his victory dispatch: "The infantry of the army was formed into

3 divisions for this expedition . . . one, composed of Major Gen. Hill's and Brig. Gen. Cameron's brigades of infantry, and a brigade of 6 pounders, under the command of Major Gen. Hill."[30] This was Wellesley's first employment of the term "division," although the use of "for this expedition" suggests a temporary measure. Within these nominally British formations, five Portuguese battalions served within regular British brigades; this experiment, though not markedly unsuccessful, was not repeated.[31] Each of the three divisions had its own artillery assets attached, as did the division-sized flanking column under Beresford and the covering forces left along the line of the Tagus under Major General John Randoll Mackenzie; this arrangement would become an important feature of the permanent divisions.

A permanent divisional system was not implemented until after the return of the army from the Oporto campaign, being announced in the General Order of June 18, 1809. Paget had been wounded at Oporto, but Sherbrooke and Hill were confirmed as divisional commanders along with Mackenzie, getting the First, Second, and Third Divisions respectively.[32] Whereas Hill took a permanent divisional command as a major general, albeit one for whom Wellesley hoped to secure a local promotion, Mackenzie's seems to have been considered a temporary appointment.[33] Accordingly, Mackenzie retained nominal command of a brigade in his division, and appears to have exercised this directly; no acting brigadier is given in any source, although Lt. Colonel William Guard of the 1/45th had clear seniority over the other battalion commanders. The Fourth Division was also initially under its senor brigadier, Alexander Campbell, until Major General Lowry Cole joined in the autumn; Lt. Colonel Sir William Myers of the 2/7th commanded Campbell's brigade whilst Campbell had the division.[34]

The army was still short of senior general officers, particularly since Paget's wounding, and Wellesley wrote to Castlereagh bemoaning the fact that Lord Bentinck and Sir Brent Spencer, now both lt. generals, had been unable to join the army as had been envisaged. Bentinck, who had commanded a brigade under Moore, declined the appointment; Spencer was sick, but did return the following year to replace Sherbrooke. Wellesley also requested that Hill and Sir Stapleton Cotton be given local rank as lt. generals, in part because Beresford, their junior, had been accorded that rank. Protest over this issue had already led to the departure of Major General John Murray, who had commanded the KGL infantry for the past year and who seems to have temporarily replaced the wounded Paget in the aftermath of Oporto. Since Murray—the same officer who would later bungle the Tarragona expedition—was a man of no obvious ability, but would, had he remained, have been eligible by seniority for a divisional command and a local lt. generalcy, his departure was perhaps no bad thing.[35]

Although the four infantry divisions would remain in existence for the rest of the war, they had not yet acquired the three-brigade structure that would largely become the norm. Instead, Sherbrooke's First Division had four brigades and the remainder two, with the KGL infantry now forming two distinct brigades. The infantry brigades were henceforth generally referred to by the name of their commander rather than by a numerical designation, thus removing what was rapidly becoming a confusing system of brigade seniority.[36] The Portuguese regiments were formed into a corps of observation under Beresford to complete their reorganization, and did not participate in the Talavera campaign. Henceforth, with a handful of exceptions, Portuguese units formed their own brigades, and were integrated into the command structure as such.[37] The British cavalry, which gained a third brigade from reinforcements, was finally united into a division under Payne.[38] The General Order establishing the permanent divisions assigned to them—or, rather, to their respective commanders—an assistant adjutant general to assist with staff duties and an assistant provost to aid in maintaining discipline. Other orders from the period refer to the existence of medical and commissary officers as part of the divisional staffs, suggesting that these appointments were already in force with the temporary divisions formed during the Oporto campaign. In either case, whether a new innovation or the codification of an existing practice, the permanent divisions would eventually come to have representatives of all the staff branches, responsible to both their divisional and departmental superiors.[39]

After the strategic failure of the Talavera campaign, with more British regiments joining the field army and with the Portuguese now fit to take their place in the line, the course of 1810 saw a steady reorganization, which, through a shuffling of existing brigades and the addition of new ones, would create the nine infantry divisions—First through Seventh, Light, and Portuguese—that would endure until the end of the war. As of February 22, 1810, Craufurd's Light Brigade was removed from the Third Division, of which Craufurd had held temporary command since Mackenzie's death at Talavera, in order to become the nucleus for the new Light Division. Completed by the assignment of two Portuguese Caçadore battalions, Craufurd's command was much smaller than the others, but it concentrated the light infantry and rifles into an elite formation. Its initial role was that of watching the frontier whilst the rest of the army completed its reorganization for the defense of Portugal.[40] Later, somewhat enlarged but still smaller than the line divisions, the Light Division became the army's shock troops, but, particularly after Craufurd was killed at Ciudad Rodrigo in January 1812, something of their original excellence was lost although the division continued to do good service under the Hanoverian Karl von Alten.[41] However, the Light Division not only held itself in considerable esteem but also contained an unusually large number of officers and men who would later write about their experiences, thus reinforcing that elite status as part of the popular image of the peninsular army.

Since the removal of Craufurd and his brigade took the Third Division down to only two and a half battalions, Lightburne's Brigade was moved across from the Fourth Division, and both the Third and Fourth Divisions then each received a brigade of Portuguese infantry to maintain them at a strength equivalent to the First and Second Divisions, which remained all-British. Major General Thomas Picton, a veteran of much controversial colonial service and more recently a brigadier in the Walcheren expedition, assumed command of the Third Division.[42] April 1810 saw the arrival of the first brigade formed out of battalions recovered from the rigors of Walcheren, and this formed the nucleus of Major General James Leith's Fifth Division along with two Portuguese brigades.[43] Two more Portuguese brigades went to form the

Portuguese Division, which operated with Hill's Second Division in what was in effect a provisional corps, and the remaining Portuguese field forces formed four independent brigades, although one of these was for a time joined with a militia brigade to form a provisional division under the Portuguese Coronel Carlos Lecor. Other than Lecor, only two other Portuguese officers, Coronel José Champalimaud and Brigadeiro A. L. da Fonseca, held senior commands after the 1810 reorganization, and both were placed directly under British officers. This unwillingness to entrust senior commands to Portuguese officers would continue throughout the war. Only Lecor and Francisco Silveira, Conde de Amarante, would obtain divisional commands in the field army, and only the former would command British troops. This was in 1813, when Lecor, by now a *marechal de campo* or major general, was the senior officer remaining when the Seventh Division's permanent commander went on leave, but his tenure as such was brief and he was soon shifted to replace Silveira, who had fallen from Wellington's favor, at the head of the Portuguese Division.[44]

This reorganization set the pattern for the organization of all seven numbered divisions of line infantry, although it would take another year before all seven could be brought to this standard. Being the basic building blocks of the army as a whole, the line divisions were each based on the model of two brigades of British troops and one of Portuguese. There were exceptions to this system. The First Division was all British—or, rather, British and KGL—as was the Second Division until 1811. During 1810 the First Division temporarily had an extra British brigade, and in 1812 the Second Division acquired a Portuguese brigade in addition to its three British, remaining as a four-brigade anomaly until the end of the war. The Sixth and Seventh Divisions both initially had only two brigades, and for a time in 1813 the First Division dropped to only two active brigades due to the detachment of the two battalions of the 1st Foot Guards that had been hit by an epidemic. Some Portuguese brigades remained independent throughout the war, albeit in reduced numbers after 1810 as some were absorbed into the expanding divisional system, but it was unusual for any British brigade to exist outside of the divisional structure. Alten's Brigade of KGL light infantry was only briefly independent in 1811 before joining the Seventh Division, whilst the brigade brought up from Cadiz by Colonel John Skerrett in 1812 was a march formation, and the Brigade of Provisional Militia in 1814 was likewise a second-line outfit. A more significant anomaly exists in the shape of Aylmer's Brigade, which always operated with the First Division but which was never formally attached to it, but this was essentially due to the formation's specialist "nursery" role for newly arrived battalions, outlined in chapter 3.

The other main anomaly in the divisional system was that of the Second Division. Judging from his correspondence, it would seem that Wellington ultimately envisaged the structure of Hill's quasi corps as comprising four British and two Portuguese brigades. Since the initial 1810 formation contained only the Second Division of three British brigades, and the two-brigade Portuguese Division, the brigades of Karl von Alten and then of Kenneth Howard were envisaged as forming the fourth British brigade. Both of these brigades included light infantry units, otherwise absent from Hill's command, and would have been posted to the Second Division in lieu of Hoghton's Brigade.[45] Although Alten's Brigade was present at Albuera, it ultimately joined the Seventh Division instead of the Second; meanwhile, the losses sustained in that battle caused the Second Division's original three brigades to be condensed into two, such that when Howard's troops did transfer from the First Division, they served only to return the brigade structure of the Second Division to the status quo ante. Accordingly, the sixth brigade for Hill's command ultimately came in the shape of more Portuguese troops.[46] Because of the inefficiency of the Portuguese commissariat, Wellington did not consider it viable to maintain large formations of Portuguese troops or to have divisions containing more Portuguese troops than British, which explains why, when Ashworth's Brigade of Portuguese troops, previously independent, joined Hill, it became part of the Second Division rather than the Portuguese, even though this created two divisions of abnormal size.[47]

The other anomaly that needs addressing is the conception that the Seventh Division was intended by Wellington to be a second Light Division. The case in favor of this is summed up by Alistair Nichols in his history of the Chasseurs Britanniques, who reminds his readers, "It was formed with a preponderance of light infantry battalions and also had attached to it horse, rather than foot, artillery."[48] Yet this is essentially the sum total of the evidence, and even this does not stand up to great scrutiny. As Nichols himself points out, the Chasseurs Britanniques, name notwithstanding, were never formally trained as light infantry, and retained a standard line organization with flank and center companies.[49] This being so, only six of the original eleven battalions were lights, which, whilst a higher proportion than in the other numbered divisions, was hardly an overwhelming preponderance. Furthermore, if this surmise of Wellington's intentions is correct, then it is hard to see why the KGL light infantry was originally slated to join Hill's command, or the Brunswick Oels Light Infantry, which ultimately formed part of Alten's Brigade, was originally posted to the Light Division; only later were these battalions redirected to form part of the new Seventh. Since the Seventh Division was formed almost entirely from reinforcements, one can perceive a triumph of necessity over choice in the assignment of battalions, which in turn suggests that the division got Captain Alexander MacDonald's E Troop, RHA, as its divisional battery simply because that unit was also newly arrived and unattached.[50]

Wellington had realized at an early date the utility of providing each division with an organic contingent of light infantry, and this was certainly apparent by early 1811, when the Seventh Division was being assembled, as it also played a part in the choice of reinforcements for Hill's command. This realization was also reflected by the removal of the Brunswickers from the Light Division and the distribution of their Jäger companies amongst the army's infantry brigades in the same manner of those of the 5/60th—a measure that ultimately ensured that every British brigade included a company of riflemen.[51] This wider distribution of the available light infantry units not only argues against a desire to duplicate the existing Light Division but also explains why the 1813 reorganization further reduced the number of light battalions in the Seventh Division by transferring two of them to the First Division—the only formation lacking such units by this stage in the war—replacing them by battalions of the line.[52] If Wellington did initially consider establishing the Seventh Division as a predominately light formation, this thinking was soon dispensed with. Nevertheless, this confused and intermediate status may go some way toward explaining the poor light in which the Seventh Division was viewed by members of other divisions, to the extent that it has been suggested that the Seventh Division's changing organization and slow gestation were responsible for an identity crisis that would dog the formation throughout its existence.[53]

At the same time as more infantry divisions were being formed, Wellington's cavalry was also expanding in numbers, although throughout the war the brigade remained the basic operational unit. The single initial division spawned a second in 1811 to encompass the cavalry serving in the southern theater, although this was only formally designated as the Second Cavalry Division on June 19 of that year.[54] Cotton, who replaced Payne at the head of the cavalry in 1810, retained administrative control over both cavalry divisions whilst commanding the First in person, and reabsorbed the entire cavalry into a single division when the army was reunited for the 1813 campaign. The increasing distribution of the army as it fanned out into southern France in 1814 therefore made Cotton's job far more an administrative post than a combat one; indeed, the cavalry seems to have functioned perfectly adequately in the Vitoria campaign with no divisional commander at all, Cotton then being on leave. During the last eighteen months of the Peninsular War, cavalry brigades were parceled out to the various wings and columns of the army, with the more senior amongst their commanders on occasion assuming command of more than one brigade. Most of these organizations were temporary and ephemeral, but as of March 1814 Major General the Hon. Henry Fane formally assumed command of all the cavalry of Hill's Right Wing. Fane henceforth led not only his own brigade, command of which passed to Lt. Colonel Arthur Clifton of the 1st Dragoons, but also that previously under Major General Hussey Vivian, who had gone home and left it under Lt. Colonel Patrick Doherty of the 13th Light Dragoons.[55] Although the shifting nature of the war during its final months speaks against any need to formalize the organization of the cavalry brigades, there remains a lingering suspicion that the failure to re-create multiple divisions of cavalry may have been due in part to the lack of suitable officers to command them; other than Fane, it is hard to see who else could have been trusted with more than a brigade even by 1814. Even when there had been two cavalry divisions, the Second was commanded for most of its existence by the terminally incompetent Lt. General Sir William Erskine, who was assigned to it largely as a means of keeping him out of greater trouble elsewhere.[56]

During 1810 and into early 1811, Wellington's peninsular command was reinforced by a steady stream of new troops, which allowed him to complete the divisional system outlined above. However, for the most part these reinforcements were integrated into the existing order of battle rather than used in their entirety to create new formations. The only division formed entirely from scratch was the Seventh, the organization of which was not completed until well after the 1811 campaign was under way.[57] Where possible, however, new divisions were created by mixing new and seasoned troops. Thus, when two brigades of British reinforcements arrived in Lisbon over the winter of 1810–11, they did not simply form a single new command. One new brigade—Dunlop's—was assigned to the existing Fifth Division, releasing one of the Fifth Division's two Portuguese brigades to form part of the new Sixth. Logically, one might have expected the Portuguese to be combined with the other new British brigade, Erskine's, but instead a more sophisticated approach was chosen. In order to spread experienced troops more widely across the army, Erskine's Brigade was posted to the First Division, displacing Pakenham's Brigade which went in turn to the Fourth Division so that Major General Alexander Campbell's veteran British brigade could be transferred to the new Sixth. Here, it would be joined in the new year by another new British brigade under Colonel Robert Burne. The new division therefore had one British brigade of veterans and one of new troops, plus a battle-tested Portuguese brigade. Campbell, who had had acting command of the Fourth Division during the Talavera campaign, was promoted to lead the new formation on a permanent basis.[58] The favoring of the double-transfer, rather than simply posting Pakenham's Brigade to the Sixth Division, was no doubt to ensure that Campbell would have troops he knew and trusted to become the nucleus of his expanded command.

This combining of new and experienced troops was a key element in Wellington's success, and was continued throughout the war. No new division was formed after early 1811, and, except in the cavalry, which continued to expand to 1813, very few new brigades were formed either. Up until 1811, reinforcements frequently arrived as organized brigades, complete with commanders, which could then be fitted directly into the expanding structure as was done during the reorganizations that created the Sixth Division.[59] Thereafter, new infantry regiments came out individually and were added to existing brigades as reinforcements or to replace units that had been sent home. New cavalry brigades did continue to arrive as fully organized formations into the early spring of 1813, but most of these were soon reorganized to acquire at least one veteran regiment. When Le Marchant's Brigade of heavy cavalry was sent out over the winter of 1811–12, for example, the raw 4th Dragoon Guards were soon replaced with the veteran 4th Dragoons.[60] Similarly, in 1813 Colonel Colquhoun Grant's Hussar Brigade had its worst performing regiment taken out and combined into a new brigade with the veteran 1st KGL Hussars. The new brigade commander, Major General Victor Alten, made it explicit to the officers of the offending 18th Hussars that this had been done for the benefit of their regiment: "The duties of hussars in the field are so various & require so much practice & experience that too many opportunities cannot be taken, even in cantonments to instruct the men in them, and the Major General will find great pleasure in giving that assistance which his experience may enable him to do."[61] The reaction of the officers of the 18th was mixed, but the reorganization had the desired effect.

After mid-1811, the continued arrival of infantry reinforcements a battalion at a time enabled Wellington to maintain the divisions as balanced entities, both with regard to their strength and to the balance of unit experience within them. Since large-scale reorganizations had to be avoided to maintain cohesion and morale, significant changes to the army's order of battle had to be delayed until active campaigning had finished for the winter.[62] Because, in practice, Wellington's forces had few extended breaks from active service, a full redistribution of forces in order to achieve a completely uniform organization was never possible; nevertheless, the lengthy operational hiatus after the retreat from Burgos allowed for extensive changes that brought the infantry divisions very close to what seems to have been Wellington's ideal. The peninsular "Marching Strength" of May 25, 1813, detailing British and Portuguese forces, demonstrates that this ideal had very nearly been achieved by the outset of the Vitoria campaign.[63] All the three-brigade divisions have a total Anglo-Portuguese strength of around 7,000 all ranks whilst the First, Light, and Portuguese Divisions all muster around 5,000 in their two brigades; these figures do not include Lambert's absent brigade, which, if added, would raise the First Division to the ca. 7,000-man standard. For the Second Division, on the other hand, the four-brigade organization gives a grand total of no less than 10,834 all ranks. The line divisions were therefore largely uniform in size, with the disparity of the large Second Division making sense when understood in conjunction with its operational pairing with the smaller Portuguese Division.

Something approaching a balance had also been achieved insofar as the level of experience of the various divisions was concerned, although here a greater disparity remained. Looking at British units only, the progression may be tracked throughout the war by taking an average of how many months each component battalion had served in the theater. Data in table 8 is for the month of April, which generally marked the end of reorganizations and the beginning of active service, although this choice of date has the added advantage of also catching the army in the final month of its existence. The second, bracketed, figure for the First Division in 1814 includes in its total the five battalions of Aylmer's Brigade, which, though nominally independent, operated in practice as part of the division.

As the table demonstrates, although divisional strengths had become roughly equalized by the beginning of 1813, a distinction in terms of experience remained between the original five divisions and the Fifth, Sixth, and Seventh Divisions raised later. This disparity may largely be put down to the fact that the Seventh, and to a lesser extent the Fifth, were initially formed out of what was available, being composed largely of battalions posted back to the peninsula after service in Walcheren or of weak second

Table 8.   Average Unit Experience (Months) of Peninsular Divisions

	1810	1811	1812	1813	1814
First Division	16	22	34	37	49 [36]
Second Division	13	25	25	37	49
Third Division	11	20	31	37	49
Fourth Division	16	23	34	39	51
Fifth Division	–	9	21	22	31
Sixth Division	–	15	24	27	29
Seventh Division	–	3	15	24	36
Light Division	11	18	29	35	47

Source: Figures refer to months since arrival in the peninsula; data largely derived from McGuigan, "Origin of Wellington's Peninsular Army," 39–70; and Oman, Wellington's Army, 343–373.

battalions. By contrast, the Sixth Division began life with a good proportion of veteran units, as we have seen, but lost out in the 1813 reorganizations.

The need to acclimatize these newer divisions, and build up their strength, explains their limited commitment to serious combat prior to 1812. Wellington's deployment of his army at Fuentes de Oñoro makes this need particularly explicit, with the three new divisions out on the extremities of the line and the tested First, Third, and Light Divisions holding the vulnerable center. Although this deployment did not save the Seventh Division from heavy fighting as a result of Masséna's flank attack, its intention is plain. In like fashion, the newer divisions were largely held back from the horrors of the storming of Ciudad Rodrigo and Badajoz, and the Fifth and Sixth Divisions would only receive their real baptism of fire at Salamanca.[64] Only from 1813 did matters appreciably change; after the reorganizations of the previous winter, the distribution of veteran and untried regiments across the army was as balanced as it would ever be, and the distinction between new and old divisions less marked. By this stage of the war, it also needs to be understood that the value of the monthly experience figures used here begins to deteriorate as many of the longest serving units had by now absorbed substantial drafts and transfers of manpower. Conversely, many of the nominally inexperienced first battalions had in fact taken in a core of veterans from departing junior battalions. Accordingly, the qualitative discrepancy implied by these figures may overstate the case a little.

One could well argue that a reorganization to create this state of affairs might ideally have been implemented earlier, but this argument begs the question of at what other point the strategic situation would have permitted a large-scale overhaul. As it was, a rearrangement of some brigades when the Fifth and Sixth Divisions were created, and minor changes thereafter, created a system that was workable for the campaigning of 1810–12 but that favored the senior divisions and in turn forced a greater reliance on them. Later, with time available, larger changes enabled a more balanced organization to be implemented. The structural development of the divisional system should therefore be read as being a steady progression toward an ultimate organizational goal, which had largely been achieved by the commencement of the march out of Portugal that would lead to Vitoria and ultimately to Toulouse.

The slow evolution of the system of command and control that has been outlined above was somewhat unusual for the era, where divisional assignments increasingly remained unchanged for years on end in most continental armies, most obviously that of France. This slow growth also had an effect on the choices of commanders for higher formations in the peninsula, complicated by the fact that the few officers who possessed the requisite experience in 1809 were unable to serve as they were senior to Wellesley. In the long term, the knock-on effect of this lack of senior commanders was to advance the careers of younger, more junior, officers who found themselves serving in posts typically filled by senior men. One only has to compare the two main field armies in 1809 to see the extent that this was the case. Chatham had a lt. general as his second-in-command, and all bar one of his divisional commanders also held that rank; the exception—Thomas Graham—was filling in for the absent Lt. General Sir John Cradock. Conversely, Wellesley's army at Talavera contained only three officers of lt. general's rank, all of whom, including Wellesley himself, were junior to the men commanding Chatham's divisions. Almost all of

Chatham's brigades had a major general in command, but the same could be said of only two of Wellesley's, who did, however, have two more officers of that rank commanding divisions whilst Alexander Campbell filled the last divisional vacancy as a brigadier general. Of the two major generals left at brigade level, one was Cotton, a cavalryman, and the other was Christopher Tilson, an incompetent. A small number of more senior officers did come out to the peninsula early on, expressly as divisional commanders—notably Sherbrooke and Payne, who completed the Talavera trio of lt. generals, and also Sir Brent Spencer, who returned in time to command the First Division for most of 1810 and 1811. However, by the time the divisional system was complete in mid-1811, all divisional commanders, with the exception of Spencer who was also the nominal second-in-command of the whole army, had risen from having previously held brigade commands either in the peninsula, on Walcheren, or both.

In the absence of an assigned divisional commander, competent brigadiers could be given acting divisional commands to fill a temporary vacancy, a policy that also helped create a pool of experienced men who could step up to a more permanent command if one were to become available. By the end of the Peninsular War, there were several senior major generals, notably Charles Colville, Edward Pakenham, and Denis Pack, who had held such acting commands for a worthwhile time, and, indeed, Colville got a permanent divisional command for the Waterloo campaign and Pakenham an equivalent command in America.[65] Henry Fane was also considered for an infantry division toward the end of the Peninsular War, and for a cavalry division in 1815.[66] Earlier, having proved his fitness in temporary command of the Fourth Division, Alexander Campbell was given the Sixth upon its formation and led it for a year before being given a new posting in India. Those who criticize Wellington for a failure to create good subordinates fail to see the way in which he worked the system to bring on men of ability; it is true that this was done for the good of the army rather than with the direct intention of benefiting those concerned, but benefit them it nevertheless did.

By the latter stages of the war, the best of the divisional commanders had proven themselves capable of more responsible roles, and, whilst Dalhousie, Picton, and Cole were clearly at their best leading single divisions, the use of Rowland Hill and Thomas Graham as commanders of quasi corps was increasingly extended and formalized. Prior to 1813, both these men had commanded multiple divisions, with Hill having responsibility for those forces deployed in the secondary southern theater whilst Graham—and, before him, Spencer—took responsibility for those in the main northern theater on those occasions when Wellington's presence was required elsewhere. Beresford temporarily replaced Hill in the south during early 1811, but Albuera called his abilities as an independent commander into question, and although he would again hold quasi corps commands during the last months of the war, he generally did so under Wellington's immediate eye. In the final months, Lt. General Sir John Hope replaced Graham. Those divisional commanders given command of corps retained their nominal divisional appointments, but an additional divisional commander was appointed to command the division under them, freeing the senior officer for the larger command. This had, less formally, been the case for Hill for most of the time since 1810, with the Second Division's senior brigade commander, William Stewart, commanding in Hill's place. Maintaining two divisions where the commander was under close supervision by a more senior officer proved useful in the later years of the war, when Stewart, who had proved a disaster when left to his own devices in command of the First Division during the retreat from Burgos, was able to again command the Second under Hill's watchful eye, whilst the relatively inexperienced Major General Kenneth Howard could begin his command of the First Division with the veteran Graham as his mentor.[67] A similar logic may also be detected in the choice of Tilson-Chowne to command the Second Division and Erskine to command the Second Cavalry Division in the southern theater during the 1812 campaign, although the need to supervise not one but two such obvious duds must surely have taxed even the legendary patience and good nature of Rowland Hill.[68]

The move toward a corps system was formalized in 1815 for the army that fought at Waterloo, although this was for reasons that were as much political as military. Militarily, it was necessary for the organization of the army to achieve a balance not only within the Anglo-Hanoverian divisions but also in the force as a whole. Politically, if the forces of the Kingdom of the Netherlands were to come under overall British command, the senior officers, and royalty, of that kingdom needed to be accorded commands that were commensurate to their rank rather than their experience.[69] Once the main difficulty inherent in the placing of Netherlands forces under a foreigner had been dispensed with by Wellington's appointment as a Netherlands field marshal, a corps system was formally implemented shortly after Wellington assumed command, with the First Corps being placed under the Prince of Orange and the Second under Hill. Even then, political sensibilities meant that the Netherlands troops in the Second Corps were at least nominally under Prince Frederick of the Netherlands, creating an additional organizational echelon between Hill and the subordinate commanders.[70] Whilst these corps initially contained cavalry units, the Netherlands Division in the First and a KGL brigade in the Second, this organization was subsequently dropped, and at no time did either corps have any artillery assets beyond those allocated to the divisions. The cavalry command as a whole was named as a corps, but in practice functioned as a large division of eight, later nine, brigades, with the Netherlands cavalry forming a separate division. As the force grew, a Reserve Corps was also created, but no commander was ever allocated and Wellington made it explicit that he preferred to retain personal operational control over at least some of his divisions, informing Horse Guards, "I must, besides, mention that in the Peninsula I always kept three or four divisions under my own immediate command, which, in fact, was the working part of the army, thrown, as necessary, upon one flank or the other. . . . It might be convenient to have something of the same kind now."[71] Whilst the "always" is a typical Wellington overstatement, this comment bears relation both to the conduct of a number of peninsular battles, most notably Vitoria, as well as to Wellington's circumvention of his corps commanders at Waterloo. Wellington certainly had no great enthusiasm for formalized army corps, and such formations disappeared from British military organization after 1815. Unlike the division—by 1815 a definite part of the organizational hierarchy—a corps system did not truly resurface in future campaigns until the next major continental commitment ninety-nine years later.[72]

Whilst 1815 saw the extension of the corps system, it did not follow that all aspects of the peninsular divisional system were re-created. Some elements were clearly recognized as essential and rapidly resurrected, most notably the mixing of brigades to create multinational divisions with Hanoverian troops standing in for Portuguese to re-create the old organization of two British brigades and one foreign. To create this structure required a complete breaking up of prior organizations upon Wellington taking command, in particular the redistribution of the KGL infantry—which had previously formed a division in its own right—into two brigades, which were placed into the new Second and Third Divisions.[73] This move caused great offense to Major General Heinrich von Hinüber, who was thereby deprived of his command and accordingly resigned in high dudgeon, but was expressly implemented because Wellington "deemed it necessary to organise [the army] in such manner as to bring together in the several divisions the experienced and inexperienced troops."[74] Hanoverian sensibilities were also hurt, since the troops of that nation had previously formed a distinct Subsidiary Corps of three divisions, which was now also broken up. The men of what had been the Light Division of this corps were luckier than most, since the small division of six battalions was converted into a large brigade and not only remained under its old commander, Major General Graf von Kielmansegge, but formed part of the new allied Third Division under Lt. General Karl von Alten who had previously commanded the Subsidiary Corps. Nevertheless, the breakup of old formations was hurtful to Hanoverian pride and, at least at first, a "severe blow to our morale."[75] Only later would it become apparent that the Germans of the Third Division—Colonel Christian von Ompteda's KGL battalions as well as Kielmansegge's Hanoverians—were there to provide some stiffening for the mixed bag of British battalions forming the division's remaining brigade.

Because a great deal of commonality can be seen between the divisional system of the peninsula and of Waterloo, it should not be assumed that the former model had become established in the

British Army as a one-size-fits-all system. Many of the problems faced in the peninsula were similar to those faced in 1815, so it is understandable that the solutions were also similar. This is seen most obviously in respect of balancing veteran and raw troops, although whereas in 1810 it had been a case of mixing seasoned British troops with untested Portuguese, 1815 saw a more complex requirement to match raw British battalions with German veterans in some divisions whilst others combined British veterans with new-raised Landwehr. In both their similarities and differences, the systems of 1810–14 and of 1815 show the obvious pragmatism in Wellington's approach. Where lessons from the peninsula could be applied to the new theater, then applied they were. Where peninsular practice was not appropriate, it was not resurrected: hence, for example, the lack of a re-created Light Division in 1815. At the same time, though, if the organizational practices of the peninsular army were not always fostered, the memory of its reputation most definitely was. For this, as we shall see, was perhaps the most significant and lasting way in which the peninsular divisional system came to influence the British Army.

Divisional Identity

William Tomkinson seems to have delighted in compiling lists; they litter his published diary and are frequently an invaluable record for the historian. But whilst most relate to decidedly practical matters concerning the organization and service of his regiment, one is rather more peculiar. Having detailed the composition of Wellington's forces for the 1812 campaign, Tomkinson then goes on to give a list of "cant names in the army for the Divisions."[76] As with many elements of Tomkinson's book, it is clear that a considerable amount of reworking was done prior to publication; even if he did make a list of the divisional names as they stood in March 1812, he certainly could not have then included—as his published version does—a note explaining how the Fourth Division gained a new nickname during the Battles of the Pyrenees. Nevertheless, the fact that Wellington's peninsular divisions acquired nicknames at all is something that might well be considered surprising for an army where the construct of identity was very firmly based on the regiment. Such a tendency might also seem at odds with modern constructs, based on twentieth-century warfare, where the division is too large, and its commander too remote, for it to play much of a part in a soldier's sense of identity.[77]

Nevertheless, it is clear not only that a sense of divisional identity did exist but also that it was recognized, and ultimately fostered, by those in authority. As with the nicknames bestowed on regiments, few were insulting but many contained an element of irony or self-deprecation; only the rivalry between the Third and Light Divisions created appellations that were out-and-out expressions of divisional pride. According to Tomkinson's list, the First Division was known as the "Gentlemen's Sons," thanks to its containing the army's Foot Guards battalions. The Second Division, detached in the south and therefore missing most of the battles of 1811–12, was the "Observing Division," and the Fourth, which assisted it at Albuera when likewise detached, was initially the "Supporting Division"; only later did it acquire a new name, "The Enthusiastics," after distinguished service in the Pyrenees that led Wellington to report on "the enthusiastic bravery of the 4th division."[78] The Third Division, commanded for most of the war by the quintessential fighting general Thomas Picton and in the thick of most of the major battles, was the "Fighting Division." Just as Picton and Robert Craufurd were great rivals, so too were their troops, and the men of the Light Division, resenting the implied snub in the Third's nickname all the more after it allegedly failed to support them on the Côa, simply called themselves "*The Division*" and treated their counterparts of the line with appropriate disdain. Hard work and little glory in the early years of the war saw the Fifth Division labeled the "Pioneers" and the Sixth labeled the "Marching Division," and these names stuck even after both formations distinguished themselves at Salamanca.

The only division for which Tomkinson did not record a nickname is the Seventh, his list simply repeating the rather sneering remark apparently then current that "They tell us there is a 7th but we have never seen them."[79] Others, however, called the Seventh Division the "Mongrels" in reference to its many foreign battalions.[80] Tomkinson's comment probably stemmed from the fact that, after heavy losses at Fuentes de Oñoro in 1811, the Seventh

was kept out of serious action until 1813, whilst the latter nickname—the only one for any division that could be deemed pejo-
rative—represents simple national chauvinism. In fact, the men of
the Seventh, whose fighting record was generally good, ultimately
came to take something of a perverse pride in their casting as
the army's misfits. Lt. George Wood of the 1/82nd, posted to the
Seventh Division in the reorganization of spring 1813, wrote, "We
were placed in the division that was considered the refuse of the
army; but the sequel will show whether this regiment, brigade and
division, were not to rank among the first, for the bravery and gallantry of their conduct."[81] Interestingly, Tomkinson also implies a
negative connotation to the nickname of the Second Division, asserting that it was given by the men of the formation to themselves
in disgust that Hill's orders only to watch the enemy meant that
"he never engaged but when obliged, and lost so many chances of
bringing on petty affairs."[82] If true—and Tomkinson did not serve
in the southern theater so his source for the story has to have been
secondhand—this interpretation again brings an element of self-
mockery into the concept of nicknames, parallel with similar cases
at the regimental level.

Wood's account is somewhat self-effacing, but other writers were
more forthright in their assertions as to the value of the divisions
in which they served, as may be seen in the account of the fighting at Fuentes de Oñoro by John Kincaid, then a second lt. in the
95th. Having begun by remarking upon his envy that the Seventh
Division was in action and his own formation only skirmishing,
he then describes how Wellington ordered the Light Division to
cover the retreat of the hard-pressed Seventh: "The execution of
our movement presented a magnificent military spectacle, as the
plain between us and the right of the army was by this time in pos-
session of the French cavalry, and, while we were retiring through
it with the order and precision of a common field-day, they kept
dancing around us and every instant threatening a charge without daring to execute it."[83] Kincaid presents the Light Division
as superior not only to the French but also to the other British
divisions since it alone was selected to cover the withdrawal. More
generous writers gave credit where due to other formations; Grattan's account of Salamanca, for example, clearly gives the honors

of the day to the Third Division, of which his own beloved 88th formed a part, but also readily allows the contribution made toward the allied victory as a result of the "intrepid valour of the 6th Division."[84]

All these examples come from the writings of officers, and it may be assumed that this sense of identity existed to a greater degree in the commissioned ranks, where the social circle would extend to the division as a whole rather than remain at a unit level as one would expect for the rank and file. This greater sense of commissioned identity was further reinforced by the establishment of divisional clubs, race meetings, and amateur theatricals for officers, set up when the peninsular army went into quarters for any length of time.[85] Indeed, this easy intradivisional sociability can be directly contrasted with Lt. George l'Estrange's comment that he encountered his cousin, serving in the 1/11th in a different division, only once in the entire war.[86] But if the rank-and-file soldier was less likely to think of the division in the context of its place in the wider military or social structures of the army, the soldier's own division nevertheless remained a key part of his identity on campaign. It remains less clear, however, to what extent this pattern continued with regard to seeing the peninsular army, in which a definite pride was also taken, as part either of the armed forces of the nation as a whole, or even as representative of the nation itself. The historian Linda Colley has presented the contemporary volunteer movement as being indicative of a growing sense of patriotism and national identity, but it is questionable how far this analysis can be extended to regular troops. Identification with the peninsular army, or one of its divisions, is perhaps better seen as a refinement of the counterargument put forward by historian John Cookson, who has presented the British Army as offering an alternative identity, and the regiment a surrogate family.[87]

Interestingly, this regiment-division-army pattern leaves out the brigade. There are a few cases where brigades were formed of troops with some obvious connection to one and other and developed their own distinct ethos, as with the Fusilier Brigade of the Fourth Division, or the Highland Brigade created in the Sixth Division by the 1813 reorganization, but these remain exceptions. As such, brigade identity in these instances needs primarily to be seen as an extension, and reinforcement, of existing regimental identities. For the most part, one can therefore assert that, after the primary self-identification as part of a regiment, the next, secondary stage, in the self-perception of peninsular soldiers was as a member of a division—or, more specifically, as serving under a particular divisional commander. The latter element is particularly significant since, unlike the bulk of their component brigades, the peninsular divisions generally had very few changes of commanding officer. Discounting temporary replacements when the regularly assigned officer was on leave or recovering from wounds, the First Division admittedly still had six commanders; however, the Second, Fifth, Sixth, Seventh, and Portuguese Divisions each had three, and the Third, Fourth, Light, and First and Second Cavalry Divisions had only two.[88] The case of the First Division differs to an extent inasmuch as it was generally commanded by the second-in-command of the whole army, but includes one tenure lasting only a few months cut short by sickness, and another of five weeks cut short by the capture by the French of the officer concerned; only Brent Spencer, Thomas Graham, and Kenneth Howard led it for any appreciable length of time. This level of continuity, combined with the small size of the Napoleonic-era division when compared with its more modern equivalent, enabled an officer of character to become a recognized personality and stamp his division in his own image. This can be seen in particular with the cases of Robert Craufurd and Thomas Picton, although the importance of the former's strength of character, including its legacy after his death, is unique and another factor behind the unusually developed self-regard of the members of the Light Division.[89]

More paternal commanders, such as Hill or Cole, were able to make a similar, if less dramatic, mark on their commands, whilst James Leith summed up his record in command of the Fifth Division as having been a "good manager."[90] Nevertheless, it was still possible for such an officer to acquire his detractors just as Picton and Craufurd did. We have already seen, with reference to the story of the 2nd Foot, how Henry Clinton used his position as commander of the Sixth Division in order to intervene in the internal arrangements of the battalions under his command. In that instance, the interference was to the good, but, on the other hand, Lt. Colonel George Bingham strongly resented Clinton's interference in his command of the 2/53rd, and was pleased when the battalion was transferred elsewhere.[91] Whereas Bingham could only seethe about Clinton in his letters home, William Stewart's brief command of the First Division saw him offend the sensibilities of the Guards, and the matter ended up being referred all the way back to the commander in chief. As colonel of the 1st Foot Guards, York swiftly came to the defense of "his" regiment and the prerogatives of its officers, something that may have hastened Wellington's decision to shift Stewart from the First Division to the Second.[92]

Having enough authority and permanence to make his presence felt, the divisional commander could, for officers, be the immediate manifestation of all that was irksome about a military career. Lt. Charles Crowe of the 3/27th found himself deprived of a chance to become acting brigade major as the result of an argument between Major General William Anson, commanding the brigade in question, and the divisional commander, Cole. It took some effort on the part of Crowe's commanding officer to persuade him that the appointment was not worth the risk of further confrontation.[93] On the other hand, for the rank and file, the person of the divisional commander gave higher authority a recognized human face and thus gained credit when the system was working as it should, something most memorably voiced by the men of the Light Division's Caçadore battalions who cheered for "General Crauford [*sic*], who takes care of our bellies!"[94]

Full bellies aside, there is no denying that the primary reason that Craufurd, and Picton like him, were able to stamp their divisions with their character is because they themselves possessed character in abundance. Nor was it the case that these men were simply respected, but it is evident that each was held in some affection by many of the troops he commanded. Picton arrived in the peninsula with the unfair stigma of a colonial scandal attached to his reputation, and the officers and men of his new command met him with considerable trepidation. Some, like Grattan, never entirely got over the initial bad press surrounding Picton, but others came to have a strong regard for their general. Joseph Donaldson's first encounter with Picton was hardly auspicious, since it came when the general delivered a "sermon" to the erring 94th on the perils of plundering, but soon came to realize that "in other respects he was indulgent; and, although no man could blame with more severity when occasion required, he was no niggard of his praise when it was deserved. Nothing could surpass his calm intrepidity and bravery in danger; and his presence in battle had the effect of a talisman, so much has his skill and valour gained the confidence of those under his command."[95] In many ways, these qualities are very much akin to those picked out as representative of the most successful battalion commanders, and reinforce the appreciation and respect that was earned by officers at any level who, however firm their discipline, were fair toward their men and willing to share their dangers. Having expected to find a tyrant in Picton, Donaldson instead found himself, as a member of Picton's division, identifying with his commander's fight to clear his name. Donaldson's memoirs, written nine years after Picton's death at Waterloo, pay tribute to his former chief's eventual success in that struggle.

The importance of the linking of the divisional commander with the divisional identity is highlighted by the case of the Waterloo campaign. The six numbered divisions in existence in June 1815 were mostly commanded by peninsular veterans, but with numerical designations that failed to correspond to the formations these men had led in Portugal and Spain. As we have seen, Lowry Cole asked for, and got, command of a division containing a large portion of his former peninsular command, and he went on to request that this new formation be renumbered as the Fourth Division to match his old peninsular command.[96] It seems that others raised the issue as well, since when Wellington wrote to Sir Henry Clinton on June 15 to canvas his views on a proposed renumbering of divisions across the whole army he informed Clinton that "Some of the General officers would wish very much to have the divisions numbered over again, and have their old numbers, which appears to be a very natural wish: and I should be very much obliged to you if you would let me know as soon as you can if you participate in it."[97] This organization left the First Division as it stood, but would have given Clinton, Cole, and Picton their old peninsular numbers whilst making Alten's new command the

Fifth Division and Colville's the Second. As is apparent from the dating of the letter, events soon overtook this scheme, but it is significant on multiple counts. Firstly, Wellington's reference to general officers in the plural would indicate Picton as well as Cole as being behind the suggestion; Alten may be discounted, since his peninsula command had been the unnumbered Light Division, whilst Clinton, the only other officer who had held a permanent divisional command in the peninsula, was self-evidently not behind the proposal since it was his views on it that Wellington was soliciting. Secondly, Wellington was clearly in favor of the idea, suggesting an intention to foster emulation of the peninsular esprit de corps within his new army. Indeed, Wellington told Cole that he believed renumbering the divisions would generate "a symptom of the old spirit we had amongst us, than which we cannot have a better."[98] This belief was notwithstanding the fact that, outside of Cole's command, the only unit thus acquiring continuity with its peninsular assignment would have been the 2/30th in what would have become the Fifth Division. Conversely, the 2/44th and 1/71st had numerical, though not commander, continuity under the original numbering, which would have been lost in the re-designation. This would suggest a greater desire to emulate the reputation of the formation, as linked with its commander, than to tie in with a regimental reputation containing a link to a particular division.

But if identity with a division, and its commander, allowed for the development of esprit de corps beyond the regimental level, it could also lead to direct rivalry between men from different formations. This rivalry was particularly keen between the Third and Light Divisions, and became particularly pronounced during the siege operations, where the assignment of different divisions to different avenues of assault set up a situation where they were effectively racing to obtain a single prize. This rivalry could, in itself, be beneficial, as with this episode described by Harry Smith in his relation of the preliminaries to the storming of Badajoz. Smith was then a captain in the 1/95th, supervising men of the Light Division who had been detailed as a working party to aid the Third Division in its assault on the La Picurina outworks: "The Light Division, the working party, consequently were sent to the Engineer

Park for the ladders. When they arrived, General Kempt ordered them to be planted. The boys of the 3rd Division said to our fellows, 'Come, stand out of the way;' to which our fellows replied, 'D[amn] your eyes, do you think we Light Division fetch ladders for such chaps as you to climb up? Follow us'—springing on the ladders, and many of them were knocked over."[99] Thus, men of neither division were prepared to concede that the other division was the more deserving, even if doing so caused increased danger for those who elected to uphold the reputation of their formation. Later in the same siege, the Third Division as a whole would incur heavy losses storming the castle whilst the Light Division, alongside the Fourth, struggled in the breaches and took even heavier casualties. On this occasion, the losses on all sides were so severe as to render success hollow, and even the strongly partisan Grattan could only mourn the losses of all three divisions—although this did not prevent him from belittling the achievements of Leith's Fifth Division, which, having escaladed the walls elsewhere, had arguably beaten them all to the prize.[100]

Yet whilst such rivalry could produce laudable feats of emulation, it has also been put forward as a contributing factor to one of the least savory episodes of the history of the peninsular army, namely the sack of San Sebastián in 1813. The reduction of this fortress was initially assigned to Graham's Left Wing, composed of the First and Fifth Divisions plus attached brigades. The Fifth Division, which took responsibility for the actual siege operations, was under Major General John Oswald in the absence of its regular commander, Leith, who only rejoined it on the day of the second, successful, attack. The first attack was repulsed with loss, and accordingly a second assault was planned for August 31. At this juncture Wellington asked for volunteers from other divisions to form storming parties to lead the assault, the main burden of which would fall again on the men of the Fifth, which is the point at which an element of controversy is introduced.[101] The historian Bruce Allen Watson argues that this implied snub to the men of the Fifth was responsible for their role in the subsequent sack of the town, forming a means in which they could vent their anger. Specifically, Watson suggests that this action represented a transferal of anger from Wellington, whom they could not revenge themselves upon directly and who was in any case idolized, to the unfortunate Spanish inhabitants.[102]

This neat psychological and sociological analysis is, however, largely a flawed one. Whilst Watson himself addresses many of the limitations of his case, his argument misses one significant point. Although Wellington did issue the request for volunteers, he was not responsible for the idea that the men of the Fifth Division had failed in their first attack through lack of courage; rather, this notion was largely spread by camp gossip and subsequently repeated in diaries, journals, and later memoirs.[103] Far from seeking to slight the Fifth Division, Wellington in fact praised the Fifth Division's conduct during the first attack, and was only driven to call for volunteers because of the despondency of its senior officers.[104] Wellington did make explicit the hope that the volunteers "would be enough to show the way to the breach," thereby implying a hope that the Fifth, or its commanders, might be shamed into action, but this was in private correspondence with Graham and not for public consumption.[105] Such an argument also fails to account for the fact that there was no remarked difference during the sack between the behavior of the Fifth Division men on the one hand and the surviving volunteers on the other.

Watson makes a fair point when he suggests that Wellington's neglect of operations on the San Sebastián front during the ongoing fighting in the Pyrenees may have highlighted a sense of neglect and added to confusion and demoralization amongst those who had survived the first assault, and repeats the view that siege work was in itself unpopular and demoralizing.[106] However, by treating the events at San Sebastián in isolation, Watson fails to pick up on the similarities between it and the sieges of Ciudad Rodrigo, Badajoz, and Burgos, all of which prefigured periods of mass disorder amongst the troops. The sack of San Sebastián, exaggerated by accounts of those who were not present, was no more horrific—and certainly less drawn-out—than the events seen at Badajoz. Its roots lay in the deeper problems of collective indiscipline experienced by the peninsular army after all its sieges, potentially exacerbated on this occasion by a temporary breakdown of command caused by the unusually high officer casualties in the storm.[107] Even if the men of the Fifth Division were aware, via camp gossip, of the slight on their reputation implied by the call for volunteers, it is inconceivable that this was enough to tip men over the brink into anarchy; if the perceived insult had any effect at all, it seems more likely that it spurred them on to greater endeavors in the attack. If a perceived snub to divisional honor could be so deeply felt at this stage in the war, then it is if anything a positive argument for the development of formation identity, with the causes of the sack of San Sebastián ultimately lying elsewhere.

Such negative constructions of divisional identity can therefore safely be discounted, leaving the conclusion that it remained an important and positive aspect of the self-identity evolved within the British Army on campaign. Regimental identity remained the mainstay of morale and self-image, but a secondary association with the division served to reinforce this rather than supplant it. It was possible to identify oneself with more than one level of the military hierarchy, and the impression one gets from the bulk of memoirs is that the writer's regiment was the best in its division, and that division the best in the army. This distinction was made explicit by Lt. Colonel James Campbell, in his 1840 work, *A British Army, As It Was, Is, and Ought To Be*, in which he praises the Third Division as a whole for its conduct at Fuentes de Oñoro but maintains a distinction between his own 1/45th—naturally in his eyes superior—and the remainder of Picton's battalions. Campbell's analysis does blur things a little through his acceptance of the 1/88th as a sister unit worthy of comparison with his own, but his work is, in fairness, a plea for reform rather than a personal memoir. Campbell may have been able to take a slightly broader view than most regimental officers, inasmuch as he had served for part of the time as a brigade major, and was in fact seeking to use the example of the peninsular divisional system to make a wider point regarding the utility of permanently embodied divisions, exactly because they generated cooperation and emulation between battalions enjoying prolonged association.[108]

Largely as a result of Wellington's successful peninsular application of divisional organization, the British Army of 1815 had developed a sophisticated and flexible system that facilitated the integration of units of varying capabilities into a functioning whole. This system of command made best use of a body of general officers who were, at least initially, lacking in the experience necessary for more independent commands, and fostered the creation of sense of a self-identity that supplemented, without supplanting, preexisting regimental loyalties. Wellington did not introduce the combat division to the British Army; the doctrinal inheritance from the years before 1809 is clear, and is also in evidence in the smaller parallel organizations implemented on Walcheren, in the colonies, and on the East Coast of Spain. Unlike his contemporaries, Wellington took that doctrine and built on it, expanding the role of the division and the divisional commander and turning the formation into a key building block of his command. Whilst Wellington's divisional system in its final form was a great improvement on what had gone before, it must therefore be seen as evolutionary rather than revolutionary, and as building on existing practice rather than beginning anew. Nevertheless, Wellington's need to maintain an army for a sustained campaign meant that far more care needed to be taken as to the composition and leadership of the various elements comprising his army and it is here, rather than in the organizational framework itself, that the true significance of the peninsular system lies. By delegating responsibility and creating additional staff posts at divisional level, this organization allowed a large army to function effectively, and in turn facilitated the most effective distribution and employment of manpower.

But whilst much of what Wellington did organizationally was based on a solid doctrinal inheritance, what was essentially new, and only made possible through the continuity of the system over a period of five years, was the creation of a strong sense of divisional identity based largely around the personality of the divisional commander. Although this sense of identity was achieved by accident rather than design, it is clear that once the beneficial effects of this tendency had become apparent they were deliberately fostered, both in the peninsula where incumbent divisional commanders unable to exercise their duties through sickness or wounds were only temporarily replaced, thus permitting their ultimate return to the same job, and more explicitly in 1815. This policy can be seen in Cole's assignment to a formation having considerable unit continuity with his former command, and, on a broader level, in the proposed divisional renumbering in line with peninsular command assignments. Such overt attempts to re-create a field army based on the peninsular organizational model indicate recognition of the importance of the divisional system and its establishment as a means of campaign organization. But just as important in 1815 was the need to use the divisional system as a means of re-creating the ethos and self-image of the old peninsular army. In this way, what began as a purely organizational and administrative measure also ultimately, if not intentionally, became one of the key motivational ties that helped give the men of the British Army their sense of identity.

CHAPTER 6

Strategic Consumption

When Moyle Sherer identified "sickness, suffering, and the sword" as the factors that had reduced his battalion's strength over the course of its five years in the peninsula, it is unlikely that his order of priorities was a coincidence. Unless the victim of a catastrophic tactical error, no unit was likely to be rendered unfit for service by battlefield casualties alone; indeed, only the 23rd Light Dragoons after Talavera and the 29th Foot after Albuera were sent home for this reason, although the aftermath of the latter battle also saw the amalgamation of several other badly reduced infantry battalions. To be sure, battlefield losses could be severe, even in successful actions, but most units were engaged in pitched battle on only a handful of occasions during the period, whereas sickness and suffering between these events exerted a constant daily toll on their strength.

It is easy to dismiss such losses as "wastage," but this is a deceptive term that implies the squandering of resources. A better understanding is gained if the concept of "strategic consumption" is applied instead, with manpower one of many finite resources being expended as a means toward achieving strategic goals. All deployments entailed a level of strategic consumption, to a greater or lesser degree depending on the station and the role of the unit in question. Obvious causes of manpower consumption, most notably the loss of men in the fever-ridden West Indies, could be minimized by preventative measures or the use of alternative manpower from sources that were better acclimatized, deemed expendable, or both, but for any unit on active campaign overseas a minimum level of loss had to be accepted. The role

217

of the commander in this context could only be, as Wellington realized, to ensure that losses—from *any* cause—be kept at a minimum. Just as it was necessary to hazard lives in combat in pursuit of larger goals, so too was it sometimes necessary to expose troops to greater hardships than would normally be warranted. The periodic need to force-march troops, as in the arduous operations during and between the Oporto and Talavera campaigns, or during one of the great retreats, is a key example of this. Likewise, it was occasionally necessary to station troops in unhealthy areas that had to be held for political or strategic reasons, as with Walcheren Island or the Guadiana valley.

But even if it is allowed that some losses were inevitable, it is impossible to forget that we are dealing with flesh and blood; the resources that were being consumed were the lives of the men in the ranks. A soldier who fell ill could expect a decidedly uncertain future, particularly at the beginning of the period when the British Army's hospital arrangements were at their worst. Under such circumstances, it is scarcely surprising that some soldiers, even if they survived the circumstances that led to their hospitalization, took months or even years to return to their units. Others, once detached, contrived never to return at all, and this brings us to the second—and secondary—drain on campaign manpower: desertion. The circumstances leading to an individual's decision to desert are themselves individual, but Sherer's identification of suffering as a factor tending toward the reduction in a unit's strength can certainly be considered as one of them. Hard service could bring on illness that could, in itself, take a man out of the ranks, but it could also break down the interpersonal bonds that held a unit together. Under such circumstances, absenteeism increased as men left the ranks to search for food, or fell out through exhaustion; however, since there was no intermediate category of being absent without leave, it was all too easy for the straggler to become the deserter. In order to minimize losses in this way, effective leadership was needed to keep a unit functioning cohesively through even the hardest service.

In a sense, it is hardly to be wondered at that some regiments became and remained sickly due to places that the fortunes of war took them. What is more important, so far as we are concerned, is how those varying sickness rates were managed and whether—and for how long—the units in question were able to remain in the field. At a regimental level, strategic consumption is therefore intrinsically linked to the history of a unit's service and the way in which it was led.

Health and Sickness

In the aftermath of the Peninsular War, Sir James McGrigor, inspector of hospitals under Wellington during the second half of the conflict, put his thoughts in order so as to be able to present them to the Medical and Chirurgical Society, of which McGrigor was vice president. Before joining Wellington's staff in the peninsula, McGrigor had seen extensive service encompassing the Egyptian campaign of 1801 and the Walcheren expedition. If anyone was in a position to speak with authority on the topic of illness amongst soldiers, then it was he. In his short paper, McGrigor gave a précis of the medical aspects of Wellington's 1812, 1813, and 1814 campaigns, which prefigured a discussion of the key diseases encountered and the means by which they might be better prevented in a future campaign.[1] What is particularly significant about McGrigor's analysis, however, is the amount of times in which a clear focus is made on the particular circumstances of a regiment—or, more common still, a division—as a direct contributor to its state of health. These circumstances could range from being in a particular station to having received a recent influx of recruits, but McGrigor was sure that they made a difference, and had the figures to prove it.

Losses from sickness, rather than from combat, represented the single greatest drain on British military manpower during the period of this study, with roughly three deaths from the former cause to every one of the latter during the Peninsular War.[2] The topic of sickness and mortality amongst troops in active service has accordingly merited much study, going right back to McGrigor's 1815 paper. However, the range of data obtained during the research for this study enables a far more wide-ranging analysis to be carried out, both in relation to the data of McGrigor and in comparison with more recent historiography. Most notable amongst modern writing is Dr. Martin Howard's *Wellington's Doctors*, which draws heavily from McGrigor's work but also uses Howard's own medical training to diagnose more accurately the symptoms experienced by British troops.[3] However, Howard's work makes use of McGrigor's statistics based on the records kept by the hospitals rather than those kept by the regiments, as here. Thus, whilst Howard is rightly full of praise for McGrigor's reforms of the peninsular medical services from 1812 onward, he fails to account for the fact that regimental records indicate that the ratio of sick to effective manpower remained poor throughout the last two years of the conflict. As we shall see, this discrepancy between the impression given by the hospital and regimental returns exposes a key failing in the systems of manpower management in the peninsula.

Before focusing on the specifics, however, it is important to try and obtain some idea of the scale of the question. The number of deaths in the theaters sampled for this study is staggering: some 55,088 in total.[4] This sample does not include troops in the eastern Mediterranean, the East and West Indies, or, before 1812, North America; nor does it include troops on the home station. However, by being selective this coverage encapsulates the death rate for the active component of the British Army—a body that averaged some 64,645 men across the period. The greatest proportion of these deaths—41,863 of them, or 76 percent—were in the peninsula, and the majority of them were not combat-related. Although the data, presented in graph form as figure 7, makes no distinction between deaths from battle or from illness, nor between sick and wounded when it comes to men in hospital, this fact is rendered obvious by the slight nature of the increases seen even at the times where heavy fighting was taking place. Waterloo, as the bloodiest battle in absolute terms, makes a significant impact, and the costliest elements of the War of 1812—the Niagara and New Orleans campaigns—also stand out in relation to the usual death rate for those theaters, but even Albuera and Badajoz do not rival the number of deaths seen during the worst epidemics.

Substantial though these figures are, it is important to recall that the dead did not represent the only cause of loss through sickness, as a great number of men were also sent home unfit for

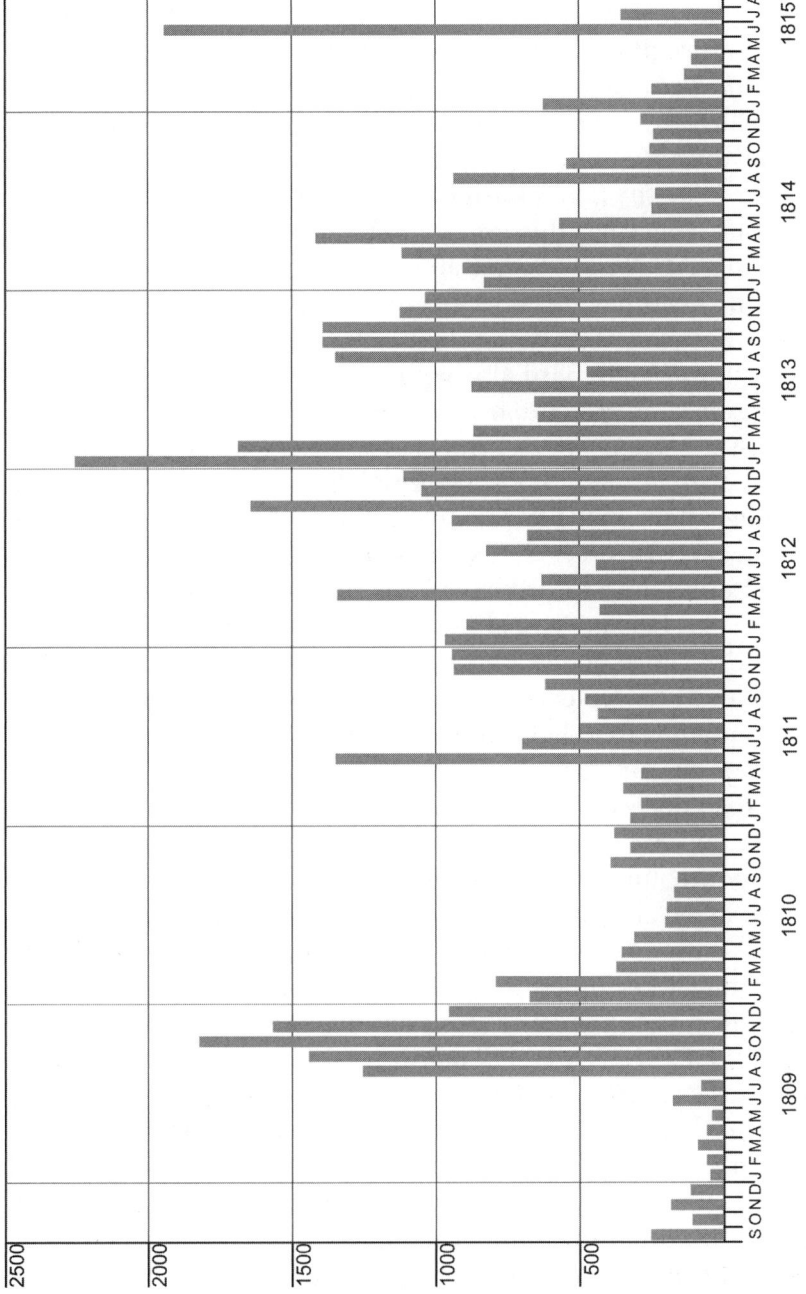

Figure 7. Deaths on Active Service, 1808–1815. Data from combined returns, as detailed in appendix 1.

active service, or else discharged from the service altogether. Data for the whole of the British Army is incomplete, but some examples at the regimental level demonstrate the point. The 1/40th began the Peninsular War with a total strength of 909 rank and file; during the course of the conflict, 967 men died whilst serving with the battalion, and a further 411 were sent home between August 1809—the first month for which this data is available—and May 1814. Together, this totals 1,378 men, and yet the battalion still mustered 829 total rank and file when it left the peninsula. Whilst the 1/40th was known as a sickly unit, and was also involved in a fair amount of heavy fighting, it is by no means too extreme an example. The 2/83rd also served throughout the conflict, beginning with 856 effective rank and file and ending with 491. During the war, it lost 626 dead and sent a further 153 men home: 779 in total. Unlike the 1/40th, the 2/83rd could not so readily call on replacement manpower and thus shrank to a lower effective strength, and this largely explains its apparently better survival rate. Whereas the 1/40th was topped up with replacements, some of whom also became casualties, the 2/83rd was allowed to shrink. The veteran core of the 1/40th is unlikely to have been any less healthy than the veterans of the 2/83rd, but the figure for the former unit is inflated by illness amongst the reinforcements. This distinction, in turn, emphasizes McGrigor's point that those fresh to active service were far more likely to fall ill than seasoned men.[5]

Whilst losses through death or permanent disability were self-evidently absolute and had to be replaced, the second drain on manpower as a result of sickness and wounds was those men temporarily unfit for duty who remained on the strength of their units, either in regimental hospitals or detached to general hospitals within the theater of war. Working as an average across the period, the figures suggest that around a quarter of all active manpower was rendered ineffective at any one time through sickness or wounds. Again, the vast majority of these men were from the forces in the peninsula. It is true that for the bulk of the period these forces represented by far the largest concentration of troops, but, even so, the proportion of ineffective troops in the peninsula is generally greater than for forces of equivalent size in Flanders in 1815 or, even, on Walcheren in 1809. However, although the

Table 9.   Sickness and Death Rates by Theater

Theater	% Sick	Deaths/1,000
Walcheren	25.8	27.6
Peninsula	22.4	12.9
American Coast	15.6	16.9
Northern Europe	9.8	7.3
East Coast Spain	9.6	4.7
Canada	7.3	6.1
Cadiz	7.0	3.3
Gibraltar	6.4	4.0
Nova Scotia	5.2	3.2

Source: Sources as per appendix 1. "Northern Europe" covers Germany, Flanders, and France, 1813–15.

peninsular figures do remain high throughout, it is worth noting that they peak in late 1812 and that thereafter the number being returned as sick declines even though, at the same time, Wellington's army was growing in size. Thus, in proportional terms, the sickness rate fell far more steeply than the absolute figures would suggest.

Considering sickness rates in proportional, rather than absolute, terms also helps clarify the situation with regard to the other stations to which the British Army was deployed. Table 9 ranks the main theaters of war in terms of their average sickness rate across the period, giving in addition the average number of deaths per thousand men. With the data presented in this way, it becomes clear that although the peninsula was exceeded only by Walcheren in terms of the levels of sickness amongst troops deployed there, the death rate was nowhere near as high. Comparing like with like, sickness levels tend to be higher in the warmer theaters, due in no small part to the greater prevalence of fevers. Figures for Cadiz and Gibraltar are higher than those for other garrison commands due to the prevalence of yellow fever in southern Spain, and, epidemics aside, the base rate for sickness and deaths was far lower; of the 1,494 deaths recorded at Gibraltar between 1808 and 1815, 362 of them were concentrated in the last four months of 1813 when there was yellow fever on the Rock.[6] Extreme cold could also pose problems, and in Canada care needed to be taken against the outbreak of contagious disease amongst troops in winter quarters.[7] To a lesser extent, contagion whilst in winter quarters was also a problem in the peninsula, whilst exposure to cold was in itself also an issue, as was also the case in Holland during Graham's campaign and—most memorably—during Moore's disastrous winter campaign in Spain. On the other hand, although the coastal United States contained unhealthy areas, particularly in the swamps around the Mississippi delta, the high incidence of both hospitalization and death there has far more to do with American bullets than American pathogens.[8]

As well as showing marked distinctions between different theaters of war, sickness levels also show marked distinctions between the various arms and subdivisions of the service, and between British troops and foreign units in British pay. Infantrymen were nearly twice as likely to be hospitalized than cavalrymen, and more than twice as likely as artillerymen. Specialist garrison and veteran units have a particularly low sickness rate, as do the Canadian provincial regiments in the War of 1812, but this is readily understandable, since these units saw little active campaigning. On the other hand, the deviation between British and foreign infantry, which also recurs in the cavalry, albeit to a lesser extent, is less readily explained. Although many of the foreign units served in secondary theaters, and in general saw less combat, this can only serve as a partial explanation for their lower levels of sickness. However, much of the foreign manpower taken into British service represented men who had already seen service in other armies, and who were accordingly better acclimatized to service. McGrigor believed that "The temperance, steadiness, and regular habits of the German legion, kept them always in a state of health."[9] These qualities speak of experience, and McGrigor's assertion is further supported by the statistical data at a regimental level. Discounting small detachments, the worst average sickness ratio in a KGL unit is that of the 5th Line Battalion, averaging 20.6 percent over eighty-one months of service in the peninsula and northern Europe; by contrast, no less than fifty-six equivalent

British units have worse ratios, with the worst having twice as many men on average sick than the KGL battalion.

Since the distinction between British and KGL sickness rates can be positively identified and attributed, at least in part, to better training and leadership, it is perhaps all the more surprising that a similar distinction cannot be made within British units. There is no obvious division between units that were notable for being either poorly or expertly commanded, nor for those that might be expected to have a strong unit identity or esprit de corps. Indeed, the two most sickly battalions—the 68th and 1/23rd—might both have maintained a claim to elite status, the one as light infantry and the other as fusiliers. One explanation for this might be found in the tendency, discussed in chapter 2, for elite units to incur heavy battlefield casualties during the course of maintaining their self-defined status, but it seems unlikely that this is the whole story. Nor does it seem sufficient to simply assert, irrespective of unit, that that the majority of British officers had little care for the well-being of their men, since there is substantial evidence to the contrary—not least the fact that many units, even in unhealthy theaters, were able to maintain a decent level of fitness.[10] Negligent officers certainly did exist, and are encountered in these pages, but it is absurd to suggest that they represented the norm. Many, it is true, could have learned from their colleagues in the KGL, but this is evidence of inexperience, not of neglect. It certainly is the case that some units known to have had bad commanders do have particularly poor sickness ratios—26.4 percent in the 2nd Foot, for example—but if inadequate leadership did contribute to higher levels of sickness, it is clear from the even poorer ratios found in well-led units like the 1/23rd that it was only a secondary factor.

Rather than try and attribute effectiveness to the qualities of individual units and their commanders, we need instead to look at the nature of the service undertaken by those units, for, as the data in table 9 suggests, it is here that the distinctions become more apparent. The main recurring factor that pushed sickness levels up at a unit level was service with one or both of the two most unhealthy commands: the Walcheren expedition or the peninsular field army. It is therefore to these two theaters, and the very different circumstances surrounding them, that we must turn our attention.

Having the highest rate of both sickness and deaths, but seeing only limited fighting of low intensity, Walcheren and its surrounding islands by far and away constituted the unhealthiest European theater of war. During the Walcheren expedition, 2,041 men died of disease and a further 1,859 after the return home, in contrast to 99 deaths through combat.[11] The cause of this disaster is, now, obvious enough—the island was a malarial swamp, although additional symptoms, including loss of appetite and an enlarged spleen, suggest that complications were involved[12]—but there are a number of important issues surrounding the campaign that have a wider impact on this study. Most obviously, the scandal resulting from the number of deaths accelerated the reform of the British Army's medical services, and indirectly led to the reforms eventually implemented in the peninsula under the aegis of McGrigor and others.[13] However, the story is by no means one-dimensional, and had its beginnings months before the expedition was even conceived. It was widely recognized at the time that men who had served on Walcheren and had been exposed to its fevers were prone to relapse, and that units that had served there were accordingly prone to develop high levels of sickness when subsequent active service rendered men's constitutions vulnerable. But whilst the future service of the Walcheren regiments forms one part of the story, it is also necessary to look at their prior careers, since in a great many cases service on Walcheren was not, in itself, uniquely devastating but, rather, represented a final straw that broke individuals—and units—that were already weakened.

This second point may seem illogical, but makes sense when one considers that the majority of the units that formed the Walcheren expedition had recently returned from the first peninsular campaign culminating in Moore's retreat to Corunna. It is therefore necessary to question the extent to which this arduous and extreme service had already rendered the units involved unfit and thus susceptible to further disease. As well as the effects of cold, hunger, and exertion, many of Moore's regiments had contracted typhus from their Spanish allies, and there had also been an outbreak of ophthalmia, rendering many men temporarily blind.[14]

Even those men who finished the march without succumbing to illness were weakened by malnutrition and exhaustion. Furthermore, the survivors returned to Britain in an awful condition with men landing in odd lots at different ports, some being retained in hospitals whilst others marched to their depots. This inevitably led to confusion as units began to reassemble, and delayed the recovery of those men who found themselves detached from their units. In the 2/59th, for example, 314 men were listed as missing in February 1809 from a rank-and-file strength of 728, though the unit's total losses on campaign had amounted to only 143. This figure, furthermore, did not take into account an additional 176 men sick in hospitals, leaving only 233 rank and file fit and present for duty and causing the unit's commander to bemoan the inaccuracies in the "supposed list of the missing."[15] The situation for the 2/59th was typical of most of Moore's battalions, and in many cases the sickness figures remained high well into 1809 as detailed in appendix 2.

As a result, many battalions embarked for Walcheren in July with a sizeable proportion of their men only just returned from hospital, and therefore likely not yet fully returned to fitness. Others were forced to make up the numbers by recruiting and drafts from the Militia; the 1/38th received 221 draftees during April and May 1809, and the 1/4th received 359 in April alone.[16] Others required the influx of large drafts from their second battalions to get them back to strength, as with the 1/26th, which received 210 men in June.[17] Given the limited time available, it is unlikely that these drafts would have been fully integrated into their units, and in any case had seen only home service and were not acclimatized to campaigning. Whilst most of the battalions that came back from Spain went on to Walcheren, three did not; these were the first battalions of the 43rd and 52nd Light Infantry and 95th Rifles, which instead returned to Portugal whilst the second battalions of the same regiments were brigaded together under Major General William Stewart as part of Chatham's command. However, before departing for Portugal, the three first battalions offloaded their ineffectives onto the junior battalions of their respective regiments. This was acceptable for the 95th, which used its new third battalion as a depository for the sick from the two senior battalions, but left the 2/43rd and 2/52nd, both under orders for Walcheren, with large numbers of unhealthy men.[18]

As a result of this combination of factors, a significant proportion of the Walcheren expedition comprised men who were either recent draftees not yet acclimatized to active service, or else veterans still not fully fit after Corunna.[19] With such widespread vulnerability, the onset of the fever, within days of the troops first landing, was sudden and in many cases overwhelming. Benjamin Harris of the 95th observed "whole parties of our Riflemen in the street shaking with a sort of ague, to such a degree that they could hardly walk; strong and fine young men who had been but a short time in the service suddenly seemed reduced to the state of infants, unable to stand upright—so great a shaking had seized upon their whole bodies from head to heel."[20] At length, by the time orders came to re-embark, only three men in Harris's company remained fit including Harris himself, but he too then succumbed to the fever during the voyage home and found himself in the hospital in Hythe along with the survivors of the other Light Brigade battalions. Having saved a substantial sum from his work as a battalion cobbler, Harris was able to purchase additional food and wine that he shared with his comrades, to which he attributed his eventual return to health—although, as it would transpire, his ordeal was not yet over.[21] Harris was at least fortunate not to have contracted the fever until already on his way back to England, where medical provision was readily available. Those who fell ill in the early stages of the campaign, however, had a hard time of it as Chatham's medical staff in the field were rapidly overwhelmed by the volume of cases. One such was John Green of the 68th, who was "amongst the first that were attacked by this disease, and laid some days in a barn, without partaking of any food whatever, and was brought so exceedingly low, that I was almost insensible to any thing that was going on amongst my comrades."[22] Green was treated in a series of barns, and then in the general hospital at Veere, but it was not until he reached the latter place that he could be provided with a decent bed and food. Eventually taken back to England, Green required a whole year to recover his health.[23]

Harris, on the other hand, initially believed that he had beaten the fever, but, like many of its victims, his troubles were just starting. Seemingly fit, he was selected to form part of a detachment from the 2/95th being prepared to go out to Cadiz, but a single day's marching on the way to embark at Portsmouth was enough to bring on a relapse, and he found himself back in hospital at Hilsea Barracks, where he was lucky enough to defeat the fever for a second time.[24] Even those of Harris's comrades who were fit enough to embark were by no means clear of the fever, since Cadiz garrison records report that on April 19, 1810, "The two companies of the 2nd battn. of the 95th Rifle Regt. marched from the Telegraph Heights to the barracks at Cadiz on account of the men suffering from a complaint[,] the remains of the Walcheren fever."[25] Harris, meanwhile, spent a further year's convalescence, the bulk of it with his family, before eventually rejoining his regiment. Being still unfit for active duty on account of the state in which the fever had left him, he was eventually posted to a veteran battalion, where he finished the war and obtained his discharge following the peace of 1814. Recalled to serve again in the Hundred Days, a recurrence of the fever meant that he was unable to do so, as a result of which he found his pension forfeit. Even thirty years later, when Harris dictated his memoirs, the fever had not entirely left him, and he still felt "the remains of it in damp weather."[26]

The long-term effects of the Walcheren campaign therefore have to be understood on two levels. The first is the number of men who were permanently rendered unfit for service, and who had either to be discharged or transferred to veteran or garrison units. The experiences of Benjamin Harris demonstrate how long and drawn-out the suffering of such men could be. This suffering not only was a personal tragedy for the individuals concerned but also cost the British Army considerable numbers of trained men, whose absence would be felt later in the war. Harris's experiences also exemplify the other problem, namely the episodic or intermittent nature of the fever, which rendered its victims liable to relapse.[27] When we look at their service over the period as a whole, we find that units that served on Walcheren have an average sickness ratio of 23 percent, as opposed to the average of 17 percent across the full range of units sampled. Death rates are also worse: thirteen per thousand for Walcheren regiments, but only ten per thousand as an overall average. When all the units sampled are ranked by sickness rate, four of the worst five, and seven of the worst ten, are units that had served on Walcheren— the worst of all being the 68th Light Infantry with an average ratio of just under 42 percent sick over three years' worth of data. Understandably, the units that suffered the most were those—including the 68th—that remained with the occupation force through the autumn. Those that left when the main force was withdrawn at the beginning of September tend to have better long-term sickness rates, having had less exposure to the fever.

As the experiences of Harris and Green exemplify, it took months, if not years, to fully shake off the Walcheren fever, and as late as February 1810 over a third of the expedition's survivors were still hospitalized.[28] Nevertheless, most of the units that served with Chatham's ill-fated army did ultimately return to active service. Of the fifty-one units, including detachments, that served in the Walcheren expedition, thirty-seven were subsequently redeployed to the peninsula, joining Wellington's command between 1810 and 1814, but their effectiveness continued to be undermined by their earlier service. The 1/26th was eventually sent away as being irredeemably sickly, whilst the 77th spent much of the war in garrison for the same reason. In the case of the 1/26th, the battalion seemed at first to have recovered itself after a period of duty on Jersey, but rapidly became sickly on being posted to the peninsula. Since Harris found that even a route-march in England brought about a relapse, it is scarcely to be wondered that the altogether more arduous circumstances of active service had the same results for the men of the 1/26th. The battalion was eventually transferred to Gibraltar, but as late as 1813 was still not entirely recovered, as repeated inspection reports testify.[29] Major General David Widdrington spoke extremely scathingly of the drafting of unfit men to the unit, who then readily succumbed to disease.[30] Losses due to the fever could also have an adverse effect on the internal economy of a unit, reducing its effectiveness even if the men were physically fit. Major General Robert Long believed that, at least in part, the poor state of the 9th Light Dragoons in early 1813 was due to "the loss sustained by the Regt. in [NCOs], as well as in some of its best men in the Walcheren expedition"[31]—a telling indictment, considering that Long was writing over three years after the end of Chatham's campaign.

Even those Walcheren regiments that were fit enough to serve with the main army in the field still required special measures to keep them in a fit state. Although a significant reinforcement joined Wellington in time for Salamanca, largely composed of battalions that had previously served on Walcheren, it proved necessary to leave the new units behind when the army pushed on to Burgos, lest the continued exertion ruin them completely.[32] Over and above the growing awareness that the army as a whole required more effective medical provision, particularly close attention had to be paid to the standard of care available in these units, forcing McGrigor to intervene in the organization of at least one of the battalions in question. Writing in April 1813, McGrigor was keen to show Wellington that he was aware of, and was dealing with, the lingering effects of the Walcheren fever in the 1/5th, which had joined as part of the 1812 reinforcement:

> In January last finding that from the age and infirmity of Surgeon Lease he was unequal to the charge of a sickly Regiment I forwarded thru' your Lordship a recommendation for his being transferred to the 13th R. Veteran Battalion, and that an active & able surgeon should be appointed to the 5th Regt. in consequence that Mr Carter is appointed to that corps. But in order that the Regiment should not suffer in the mean time I directed one of the best and most active Surgeons in the Army to take charge of the medical concerns of the 5th Regiment, and Staff Surgeon Arthur remained with that Regt. and did the duty of Surgeon during the most sickly period of the Corps.[33]

More than three years after the failure of the Grand Expedition, its aftereffects were still causing concerns for the chief medical expert with the main British Army in the field. As such, however, they cease to become an independent issue and instead become intertwined with the wider problems surrounding the health of the peninsular army, and need to be addressed as part of that analysis.

Up to a point, it is relatively easy to assess the medical situation in the peninsula in an informed manner thanks to the analysis provided by McGrigor. However, that officer only arrived in Portugal in December 1811, meaning that detailed information on the earlier years is less readily available, although it is possible to a degree to identify common factors between earlier events and those from later in the war. Through McGrigor's analysis, it is possible to appreciate the seasonal nature of many of the illnesses affecting the troops, and this can then be more widely applied to aid understanding of comparable circumstances in other theaters. Many of these seasonal trends are logical when considered in relation to the weather conditions, as with the prevalence of rheumatic and respiratory problems in the cold and wet weather, which McGrigor particularly noted during the 1812 operations against Ciudad Rodrigo and Badajoz.[34] Winter conditions also contributed to incidence of typhus and typhoid fevers, as climatic conditions forced troops to spend long periods of time in overcrowded and potentially insanitary winter quarters.

McGrigor divided his service into four periods, each representing different sets of circumstances, and this approach can also be extended to earlier periods. Figure 8 superimposes this division onto a graph plotting sickness and death rates; in order to avoid undue distortion of the figures by losses attributable to combat, the data has been averaged out on a quarterly basis. This creates a total of eight phases—those identified by McGrigor are numbered one though four, in order to correspond with his own designations, whilst those for the first half of the period are designated A through D.

Phase A represents a period during which the army was inactive in Portugal, but although the sickness ratio is at its lowest point, it is by no means as low as one might expect. This can partly be attributed to wounded from Roliça and Vimeiro still in the hospitals, but a significant proportion of these casualties belonged to regiments that marched into Spain with Moore, and, after October 1808, these are no longer included in the returns from which this data is assembled. Evidently, a significant proportion

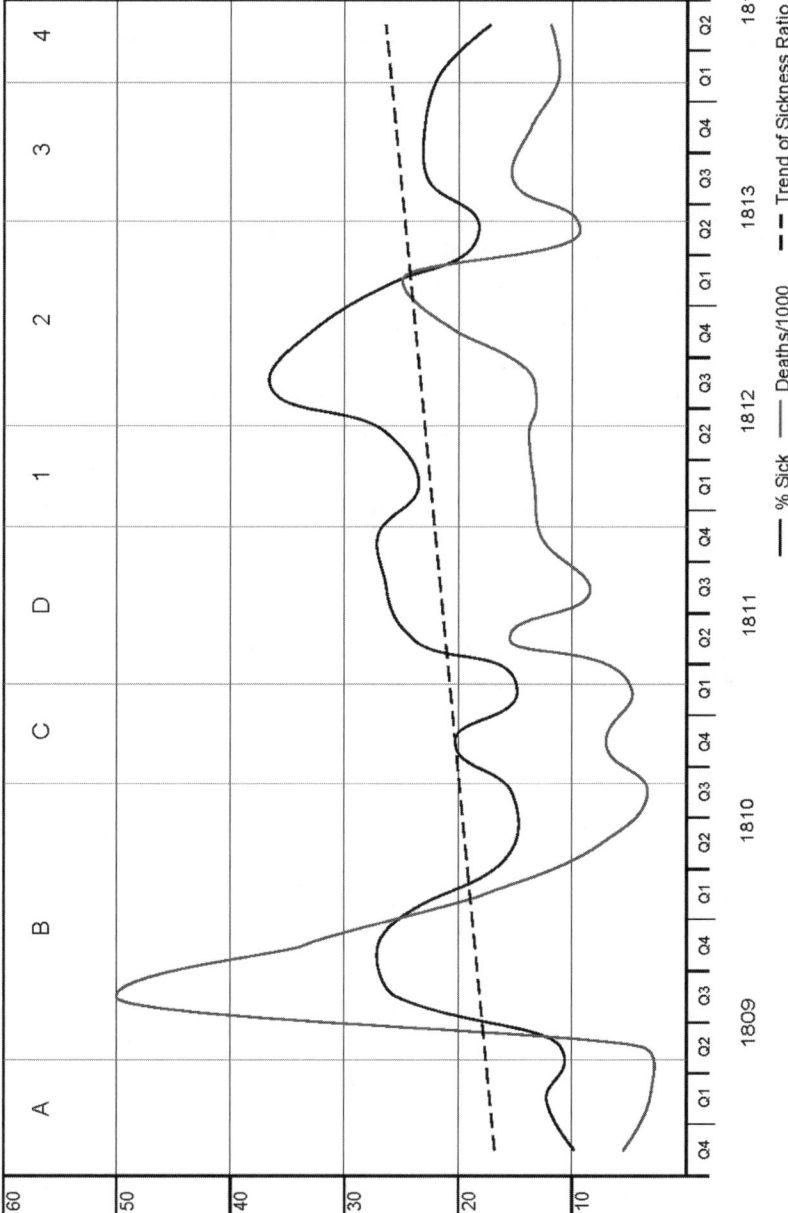

Figure 8. Sickness and Death Rates in the Peninsula. Data from Monthly Returns, TNA, WO17/2464–2465, 2467–2476.

of manpower was hospitalized through illness during this period, although this is in one sense unsurprising since we have already seen how high levels of sickness prevented units like the 29th and 1/40th joining Moore. Furthermore, time was needed for the men to become used to a foreign campaigning environment, and in 1808 this applied to everyone, whereas later in the war the troops going through this process represented a small proportion of the total. This acclimatization was exacerbated by the fact that, with the force as a whole being inexperienced, mistakes were made that would simply not have happened in later years. Commissary officer August Schaumann remarked of his service in Portugal at this time that the offal from the bullocks slaughtered for food built up around the camp to the detriment of health and hygiene, but it was only when this accumulation "became putrid in the great heat," and infested with flies, that it occurred to him to bury the mass.[35] This ignorance of the need for basic sanitary precautions stands as a marked contrast with the organization of a more experienced force, as evidenced in contemporary French practice.[36]

The other factor contributing to the higher-than-expected levels of sickness during this time is the fact that several of the units concerned had come to Portugal after several months aboard transport ships, where poor conditions and lack of fresh food, sunlight, and exercise all had a deleterious effect on health.[37] This factor was particularly noteworthy amongst the four battalions that had formed Spencer's initial command in the summer of 1808, which had spent the hottest months of the year cooped up aboard transports. These units lost thirty-five dead between them whilst embarked, and all had well in excess of one hundred men each sick by October 1.[38] At this same point, before Moore took the bulk of the troops into Spain, the average rate of sickness in the whole force was 10.4 percent; in contrast, Spencer's four original battalions averaged 17.8 percent.[39] The 1/40th, with the second highest sickness rate, had also spent much of its immediate prior service aboard transports during its passages to and from South America.[40] The worst sickness ratio of all was in the 2/43rd, which had not previously served outside the British Isles at all, but this battalion, being particularly inexperienced, suffered heavily with dysentery—a condition which, in itself, reinforces the harm done by ignorance of camp sanitation practices.[41]

The issue of acclimatization must also be seen as having a considerable part to play in the rapid rise in the sickness and death rates encountered at the outset of Phase B, which represents the Oporto and Talavera campaigns and the subsequent withdrawal, via the Guadiana, to Portugal. Troops that were not yet fully acclimatized to the theater inevitably suffered during the long and arduous marching involved in these campaigns, a trend also noted amongst those regiments that marched into Spain with Moore the previous autumn.[42] The pursuit of Soult after Oporto was forced through poor terrain in bad weather, and this same combination of forced marches and shortages of supplies was also experienced in the Talavera campaign. By then, heat had replaced rain as the main climatic problem, but so far as the sickness ratio was concerned, these extremes caused a continuation of what had already been experienced in the march to Oporto and back. The Talavera campaign saw a prolonged shortage of rations, and after the battle the supply system collapsed completely, to the extent that the men's pay stoppage for food was reduced to three pence in recognition of the fact that they had "not received their rations regularly since the 22nd July."[43]

The result of this combination of factors was that when the army, forced to remain in Spain to show continued support for Britain's ally, went in to cantonments in the marshy Guadiana valley, many of the men were already weakened. The similarities with the ongoing ordeal of the troops on Walcheren are marked, and the results were almost as severe as men began to rapidly succumb to fevers. The subsequent epidemic was, in relative terms, the single most destructive event in the Peninsular War so far as the British Army was concerned. Total deaths in November 1809, when the fever was at its height, were comparable with the casualties at Albuera or the storm of Badajoz. Only the even more virulent, but thankfully localized, epidemic that decimated the First Guards Brigade during the winter of 1812–13 led to more deaths through sickness. What was more, the death rate remained high over a period of six months; the peak came with 1,186 deaths in November, but the death rate remained continually in excess of 500 per month from

September 1809 to February 1810.[44] This figure does not take into account the number of men either temporarily or permanently weakened by illness, whose numbers were such that some battalions were reporting over half their strength as sick.[45]

What makes understanding the Guadiana epidemic harder is the arbitrary way in which units suffered. Those battalions that had undergone the rigors of the Oporto and Talavera campaigns generally returned higher numbers of sick, although a proportion of combat casualties must be factored in. Yet the 2/39th and 2/42nd, newly out from home, both fell heavily sick whilst the 5/60th emerged largely unscathed despite considerable arduous prior service.[46] To an extent, this discrepancy mirrors at a regimental level the problem seen within units during the Walcheren expedition. Some men succumbed because they were not acclimatized; others, though acclimatized, succumbed because they were worn out. However, like the Walcheren fever, the circumstances of the illness are confusing. Whilst the marshy Guadiana was notoriously malarial, the time of year makes this diagnosis problematic, and similarities with the better-documented epidemic of 1813 suggest that the 1809 epidemic was also a typhoid fever, or, at least, a combination of typhoid and malaria as may also have been the case on Walcheren.[47] Whatever its cause, the contagion was certainly specific to the locality, and battalions leaving the field army, such as the 2/83rd and 2/87th, showed a marked improvement once in a healthier clime.

Even before the epidemic, Wellington had a considerable problem by virtue of the number of Talavera casualties. This circumstance did, however, ensure that when sickness broke out steps had already been taken to organize hospitals, notably in the fortress of Elvas. Command of the sick was given to Lt. Colonel Henry Mackinnon, who oversaw the evacuation of the wounded back to Portugal and subsequently took command at Elvas, working alongside James Franck, McGrigor's predecessor as inspector. But, even with these provisions, the numbers of sick swamped the system, and there was little Mackinnon could do so far as the actual medical care was concerned.[48] Provision in this regard was sadly lacking at this stage in the war, and though the Elvas hospital was organized with a main section in a convent and a convalescent hospital in one of the barracks, overcrowding and lack of equipment and trained personnel ensured that conditions were hellish and recovery rates low.[49] Cooper of the 2/7th described conditions in the "Bomb proof barracks" at Elvas: "No ventilation, twenty men sick in the room, of whom about eighteen died. In this place there were [*sic*] one door, and one chimney, but no windows. Relapse again; deaf as a post; shirt unchanged and sticking to my sore back; ears running stinking matter; a man lying close on my right hand with both his legs mortified nearly to the knees, and dying. A little sympathy would have soothed, but sympathy there was none."[50] Unsurprisingly, those who had the choice preferred to remain with their units, one such being Lt. George Simmons of the 1/95th, who, despite three recurring bouts of fever—which he attributed to overexertion in the heat—was "allowed to proceed, by my own wish, with my regiment."[51] No matter how well organized, military hospitals were places to be avoided if at all possible, but for the rank and file the opportunities for private convalescence available to an officer like Simmons did not exist. In November, 5,740 rank and file from the infantry were sick in hospital, as opposed to 2,268 remaining with their battalions, but conditions for the latter can hardly have been improved by the fact that many battalion surgeons were taken from their units to make up for shortages at Elvas.[52] Although the eventual move away from the Guadiana was marked by a decline in death rates, it would take far longer for the men on the sick lists to return to their units, as is demonstrated by the lack of correlation between sickness and death rates during the epidemic.

Whilst the Guadiana epidemic was a one-off case insofar as its affecting the entire force was concerned, the trend seen during the initial months of Phase B, of an increase in the sick rate coinciding with the opening of active operations, continued throughout the war. As the army became more acclimatized, the levels of sick became lower, but each break from settled quarters saw a distinct increase. This pattern can be seen in Phase C of figure 8, representing the Busaço campaign; even in this short time, the sickness ratio climbs markedly before dropping off again when the army went into winter quarters behind the Lines of Torres Vedras. It also recurred in Phase D, beginning with the pursuit of

Masséna in March 1811 and continuing through to the following December, when the army was again in winter quarters. During this latter period, the army undertook a series of lengthy marches, with elements of the force being shifted between the northern and southern theaters. The debilitating effect of these marches, along with heavy combat casualties, ensured that sickness and death rates remained high. The contribution of attrition through strategic consumption is made clear by the fact that whilst the death rate drops after being forced up by the bloody actions of spring 1811, the sickness ratio continues to climb until active operations ceased.

Whilst the coincidence of active operations and a higher sickness ratio is understandable, with physical exertion and exposure to the elements both being recognized at the time as major contributors, a recent study has suggested an additional cause. The historian Edward Coss has demonstrated that the standard British ration issue during the Napoleonic Wars, composing biscuit, meat, and wine or spirits, compared unfavorably not only with that of the British soldier at earlier points in history but even with the food supplied to Renaissance galley slaves. Quite apart from occurrences of commissariat failures, which naturally made the situation worse still, the rations, even when delivered, are shown to have surprisingly low nutritional value. In particular, Coss's comparison between the standard ration and modern medical guidelines indicates a very low level of protein being provided. The standard ration was further deficient, wholly or in part, in fifteen of twenty-one major vitamins or macronutrients, with accordingly negative effects. Whilst the collective contribution of these dietary deficiencies was debilitating in itself, they also impeded recovery from other illnesses and wounds. Significantly, for example, the ration contained only 4 percent of what is now the recommended allowance of vitamin K, the absence of which can impede blood clotting and therefore increase both bruising and bleeding from wounds.[53]

Whilst Coss's calculations give a valuable insight into the nature of the nutritional problems faced by British soldiers, it should be kept in mind that the sparse nature of the basic ration was essentially down to financial and logistical constraints rather than medical ignorance. Nutritional theory in the early nineteenth century may not have progressed beyond the concept of the "universal aliment," believed to be present in equal proportion in all foods, but the doctors who treated the British solider of the Napoleonic Wars were fully aware of the advantages to be obtained through a healthy and regular diet, even if terms like "macronutrient" would have meant nothing to them. McGrigor summed up the situation as understood in 1815, asserting,

> The health of an army depends, in no slender degree, on the quality of the provisions and on the regular supply of them. Some of the divisions of the army appeared to derive a superior degree of health from attention to these circumstances. Some of them were always supplied with abundance of good meat, wholesome wine, and excellent bread; while others complained of their meat, got spirits instead of wine, biscuit instead of bread, or sometimes had neither bread nor biscuit, receiving in lieu of it a portion of flour, or an additional quantity of meat. It was the duty of the superintending medical officer of a division to see these things, and to report to me whenever they were complained of, or were equal to the production of disease. This was done for my satisfaction: at the same time I must state, that generals of division were usually paternal in their attention to the soldier, as well as most commanding officers of regiments.[54]

Further emphasizing the importance of effective leadership as essential to a good level of fitness in a unit, and reiterating too the paternal nature of that leadership, McGrigor went on to note, "If left to himself, the soldier would broil his modicum of meat and eat it at one meal, drinking his allowance of wine or spirits at a draught. It is needless to say, how hurtful this must be to a man undergoing great fatigue and requiring much nutriment. The orders of the Duke of Wellington were, that whether in the field or in quarters, the men should be divided into messes, have regular meals, their meat be well boiled, with a portion of vegetables and salt (whenever they could be procured): and under the inspection of their officers."[55] The fact that efforts were clearly being taken to supplement the standard ration, even if those efforts were not always successful, affirms that Wellington, his subordinate commanders, and his medical staff were fully aware of the shortcomings of the standard ration and sought to increase it where possible. This was also the case in other theaters, with Graham in January 1814 requesting that Horse Guards consider the viability of "the sending out of a large quantity of Cocoa ready pounded" in the hope that a warm breakfast would help soldiers retain their health in the viciously cold Dutch winter.[56]

Despite such measures, it is nevertheless clear that nutritional factors played a considerable role in the rapid increase in sickness figures during active operations. When the army was in billets, less energy was expended and food was more readily available, allowing soldiers to recover their health before another spell of active operations, such that the sickness rate only rose when the army was engaged in sustained active operations over an extended period of time. Attempts were made during quiet moments to keep the men hardened by route marches, but whilst the practice would have been beneficial in keeping feet and backs used to the strains of long-distance marching, and thereby limit the numbers of men who might fall out as footsore once operations recommenced, it did nothing about the problem of insufficient nutrition once the army was on the move again.[57] Not only were men then far more physically active, and lacking in ready access to shelter, but a failure or hiatus in supply was far more probable and fewer opportunities existed for obtaining food legitimately from other sources. It may well be that this led to men seeking to obtain food by illicit means, but, equally, the opportunities to do so whilst on active operations would not always have been available. Nor, in well-led units where rations were supplemented where possible, and their consumption supervised, should such recourse have been necessary under ordinary conditions. This is not to deny that plundering and illicit foraging occurred—examples aplenty have already been encountered in these pages—but such occurrences seem to have generally been crimes of opportunity and, crucially, to have taken place when soldiers had the time and energy to engage in them. Large-scale foraging in defiance of orders whilst on campaign is only to be encountered on those occasions when the supply system collapsed under pressure, as in the aftermath of Talavera or during the retreat from Burgos.[58]

Whilst McGrigor appreciated the problems caused by poor nutrition, there were limits as to what he could achieve in this area. On the other hand, he did have primary responsibility for the peninsular army's hospitals and is primarily remembered for his reforms of their arrangements, something that was certainly needed as accounts such as Cooper's recollection of his experiences at Elvas testify. The success of these measures is readily apparent in Phase 2 of figure 8. Here, after being pushed to its highest point by the conditions experienced during and after the Burgos retreat, the ratio of sick drops hugely during the first half of 1813. In part, this decrease was because the army had the opportunity to spend a good six months in relatively settled quarters, but there is no denying that men were now getting better medical care than had been the case three years previously. McGrigor's primary success was in increasing the role of the regimental hospitals, leaving only the worst cases to be treated in the larger general hospitals; this change was of course extremely beneficial when the army cut loose from its established Portuguese bases prior to the Vitoria campaign. New general hospitals were established as the army advanced, notably at Santander and Pasajes, and, both in these and the older established hospitals, greater care was taken to ensure cleanliness and an adequate supply of food, medicine, and clothing.[59] The impact of these reforms is, however, masked in part by the fact that the winter of 1812–13 saw the recurrence of epidemic fever. As we have seen, swift quarantine largely confined this contagion to the First Guards Brigade, but the Guards battalions were not the only units to contract the disease during that winter, and it is a tribute to the hospital reforms at both regimental and army level that, unlike in 1809, only an isolated part of the total force was seriously afflicted.[60]

In the final winter of the war, the provision of tents for the infantry meant that the problems encountered in 1809–10 and 1812–13, with confined conditions leading to diseases such as typhoid, did not arise. Tents not only provided basic protection from the elements, but also kept the troops out of potentially insanitary billets. Each infantry company was allocated one tent for its officers and a further three for its NCOs and men; known as "Flanders Tents," these were of a large bell configuration with a central pole. In addition, one tent was allocated to each field officer in a battalion, and to the paymaster, with an additional tent shared by the adjutant and quartermaster and another by the medical staff. Transport space was obtained by replacing the old iron camp kettles with small tin ones that could be carried by individual soldiers; the mules that had previously carried the large kettles were then able to carry the tents.[61]

That it was possible to issue tents for use in the field is in itself testament to the improvements in logistical organization during the course of the conflict—back in 1810, it had not even been possible to provide sufficient tentage for the troops within the Lines of Torres Vedras—although a shortfall did develop by early 1814 that required the shipping of a further 1,500 tents.[62] Poor storage did also cause problems, as George l'Estrange recalled, "[The tents] had come out from England at the beginning of the year; they had therefore been a long time in store, and consequently the tent cloths were rather degraded, which was unpleasant indeed, for when the snow begins to fall on the top of the Pyrenees it comes with a vengeance."[63] Ultimately, matters reached an inevitable conclusion, and the rotten canvas of l'Estrange's tent gave way under the weight of the snowfall. Such difficulties aside, the provision of tents—in conjunction with efforts at a unit level to construct huts or adapt abandoned civilian dwellings—certainly contributed to the reduction of disease, and this would in turn reinforce McGrigor's view that camp discipline was being well enforced at a unit level by this stage in the war. Indeed, McGrigor noted that it was amongst the cavalry regiments, billeted in villages well to the rear and thus still exposed to the contagions from which the tented infantry had escaped, that the worst levels of sickness were to be found during the winter of 1813–14.[64] The Historian John Elting suggests that for troops not sufficiently "housebroken," tents could in fact increase the likelihood of disease, but this does not seem to have posed a problem in Wellington's regiments.[65]

In his account of the closing months of the war, McGrigor implies that he had all but eradicated serious sickness amongst the troops, but this does not tally with the impression given by the data from the regimental returns, which show that the sickness rate, though falling, remained above 20 percent while active operations continued. However, it needs to be kept in mind that McGrigor's account was based on the figures kept by his own department and listed the numbers of troops recorded as present in the various hospitals. By contrast, the figures used for this analysis are those kept by the units themselves, and at this stage in the war a distinction is no longer made between those sick men being treated in regimental hospitals and those detached to general hospitals. Alongside this point, two additional considerations serve to reconcile McGrigor's analysis with the data from the unit monthly returns. Firstly, as we have seen, the quality of reinforcements had deteriorated considerably, leading to high instances of sickness among them due to unfit manpower being sent out without time to fully acclimatize—thus, there was a concentration of sickness in a handful of units. The second issue is more significant and can in fact be applied to the war as a whole, with its effects during the last eighteen months being exacerbated by the movement of the army away from its established depots. This was the problem of getting convalescent manpower back to the front from the base hospitals, which is the point at which strategic consumption through sickness and strategic consumption through absenteeism become linked.

As early as October 1810, regimental returns indicated that there were twice as many men in the base hospital at Belem than the hospital's own records showed.[66] These "Belemites," or "Belem Rangers," numbered many malingerers but also included men awaiting the opportunity to return to their units.[67] Considering that this problem was already pronounced in 1810 when the theater of war was centered on Lisbon, it is readily apparent how much more complex and problematic matters became as the army moved further from its bases. Convalescents had to be grouped together—usually by increments of twenty—and officers found to oversee their movements.[68] Not only was the rate of progress slow, but there was frequently a shortage of superintendence, since many convalescent officers preferred to make their own way back to their units, frequently at a pace to suit themselves; that said, at least one officer landed himself before a court-martial after an out-and-out refusal to superintend convalescents.[69] The ease with which it was possible for men to slip out of the system during transit is emphasized by the fact that in March 1814 it was necessary to strike from the records a total of 1,815 rank and file, eleven sergeants, and eleven drummers who had disappeared between the hospitals and their regiments.[70]

McGrigor argued that improvements to regimental hospitals by 1813 meant that far more cases could be treated without the sick man leaving the army; this cannot be proven or disproven using the unit returns, but McGrigor's own analysis suggests that it was so. If McGrigor was indeed correct, then a significant portion of the men being returned as sick by their units after mid-1813 were in fact convalescents in the process of returning to the front, and McGrigor's statement that "we suffered very little from disease, our hospitals containing few besides the wounded"[71] is closer to the truth than the data might imply. Irrespective of these points, additional proof of the efficacy of McGrigor's policies can be seen in the fact that whilst the overall trend in sickness rates—shown in figure 8 as a dotted line—is upward, the actual figures for the final year of the war remain below the level that the trend line would suggest as typical.

It is in this context that McGrigor's supporting argument, that his figures were artificially inflated by the fact that sickness remained high in the cavalry, which is otherwise impossible to reconcile with the data from the monthly returns, makes sense. McGrigor's analysis, being based on hospital records, accordingly ignores the absent convalescents who were artificially inflating the sickness ratios in the infantry. So far as the hospitals were concerned, these convalescents were fit and on their way back to the front, but since they had not yet rejoined their regiments the latter continued to record them as sick. This in turn suggests that, outside of the cavalry where some sickness still needed to be eradicated, many of the men being returned as sick by their regiments were in fact lost in the system, by accident or by their own design. If McGrigor is correct about the success of his reforms, then the primary problem at this stage of the war was not healing the soldier—vital as that remained—but getting him back to his regiment.

Straggling, Absenteeism, and Desertion

In marked contrast to the attention paid to sickness both by contemporaries and historians, very little has been written about desertion on campaign. Not only are personal accounts understandably scant, but there is also only limited coverage in official papers. Furthermore, for the most part such official correspondence as there is deals largely with the courts-martial of individual deserters; rarely is the problem discussed in any wider context. This lack of coverage stands in marked contrast to the situation for troops stationed in the British Isles, where desertion was a far more serious problem and one to which considerable effort was devoted by the authorities in an attempt either to stem the tide or, at least, recover the services of men who had absconded.[72] In any case, desertion on active service generally represented a set of circumstances distinct from those encountered at home. Whereas a typical case of desertion in the British Isles can be broadly characterized as a desire to make a permanent break with military service, the situation on active service could be far more fluid with a sizeable grey area tending from the extreme of desertion, to absenteeism—deliberate or inadvertent—to simple straggling. Since, however, there was no category in the monthly returns akin to being absent without leave—at least, not for the rank and file—desertion covered the lot so far as the official record was concerned.[73]

Working with averages, calculated in desertions per thousand, it is possible to make a comparison between the various theaters of war. However, whilst a ranking of the theaters in order of relative levels of desertion does produce some clear distinctions, the most obvious conclusion to be drawn from such figures is the relatively low instance of desertion as a contributor to strategic consumption for the British Army on service overseas. Within table 10, most trends are logical enough, particularly with respect to those theaters with the lowest rates of desertion. It would take a desperate man indeed to voluntarily remain on fever-ridden Walcheren, or to risk his lot along the coast of the United States when his uniform immediately identified him as one of those who had been engaged in attacking its citizens. The desirability of remaining in Cadiz or Gibraltar, with their exposure to periodic fever

Table 10. Average Desertion Rates by Theater

Theater	Total Desertions	Monthly Average	Average per 1,000 Men
Northern Europe	2,044	82	4.95
Canada	2,586	60	4.35
East Coast Spain	369	28	2.62
Nova Scotia	406	9	1.98
Peninsula	4,793	68	1.47
Cadiz	341	7	1.31
American Coast	34	4	0.72
Gibraltar	246	3	0.66
Walcheren	23	6	0.32

Source: Sources as per appendix 1. "Northern Europe" covers Germany, Flanders, and France, 1813–15.

epidemics, is also questionable, but here there is also the issue of where one might desert to, particularly in Cadiz under siege. Short of being able to obtain a clandestine sea passage, the only alternative would be to go over to the enemy, which may be assumed to have been the recourse taken by the handful of men who ran whilst on Walcheren or in the United States.

With regard to those theaters with the worst rates of desertion, an element of distortion is admittedly involved. This is in part due to the large numbers of foreign troops—justly noted at the time as being far more likely to desert—in both Flanders and Canada, which artificially raise the average for the theater. This distinction also helps explain why the force in eastern Spain, containing many Italian, German, and émigré units, has a higher incidence of desertion than the main peninsular army under Wellington. In the case of Canada, it should also be remembered that the English-speaking colonial population provided a society into which it would be far easier for a deserter to find a new life than in many of the other seats of war—a situation analogous to eighteenth-century experiences in North America.[74] Data for Canada may also be further distorted by the fact that data for this

Table 11.   Desertions on Active Service, 1808–1815

Year	1808	1809	1810	1811	1812	1813	1814	1815
Desertions	236	479	481	641	1,441	2,185	3,685	1,754

Note: Sources as per appendix 1.

theater is for the years 1812–15 and therefore reflects in part the general trend toward a higher incidence of desertion toward the end of the period, something that is also apparent in the data for northern Europe. The extent by which desertion levels increased toward the end of the war is emphasized by the data in table 11 where the numbers of troops deserting whilst on active service are given year by year.

Clearly, the increased number of desertions is in part due to the greater number of troops on active service, but, even so, it is clear that the rate of desertions did increase to a considerable degree in relative as well as absolute terms during the last third of the period. Yet even in the worst cases, five men in a thousand is by no means an immense drain, and this is reflected when one makes comparisons at a unit level with losses through other forms of wastage. We have seen, for example, that the 1/40th lost 1,378 men dead as a result of sickness or wounds during its peninsular service, and the 2/83rd lost 779; in comparison, the numbers lost through desertion total 14 and 53 respectively. Furthermore, desertion figures from the monthly returns give a deceptively high total, for they make no account of deserters who subsequently gave themselves up or were forcibly returned to their units; when these men are taken into account, the totals are reduced to a considerable degree.

Unless aggravated by the deserter having taken service with the enemy, or compounded with other crimes—in which cases the death sentence usually resulted—recaptured deserters would generally end up back in the ranks.[75] Many who were listed as deserters may well have never had any intention to permanently desert at all, but the absence of an intermediate "absent without leave" status caused all cases to be tried as desertion. This for example, was the fate of Private Thomas Jones of the 2/53rd, brought before a court-martial on August 26, 1811, but Jones was able to satisfy the court that he had had no intention to desert but, rather, had gone "absent from his regiment, without leave, in search of wine." Jones thereby escaped the death penalty, which the court had already handed down once that day to a previous defendant facing the same charge, and was sentenced to the lesser punishment of four hundred lashes. Since Jones had already spent some months as a French prisoner as a result of his ill-judged efforts to quench his thirst, Wellington evidently deemed that he had suffered enough and granted a pardon that permitted Jones to return to his unit.[76] Cases like Jones's demonstrate that there was considerable flexibility and understanding within the British Army's legal processes, but so far as the keeping of returns was concerned it led to men first being struck off as deserters and subsequently reentered onto their unit's strength. Furthermore, particularly when a unit was split up or men were serving detached, it was relatively simple for error to creep in and for men who had legitimate reason to be away from their units to find themselves wrongly listed as deserters. A particularly extreme case of this occurred early in the Peninsular War, when the 5/60th was serving for some months as two independent wings attached to different brigades. When the battalion was reassembled in March 1809, it proved necessary to reenter onto the strength no less than 110 individuals wrongly assumed to have deserted.[77]

Because of the detailed nature of the notes accompanying the army-level monthly returns for the peninsula, it is possible to track the number of deserters who did return to their units. Details of the data sample are shown in appendix 3. By looking at return rates for the years of 1810 and 1812, it is possible not only to calculate a representative average that can be applied to the whole conflict but also to identify changes between the early and late period of the war. In 1810, 296 British troops deserted along with 126 foreigners; over the same period, 127 Britons and 29 foreigners came back. This gives a return rate of 43 percent for British deserters and 23 percent for foreign deserters. In 1812, 546 Britons deserted and 128 came back, in comparison with 458 foreigners of whom 46 came back. These figures produce lower return rates than those for 1810: 23 percent of Britons and 10 percent of foreigners. To some extent, the lower return rates for 1812 may be accounted for by the far more rigorous nature of campaigning in that year as opposed to the conditions of 1810, but the deterioration of manpower quality, particularly in the foreign units, also plays a part. Taking an average of both years, the return rates come out as 30 percent for British deserters and 13 percent for deserters from foreign units. Extrapolation from this would therefore suggest that, in terms of men permanently lost to the service, the total figure of British desertions during the Peninsular War should be corrected from 2,719 to ca. 1,900, and that for foreigners from 2,052 to ca. 1,800. Regrettably, data for other major theaters is less readily available due to the far less detailed nature of the notes accompanying the returns; information for Canada, in particular, is practically nonexistent. The 30 percent return rate for British deserters may be considered as representative, at least for continental Europe, but the wide fluctuation in the nature and quality of foreign manpower in the various theaters prevents any meaningful application of the peninsular figures elsewhere. The primary point remains, however, not only that desertion on active service was low for the bulk of our period but that this low incidence was further reduced by the sizable number of men who came back. This then begs the question of why this was so.

One key point, already mentioned, is the blurred distinction between straggling, absenteeism, and desertion. The taxing effects of sustained active service have already been analyzed with respect to their effect on health, and under such circumstances straggling could then become a serious issue. However, whilst becoming detached from one's unit in this way provided the means and opportunity for desertion, it should not be taken as implying the presence of motive. The breakdown of unit integrity during the great retreats to Corunna and from Burgos is well noted, but similar problems were encountered when the army had to be force-marched during an advance. This was particularly apparent during the attempt to cut Soult's corps off after Oporto, which represented the first sustained active service for many of the men involved. Conditions for Beresford's flanking column, pushing into the mountains through heavy weather and on poor roads, took a particularly heavy toll on the battalions of Tilson's Third

Brigade. The ordeal was related by Hugh Gough, then a major and in command of the 2/87th: "In the dark many men lost the column, several fell into pits excavated by the falling water: many lay down in the road from fatigue and hunger, and the greater part lost their shoes. . . . We pursued our melancholy march at five o'clock, the men nearly fainting with hunger: about twelve we fell in with some carts of bread belonging to a Portuguese division, which General Tilson pressed for the men; this (with some wine) enabled us to proceed. . . . Part of the officers and nearly all the men had their feet cut to the bone for want of shoes."[78] Gough further notes that the brigade was on the march in these conditions for three days, with only twelve hours' rest, by the end of which its other battalion, the 1/88th, had only 150 men still present with the colors. Yet, despite this disintegration, the men who had fallen out were nearly all able to rejoin, and the 1/88th recorded only one desertion during the whole of May, with none at all in Gough's own battalion.[79] Under ordinary circumstances, steps were taken to prevent straggling either through extreme disciplinary measures, as practiced in particular by Robert Craufurd, or—more typically and humanely—by making provision for men who fell out to be brought along after the column.[80] Rank-and-file memoirs also give the impression that whilst falling out during combat was bad form, it was considered acceptable to be absent without leave on the march, or even when in billets; Lawrence, as we have seen, certainly felt extremely aggrieved to have been sentenced to four hundred lashes for a twenty-four-hour absence.

The lengths to which some men went to get back to their units emphasize that such behavior did not inevitably imply an intention to desert. This was seen on a wide scale during the Corunna retreat, although in this instance the unthinkable nature of any of the alternatives must have played some part in motivating the groups of men left behind to keep moving. It was quite one thing to fall behind during an advance, when the opportunity existed to rejoin, but quite another to do so during a retreat when death or capture were the only likely outcomes. Men who fell out on the road to Corunna and Vigo were men who were too exhausted, mentally and physically, to keep moving.[81] But even if the Corunna retreat is discounted as an exceptional circumstance, there are numerous other accounts of individuals making stringent efforts to get back to their units, or, failing that, to another British outpost. In February 1811, for example, an amendment had to be made to the strength of the KGL Garrison Company, then stationed at Lisbon, as explained in an accompanying note: "1 Rank and File was in former Returns struck off the strength as deserted, but however it appears that on the 1st Oct when on the Retreat from Coimbra, he was taken Prisoner from which he Effected his Escape and Joined the Company."[82] Yet if the experiences of this unnamed KGL private were remarkable enough, they nevertheless pale somewhat in comparison with another instance recorded the same month. When Brigadier General Daniel Hoghton was inspecting the rifle detachments at Cadiz, he uncovered a story that hints at a quite remarkable adventure: "Bugler Bryan McGrath of the 1st Battn. 95th was taken prisoner in Spain on the 3rd of Jan. 1809, on the retreat of the British Army to Corunna, he made his escape and joined at Gibraltar on the 1st of March 1810 from whence he was sent to Cadiz and joined this Detachment on the 2nd of May."[83] No further details are provided of McGrath's journey, but even the bare-bones account sounds like something from the pages of C. S. Forester.

Even in northern Europe, men turned up from surprising locations. Of seventeen detached personnel temporarily attached to the 2/91st at Stralsund in November 1813, two were escaped prisoners of war. Private Bennett of the 20th could conceivably have been taken during the Walcheren expedition, but Private Cunningham of the 87th could only have come from Spain.[84] On the other hand, confusion could also ensue from such cases, as seems to have been the case judging from a rather bizarre note accompanying the June 1810 monthly return for the 2/53rd, stating that "the man returned Discharged is a man that was supposed to belong to the Regt. who had made his escape from a French Prison, but it appears that he does not either belong to this nor to any other Regt. and is therefore struck off the strength in consequence."[85] Sadly, no further clue is given as to the man's true identity, or his subsequent fate.

In many cases of this nature, there is no way of knowing if the individual concerned had truly been taken prisoner, or whether he had taken service with the French. The practice of joining the enemy with a view to getting back to one's own side was common enough during the period, most notably in the case of the Portuguese garrison of Almeida, but whilst Wellington condoned that example he was less happy about the potential result of British troops who had made this recourse subsequently rejoining their regiments.[86] In early 1811, perceiving an increase in desertion, he aired his views on the matter to Lord Liverpool, producing some startling conclusions:

> It is difficult to account for the prevalence of this crime, particularly in this army lately. The British Soldiers . . . know and feel that they are suffering no hardship and distress; that there is not an Article of food or clothing which can contribute to their health and comfort, that is not provided for them; that they are well lodged and taken care of in every respect, & not fatigued by work or duty, and having every prospect of success.
>
> The deserters from the British Regts. are primarily Irishmen; and I attribute the prevalence of the Crime very much to the bad description of Men in all the Regiments which are drafted from the Irish Militia; and also to the irregular habits which many Soldiers had acquired and had communicated to others in the Retreat of the Army through the North of Spain in the Winter 1808–9, and in their subsequent service in the French Army and in their wandering through the Country back again into Portugal.[87]

Lack of suitable data renders difficult a judgment of the truth of Wellington's assertions about Irishmen, but coming from a member of the Protestant ascendancy there is clearly a strong streak of prejudice implicit in them. Yet more remarkable, however, is the insistence that conditions were such that no man ought be dissatisfied with them and desert as a result; to say the least, this portrayal of life in the ranks is decidedly rose tinted. Nevertheless, Wellington's expression of surprise that men would have any cause to desert from the British service is a recurring theme on those occasions when the issue comes up in official correspondence.

Wellington's concerns reflect an understandable fear of his army being contaminated with the ethos that led men to accept service in the pay of any nation as an alternative to an existence as a prisoner. Two months before he had written to Liverpool, Wellington had explicitly expressed this concern in a General Order relating to the courts-martial of Private Henry Davy of the 52nd and Henry Vandebruck of the 97th, which took place on the same day as that of Jones of the 2/53rd. Like Jones, Davy and Vandebruck had been taken prisoner whilst absent from their units looking for wine; unlike Jones, they had elected to take service with the enemy, as the court-martial recorded it, "in order, if possible, to make their escape." This they managed to do, but having been recaptured in enemy uniform they clearly had a lot of explaining to do. It eventually took a pardon by Wellington himself to save the two men from the firing squad. In the General Order outlining the case, Wellington made it clear that theirs was not an expedient to be recommended, not least because it was "very possible that those who attempt it may find themselves engaged with their countrymen and comrades before they can carry this design into execution."[88]

It is telling that one of the men involved in this case can be assumed—bearing in mind his name and regiment—to have been a foreigner. Although Wellington's assertions with regard to the tendency of British troops to happily seek service with the enemy, as expressed in his letter to Liverpool, seem to be an exaggeration, he was certainly aware of the high desertion rates in the foreign units under his command, and it is hardly to be wondered at that men like Vandebruck cared little which army's coat they wore.

Even without the additional distortion caused by the fact that far fewer of these men were ever recovered, the high rate of desertions by foreigners serves to considerably distort the data in this sample as a whole. Across the period as a whole, in all the theaters from which data was sampled, the average loss rate through desertion was two men in every thousand. However, the figures for foreigners in British pay comes out as seven men in every thousand and that for Canadian Provincial units as ten men in every thousand. When British troops are considered alone, the average figure comes out as 1.2 desertions per 1,000 men. These figures also indicate a distinction between arms of service, with a far greater incidence of desertion to be found in the infantry than in the cavalry and, in particular, in the Ordnance services. It may be suggested that the better levels of health in these two arms led to less straggling and absenteeism than in the infantry and therefore to less desertion as a result. That said, it is important not to overstress this point, as there is no obvious correlation between sickness and desertion rates when these are plotted together at a unit level.

The far higher rate of desertion in Canadian Provincial units is understandable in light of the fact that these units were stationed in the same territories in which they had been raised and—with a few exceptions such as the Canadian Voltigeurs and Glengarry Light Infantry Fencibles, which took the pick of the manpower— were generally employed in garrison roles from which it was relatively easy to slip back into the civilian population.[89] In effect, their circumstances were more akin to British units in the British Isles, where desertion was far more of a problem, than to those of British units on active service. For the foreign units proper, the dubious nature of much of the manpower utilized, particularly in later years, must chiefly account for the greater tendency to desert. If additional reinforcement for this assertion is needed, it should be noted that the two British units with the worst relative records in terms of desertions were the 33rd and 2/91st, both of which had recruited extensively amongst foreigners whilst in north Germany. On other occasions, what seem at first to be large-scale desertions from other British units turn out to be down to foreign recruits taking the opportunity to abscond; in October 1810, for example, the 1/3rd lost eight men, but "Seven of the men that deserted were Dutchmen who volunteered from Prisoners of War."[90] Whilst recruits of this description might remain with a unit if doing so represented the best alternative for the present, likewise, they were not averse either to taking a chance to desert when the opportunity arose. Thus, substantial numbers of Spaniards served with distinction in the Chasseurs Britanniques in various Mediterranean postings, but absconded en masse when posted to the peninsula.[91] Similarly, there was a flood of deserters from the KGL units stationed in Flanders during 1814, although these were exceptional circumstances due to the fact that the Legion was in the process of being run down prior to its disbandment. Ordinarily, the KGL generally represented something of an exception to the rule that foreigners had no commitment to the service and were liable to desert, and it was only when the initial genuine volunteers began to be replaced by manpower from more dubious sources that the desertion rate in KGL units began to climb. Even then, a continued prioritization of the best foreign manpower for the KGL prevented circumstances in the Legion from becoming as bad as those in the other foreign corps.[92]

Yet even allowing for a relatively low instance of desertion, a substantial return rate, and a distortion of the figures by an increased use of unreliable foreign manpower, there is still no escaping the fact that desertion rose sharply toward the end of our period. What was more, the closing months of the Napoleonic Wars, and the brief peace of 1814, saw several cases of mass desertion. Most extreme of all these cases was the 1/27th in Canada, which lost no less than seventy men in the single month of September 1814.[93] The timing and location of that episode are in themselves significant, and will be returned to, but the same tendency occurred in all theaters, even whilst the fighting was still under way, and to an extent that was remarked upon at the time. When Major General John Byng inspected the 1/3rd in October 1813, for example, he noted that he had "endeavoured to ascertain the cause for so remarkable an occurrence as the desertion of twelve men in ten days—but without success. One man has since returned and surrendered himself. The others were bad characters and there is little doubt that they have gone over to the enemy."[94] As Wellington had done three years earlier in his letter to Liverpool, Byng expressed some surprise that desertions were occurring in such numbers, but writing off such occurrences as being down to "bad characters" seems somewhat naïve. Rather, these cases should be seen as indicative of a growing war-weariness, and a desire to escape from military life.

This trend manifested itself in the high levels of desertion during the very final months of the peninsular army's existence, after the announcement of peace, with 251 desertions in May 1814 and 377 in June.[95] In part these desertions represented men who had built themselves a family during their years in the peninsula, and preferred to make a new life there, but there was also an understandable reluctance to be shipped across the Atlantic to fight a new war. William Grattan wrote eloquently of the fate of the "poor faithful Portuguese and Spanish women, hundreds of whom had married or attached themselves to our soldiers, and who had accompanied them through all their fatigues and dangers . . . had lived with our men for years, and had borne them children . . . had staunched their wounds with their tattered garments, or moistened their parched lips, when without such care death would have been certain; they had, when such aid was not required, devoted days and nights in rendering those attentions which only they who have witnessed them can justly appreciate."[96] The same restrictions that prevented many men taking their wives on campaign in the first instance now prevented many of the wives who had been acquired on campaign from continuing with their husbands; under such circumstances, and with France defeated, it is scarcely surprising that so many men preferred the tie of a family over the tie of a regiment, to the extent that they ran the risk of desertion in order to remain with their womenfolk and children.

In similar fashion, late 1814 and early 1815 saw growing numbers of desertions in Canada, representing men not wanting to leave North America due either to a wish to settle or to an unwillingness to face yet another war in Europe; the mass exodus from the 1/27th forms part of this pattern. Desertions of this nature need to be read not as part of the ordinary wastage of campaigning, nor as the actions of disgruntled bad characters, but as a reaction to perceived rights and expectations being interfered with. In this regard, parallels may be identified with the situation in the closing years of the Seven Years' War, in which men who were content to serve in the war that they had signed up for—in that case against France in North America—actively resisted attempts to send them on further service in the Caribbean or to withhold their discharges.[97] Even when it did not develop into a resolve to desert the service, the feeling that hard-earned rights were being interfered with, and liberties taken with the terms under which men had enlisted, led to widespread disaffection. John Cooper had signed up for seven years' service, and when he returned from the peninsula with the 1/7th—Sergeant Cooper, by now—he had only days to serve, and every hope of a discharge, when his battalion received orders to embark for the Americas. Even as the force lay at anchor off Portsmouth, seven-year men from the 1/43rd were discharged and sent ashore, but Cooper's commanding officer was absent and could not sign the papers to release their counterparts from the 1/7th. By the time that the fusiliers were preparing to attack Major General Andrew Jackson's breast-works outside New Orleans, the discharge date was long past, and Cooper was one of several who protested their case to their commanding officer Colonel Edward Blakeney, only to be told that "it could not be helped" and that they must join the attack. Only upon their return to England were the surviving seven-year men given their discharge, although news of Napoleon's return led to speculation that they would be retained still longer. There is a decided bitterness in Cooper's telling of the conclusion of his service; he had fulfilled his part of the contract he had entered into, only for the British Army to fail to reciprocate. Cooper stuck it out, but it is easy to see why others saw themselves as within their rights to leave an army that had broken its promises to them.[98]

Although the closing months of the period saw the most pronounced instances of desertion as a form of protest, the trend was not without some parallels earlier in the period. Unusually hard conditions also led to cases of desertion as a matter of individual protest and escape, as in the hard winter of 1811 when even so harsh a commander as Robert Craufurd was prepared to intervene with Wellington and present these circumstances as explanation, if not justification, for a spate of Light Division desertions. Wellington was not convinced, however, and when Private Joseph Almond of the 1/95th was recaptured the following year, his plea of hardship did not save him from the firing squad.[99] But if some instances of desertion are to be understood in the context of protest, this seems to have been the case only at an abstract level. There is little or no correlation, for example, between units known for heavy punishments and units with a high incidence of desertion. Indeed, the 1/40th, commanded for much of the

Peninsular War first by James Kemmis and then by Richard Archdall—the former portrayed by Lawrence as a flogger, and the latter removed for tyranny—has one of the lowest relative desertion
rates, bettered only by the Foot Guards amongst long-serving peninsular battalions. Conversely, there does seem to be a more obvious correlation between units having a high reputation, and/or
some pretension to elite status, and a low incidence of desertion.
There was not only very little desertion from the Foot Guards on
active service but none whatsoever from the three Household Cavalry regiments. Within the regular line, the general trend of desertion in light, rifle, highland, and fusilier battalions is generally
better than the norm, serving yet again to reinforce the practical
value of their strong sense of unit identity.

If, then, we accept that, in addition to the caveats already noted,
the desertion rate for British troops is to some extent further inflated by "protest" desertions stemming from the atypical circumstances in the last months of the Napoleonic Wars, the generally
low rate for the era as a whole becomes even more remarkable and
stands out in some contrast to the situation experienced amongst
units serving at home. In contrast to the average rate of two desertions for every thousand men overseas between 1808 and 1815, the
former year saw fifty-five desertions for every thousand men at
home.[100] This, however, is not entirely a fair comparison, since not
only did home service make desertion a far more realistic possibility, with aid and concealment readily available, but additional mo-
tivations also existed with respect to the possibilities of deserting
in order to reenlist for bounty.[101] The tendency toward desertion
by new recruits also helps explain the higher incidence at home.[102]
Over and above this, however, there is a strong psychological element to add to the practical; for many men who left accounts
of their time in the ranks, their embarkation for service overseas
seems to have represented a distinct cutoff between home service
and the acceptance of active duty. Many memoir accounts place
a particular significance on the first time the writer left Britain,
causing a general reassessment of personal circumstances.[103] For
men being taken away from proximity to hearth and home, to a
potentially distant and unfamiliar theater of war, the army, and
more specifically the regiment, could readily change from an

oppressive force to a place of security and refuge, to be left only under extreme circumstances.

During the course of the period under analysis, the British Army, collectively and individually, benefited greatly from the practical experience of sustained campaigning, and this contributed to a reduced rate of strategic consumption. In this way, the connection between losses through sickness and losses through straggling and desertion is reinforced, since measures to improve the health of the army also rendered it more effective and, in conjunction with better logistical management, reduced the scope for the straggling and absenteeism that was frequently the precursor to desertion. However, on an individual level, an increase in experience also led to new sorts of problems. Thus, whilst greater experience of campaigning led to improved ratios of strategic consumption, it also led soldiers to develop a greater understanding of what they could get away with, how they could play the system, and what rights they perceived themselves as having. Whilst only a desperate or foolish man would completely abandon the security offered by the British Army on active service, the worst rigors of service could be circumvented by taking care to extend any opportunity to remain away from active duty. Such tendencies should not be thought of as universal by any means, as for the majority of the rank and file the key connection was with the regiment rather than the army as a whole, and it was deemed better to take one's chances with one's comrades than strike out alone. Nevertheless, enough men did take this approach to cause the marked discrepancies between the sickness rate as understood by the Medical Department on the one hand, and the sickness rate as understood by the regiments on the other: a discrepancy that can only be reconciled by allowing for the existence of the "Belem Rangers" and their like. Greater experience also brought with it a greater understanding of the perceived rights that had been earned through service; this understanding in turn could lead to feelings of grievance when these rights were infringed upon, culminating in the surge of desertions in 1814 and 1815. An experienced army was therefore a more effective army, but by no means a perfect one.

CHAPTER 7

Beasts of Burden

The records of cases tried by Regimental Court Martial contained within many inspection returns are, for the most part, dry and sparse, giving little clue as to the details of the stories behind them. Nevertheless, there are some cases where a picture of the events leading to the trial can be guessed at. One such case is that of Private William McKean of the 5th Dragoon Guards, who was sentenced to two hundred lashes on November 4, 1812, "For ill treatment to the Camp Kettle Mule and disobedience of orders."[1] The date immediately conjures up a rain-sodden vignette from the early stages of the Burgos retreat, and we can be sure from the circumstances that the unfortunate McKean had already, for whatever reason, lost his own horse.[2] For a cavalryman to be reduced to marching on foot was no doubt bad enough, but to be given charge of a mule—it being reasonable to assume that this animal was as contrary as the reputation of the breed suggests—no doubt added insult to injury. Add the fall in morale engendered by the weather conditions and the inevitable demoralization of a retreat, and it is all too easy to see the private taking out his frustrations first on his charge and then at whatever officer or sergeant attempted to intervene. Speculation perhaps, but if the above reconstruction approaches the truth, then it renders it abundantly clear how the suffering of the men in the ranks of the British Army was closely bound up with the suffering of the animals on which they relied, and emphasizes the vital role played by draught and pack animals in the force's continued effectiveness on campaign.

Quite apart from continual need for riding horses to maintain the mounted strength of the cavalry regiments, it is clear that without the services of thousands of horses, mules, and other draught animals, those campaigns could never have progressed beyond an initial landing. Inevitably, in a war where human lives were not infrequently lost as a direct or indirect result of famine and want, equine lives were often counted cheap; one of the few things more staggering than the number of animals needed to maintain the British forces in the peninsula is the number that died. However, the British Army—as distinct from the King's German Legion, whose care for its horses is repeatedly lauded—has received a poor press over the years for the level of provision that the private soldier made for the animals in his charge. As we shall see, however, the real situation is far more complex, and the respective reputations of the British and KGL troopers have become increasingly distorted over the years. In fact, the statistical data with regard to strategic consumption of equine resources, as with regard to manpower, suggests the existence of a distinct learning curve where, as with the manpower story, leadership and experience—not nationality—stand out as the true distinguishing factors between regiments.

The Nature of the Beasts

In addressing the general question of the British Army's requirement for animals, two points need to be emphasized: on the one hand, the variety and on the other the numbers. As with all armies until well into the twentieth century, horses were essential to the British Army of the Napoleonic Wars. Nevertheless, one cannot take a general approach to horseflesh, since there was—and is—a world of difference between the animal best suited for heavy draught work and that best suited to riding. Within the latter distinction too, a far more substantial creature was ideally required for heavy cavalry service than the lighter, lither, mounts of the light dragoons and hussars, and a creature of a far better description than any troop horse was naturally de rigueur for all but the most impecunious of mounted officers.

The range of requirements can be seen in a memorandum drawn up by Major General Herbert Taylor, intended for translation into Dutch so as to give advance notice of the animals required for purchase to facilitate the operations of Sir Thomas Graham's forces then on their way to the Low Countries. As draught animals for the artillery, Taylor sought "Geldings or Mares, stout and active, with good bone, and in good order; having been broken to draught. Age from 5 years old to Eight Years and upwards. Size from 15 hands to 15 hands three inches. To be perfectly sound in Eye, Wind, and Limb, and free from any Blemish."[3] In order to prevent any confusion over the measurements, Taylor confirmed that "The Hand is four inches English measure." Whilst that satisfied the formal military requirements, individual officers would need to purchase their own mounts, to which end Taylor went on to recommend that "it may be suggested to the Contractor or Contractors to provide 40 or 50 Riding Horses to be ready for delivery on approval by the Officers requiring them, at a price not exceeding from £30 to £40. Age from 5 to 8 years. Size from 14 [hands] 2 inches to 15 hands 2 inches."[4] As with many such attempts to obtain mounts locally, Taylor's efforts would prove largely fruitless, but his specifications do provide some idea of the variety of horses that were needed.

Horses, however, were not the only animals employed, and, particularly in the peninsula, the mule became the indispensable beast of everyday burden. Mules were employed to carry forage, supplies, camp kettles, tents, and—albeit illicitly—private baggage.[5] Under exceptional circumstances they could even be used directly in lieu of horses; faced with a complete lack of artillery horses when assuming command at Cadiz, Major General William Stewart authorized the formation of "a field brigade of Light Artillery consisting of 4 lt 3 pounders and 1 5½ inch howitzer drawn by mules . . . to be stationed at Fort Puntales."[6] The main role of mules was always as pack animals though, in which role Assistant Commissary Richard Henegan of the ordnance field train considered them "worth [their] weight in gold on a long march."[7] As a result of their indispensability, prices of mules rose from 25 Portuguese dollars in 1808 to not less than 180 dollars by 1813.[8]

Also forming a vital element of the supply system, particularly in the peninsula where the design of cart had hardly changed in centuries, were draught oxen whose presence was essential not only for moving supplies and ordnance but also, on occasion, shifting artillery behind the lines.[9] Like the bulk of the mules, however, these animals were privately owned and hired for military service. They were controlled by the Commissariat and Ordnance Departments, but were managed by Portuguese and Spanish civilians, and thus there are no official records relative to their numbers and mortality rates. Their organization, though, brought headaches enough for the officers assigned to deal with their drivers, with Henegan of the field train and August Schaumann of the Commissary Department both frequently descending into exasperation when their memoirs touch on both the animals and their equally self-willed masters; not the least of Henegan's problems was preventing the bullock drivers smoking whilst hauling powder![10]

In attempting to understand the numbers of animals directly employed by the British Army, and the resulting problems of supply and demand, the situation is further complicated by the fact that there is little useful data over and above that from the main peninsular field army. Lack of substantial cavalry deployment outside the peninsula means that it is necessary to focus more heavily on that theater in any case, but, by and large, much of what data is available from elsewhere is patchy and incomplete. Only in Flanders and France during 1815 was a mounted contingent fielded on a par with that in the peninsula, and even then the campaign was of so short a duration, and the figures so badly skewed by horses lost at Waterloo, that it is hard to draw any meaningful conclusions from this theater.

At its peak strength in April 1814, the army under Wellington had 13,887 horses on its rolls, a ratio of 2:9 in relation to its 61,535 rank and file.[11] This ratio had nearly halved from 1:8 in the earliest days of the war, in reflection of the growing number of cavalry regiments assigned to Wellington's command, but the growth rate is staggered over the course of the war, with the sharpest increases generally coinciding with preparations for a new campaign, as shown in figure 9. These totals represent only half the story, for the figures not only fail to include the privately owned mounts of officers but also omit the vast numbers of draft animals, official and unofficial, that accompanied the army. Complete data does not exist for these creatures, although in some of the smaller forces serving elsewhere the printed "horses" on returns has been amended by the addition of "and mules." We can, however, gain some insight into the scale of this unrecorded mass since Commissary General Sir John Bisset noted the presence of 8,815 commissariat mules with the peninsular field army in June 1812, along with a further 16,165 horses and mules for baggage and forage.[12] If the 567 horses listed on the strength of the Wagon Train are also included, this yields 25,547 animals, or roughly five-sevenths of the total number serving with Wellington's army, employed in support of the logistical train. Including the combat arms and comparing the grand total of 34,127 animals with the rank-and-file strength of the army for the same month, this now gives a ratio of animals to men in the region of 3:5.[13] These figures, of course, still make no consideration of the vast numbers of oxen, for which no data is available.

Nor, indeed, does that data make any inclusion of the privately owned mounts of officers, for which there was an ongoing demand. We have already seen Herbert Taylor noting his expectation that officers from Graham's command would need to mount themselves upon arriving on service, and a similar concern had earlier prompted Wellington to "beg leave to recommend that 50 or 60 Horses or mares of a superior description should be purchased at the price of 40 or 50 £ each as a Remount for the Officers of the Cavalry, who cannot find Horses in the Peninsula at present fit for the service; and would pay this price for these horses."[14] Providing sufficient high-quality mounts for officers was by no means a small requirement in itself, when one considers that Wellington himself maintained a stud of some ten horses whilst Graham began the Vitoria campaign with thirteen.[15] Even junior officers maintained a private establishment of riding and draught animals. George Gleig, as a young lieutenant still swept up in the romance of the open-air campaigning lifestyle, nevertheless accumulated kit "quite enough to load a mule,"[16] but others had rather more extensive stables. John Aitchison, as a lieutenant of the 3rd Foot Guards, also made do with a single mule for his kit, but added a pony in addition as a mount for himself on returning to the peninsula in 1812, complaining as he did so that the cost of animals had risen to exorbitant levels since he had first arrived

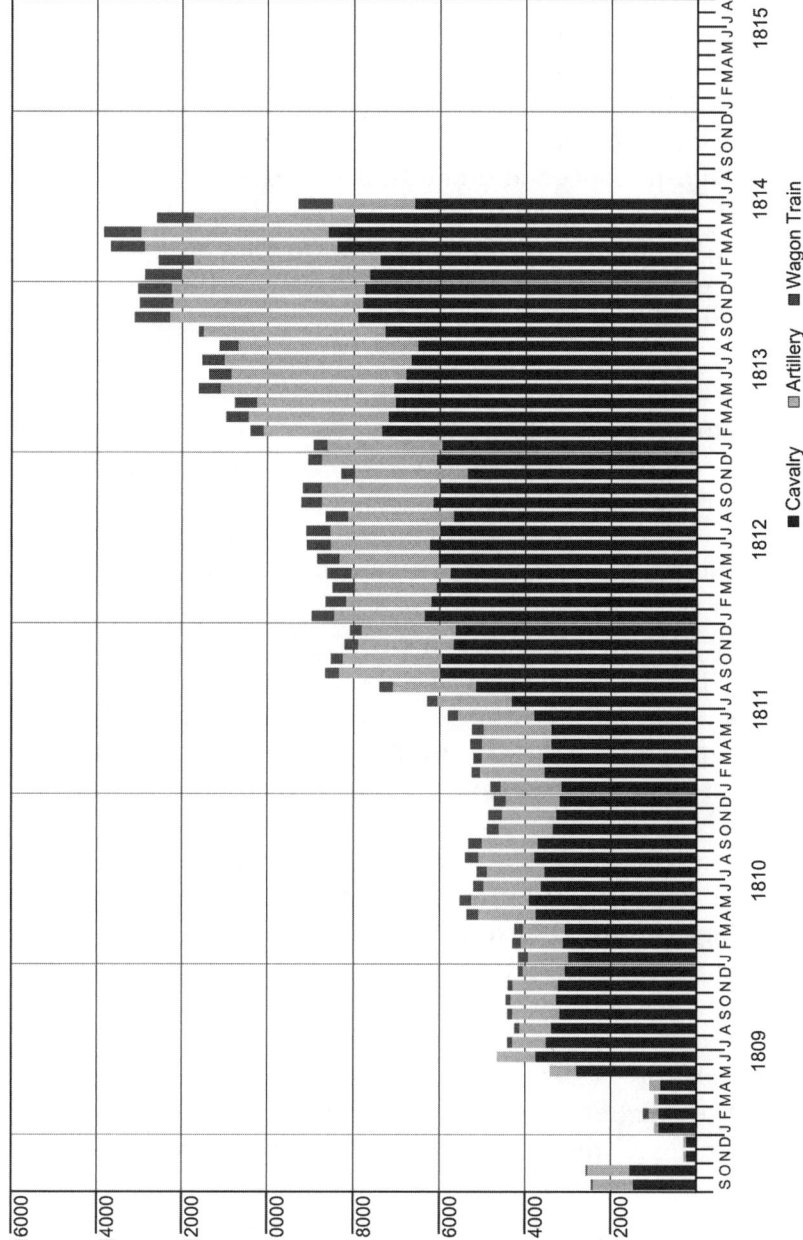

Figure 9. Peninsular Horse Strength by Month and Arm of Service. Data from Monthly Returns, TNA, WO17/2464–2465, 2467–2476.

in the theater three years previously.[17] In these provisions, as in the fact that each also maintained a brace of gundogs, Gleig and Aitchison were fairly typical subalterns. More extravagant was the sporting William Hay, who, as an ensign of the 1/52nd, had as his mount a spirited Portuguese mare, which, if his memoirs are to be believed, nevertheless repaid his investment by winning him a wager through her speed and endurance over an arduous long-distance ride from "Castel de Vieda" to Portalegre and back in less than two and a half hours.[18] Captain Neil Douglas of the 1/79th, meanwhile, preferred comfort to speed, and on arrival in the peninsula "Purchased two Mules, and a Jack Ass, also a Mess Mule for carrying prog [i.e., provisions]."[19]

After the total number of horses involved, the second figure that needs to be appreciated is the sheer scale of equine mortality, and the consequent demand for replacement animals. The British Army recorded the loss of a staggering 18,940 horses by the main body of troops in the peninsula. In addition, Sir John Moore's forces recorded a further 218 equine deaths during November and December 1808, to which must be added the horses shot at Corunna to prevent their use by the French; based on the December 1808 totals this represents a further 5,006 animals, to give a grand total of 24,164 horses dead for the war as a whole.[20] Even leaving aside Moore's losses, which nevertheless served to remove the Hussar Brigade from the roster of effective regiments for several years whilst its component units were remounted,[21] this level of losses represents an immense requirement for replacement horses to enable the army to maintain its effectiveness. The nature of these losses is tracked in figure 10.

Whilst an approximate base rate in the region of twenty to twenty-five deaths per thousand horses can be identified throughout the course of the war—the overall average works out as thirty-eight deaths per thousand horses, but that is distorted by the periods of higher mortality—there are nevertheless obvious points of far higher mortality. The worst losses in absolute terms are during the retreat from Burgos, but in relative terms the huge number of horses lost during the Talavera campaign make the month ending on August 25, 1809, the worst, with a quarter of the total equine strength being lost. Here, the already high level

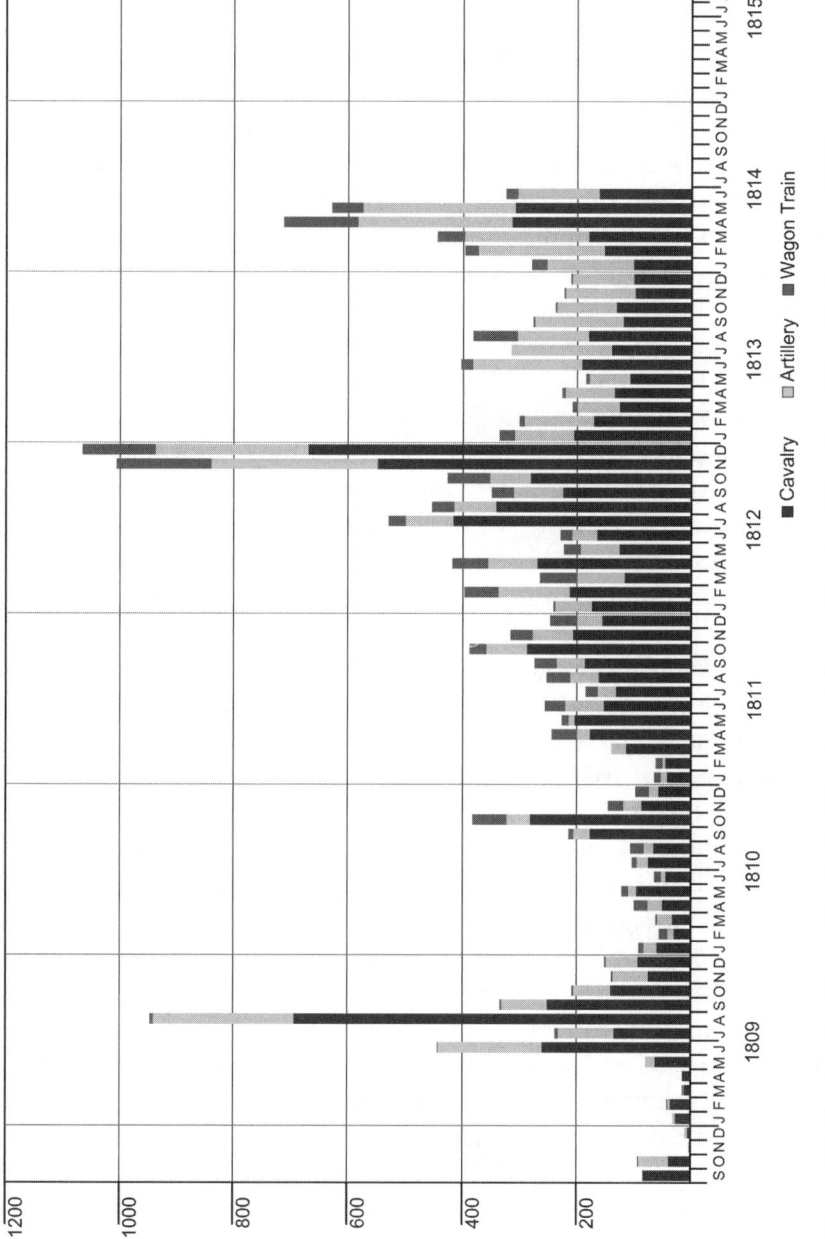

Figure 10. Peninsular Horse Deaths by Month and Arm of Service. Data from Monthly Returns, TNA, WO17/2464–2465, 2467–2476.

of strategic consumption due to the shortage of forage and the forced march south through Portugal is further increased and distorted by the heavy battlefield losses sustained by the 23rd Light Dragoons at Talavera—228 deaths out of 940 being from this regiment.[22] The smaller peaks and troughs in losses correspond either to major battles, or else to periods when the army was either unusually active or recommencing active operations after a period in cantonments—in other words, precisely the same factors as those influencing fluctuation in manpower wastage.

Of course, equine wastage was not just restricted to horses lost permanently through death. Just as men who had become permanently unfit for active duty could be drafted off to continue serving in garrison units, horses that were not fit for cavalry or artillery service could be "cast" and passed on to train units where they could still be useful. The 1812 second half-yearly inspection return for the 14th Light Dragoons, for example, contains an annex: "A Return of horses proposed to be cast, most of them able to do duty in the Wagon Train."[23] There was little point in shipping a worn-out cavalry horse back to England, whereas even a few weeks' draught work represented a continued return on the initial investment. The poor state of these cast animals, and their obviously higher mortality rate, is made clear by an incident from 1809 when, in a bizarre inversion of the normal practice, sixty-one animals from the Irish Commissariat Corps were utilized as remounts for the 14th and 16th Light Dragoons due to a complete lack of any other available horseflesh. Unfit for cavalry service—most likely unfit for any service—the entire batch perished after only a single day's work.[24]

In the same way that it was rapidly established that soldiers needed to become fully acclimatized to a theater of operations before becoming fully effective, so too did it become clear that the same was the case for horses. Even before being considered for active duty, the animals needed to be properly broken for service and become accustomed to the role they were to fulfill, a process that required one and a half to two years.[25] This, and the fact that more mature animals were considered less prone to illness, led Wellington to recommend in December 1810 "that no horses should be sent for Service to this Country which will not be Six years old in

May; and that Mares should be sent in preference to Horses, as it has been found that they bear the Work better than the Horses."[26] Once horses had arrived in theater, it was also important to ensure that they were given time to adapt to the local forage, and, shortly after reassuming command of the British forces in the peninsula, Wellesley stressed this requirement in a General Order: "Those horses of the Dragoons and Artillery, which will eat the corn and forage of the country, are to feed with that description of forage only; the commanding officers of Dragoons and Artillery will give direction that all horses may be accustomed to the corn and forage of the country, by being fed at first, in the proportion of half English and half Portuguese corn and forage; then of two-thirds Portuguese and one-third English; and lastly of the whole Portuguese."[27] By such means, in theory, suitable remounts would arrive during times of relative inactivity and would become accustomed to the climate and country before serious campaigning began; in practice, of course, nothing was that simple.

What made strategic consumption of horses so problematic was that the remount system did not permit the steady provision of replacements from home. Indeed, lack of available transports at times prevented mounted units from even embarking on service with sufficient animals, as in the 1808 campaign when some of two-fifths of the available cavalry and artillery could not be moved for lack of horseflesh.[28] The assumption had been that animals could be obtained on active service, but this rapidly proved to be a fallacy, and the dispatch of remounts had to be instituted although this took some time to be implemented and Cradock was forced to send officers to North Africa in search of horses.[29] Even then, the same overconfident assumptions about the availability of horses continued in other theaters. Lt. General Frederick Maitland, commanding the first wave of forces dispatched to eastern Spain, complained on arrival at Alicante that "The necessity of procuring mules for the Conveyance of Provisions, Ammunition & Baggage, as also to move the Field Artillery for which the number of horses with us was very insufficient, delayed us several days."[30] Yet more naïve was the assumption that Taylor would be able to obtain draught and riding animals for Graham's forces in the Netherlands at the end of 1813; considering that Europe had been scoured for horses to remount the Grande Armée after the Russian debacle, it is scarcely to be wondered that only a few animals, mostly unfit or unsuitable, could be rounded up.[31]

The relatively low ratio of cavalry to infantry of almost all British forces sent overseas is largely connected to these difficulties, which relate not so much to a shortage of horseflesh, but, rather, a shortage of means to get it to where it was required. This problem played an important role in dictating to what extent, and in what quantity, cavalry regiments could be deployed. For example, when Wellesley's force set out for Portugal in 1808, it required twenty-nine transports for the nine battalions of infantry, but a further twenty-one horse transports for the animals required by three artillery batteries and two squadrons of cavalry.[32] To place this requirement in context, in 1810 the Transport Board could deploy 980 vessels totaling 250,000 tons burthen, of which 320 were eventually assigned to serving the forces in the peninsula.[33] Shipping remounts alone made a substantial inroad into that force, with the need to deploy a new cavalry regiment or brigade representing a substantial investment in shipping space, not to forget the need for escorts, the absence of which in 1813 saw the best part of a whole troop of the 18th Hussars—forty-five men and sixty horses—captured and ransomed by an American privateer.[34] There was also the issue of cost, with Wellington observing that, over and above a purchase price of twenty-five guineas, it cost ten pounds per horse to ship remounts to Portugal.[35]

Like the men themselves, horses did not respond well to prolonged spells aboard transports. Indeed, some objected to the whole shipping process and responded accordingly; William Tomkinson's bay horse Bob twice escaped from the slings by which he was being hoisted aboard a transport at Falmouth, and then kicked out again "and nearly killed the second mate of the vessel by knocking him overboard."[36] One of the regiment's own troop horses subsequently followed the unfortunate mate over the rails, having been ordered to be shot after exhibiting symptoms indicating the onset of what Tomkinson referred to as "farcy."[37] More commonly known as glanders, this contagious disease, transmittable through contaminated food or water but also by the nasal discharges of the afflicted animals, would have likely infected the whole shipment had drastic action not been taken. Similar precautions were necessary on campaign, particularly if occupying quarters that had previously been used by the French, whose cavalry mounts were believed to be poorly cared for and thus liable to spread infection.[38] Even if serious contagion was avoided, a sea voyage did nothing for the health of horses cooped up aboard ship, particularly if sanitary conditions were poor, and animals inevitably fell off in condition.[39] Others contrived to harm themselves or their counterparts by following the example of Tomkinson's Bob, whilst a rough passage led almost inevitably to broken legs and the consequent destruction of the injured animals.[40]

The remount situation was further complicated by the restrictions imposed on it by the regimental system, which meant that, rather than being organized centrally, remounts were obtained individually by each unit.[41] This problem, relating to the retention by regiments at home of mature, campaign-ready horses whilst those on active service had to use young animals not accustomed to the rigors of service, was one of several associated issues raised by Wellington in correspondence with Bathurst as preparations for the 1813 campaign got under way. Wellington also expressed his dislike of the need, again brought on by the confines of the system, to receive whole new regiments rather than simply acquire their horses for the use of existing units. The same round of correspondence notes the lack of horses to haul the guns, which kept a portion of the artillery confined to garrison duty at Lisbon.[42] By extension, the employment of a significant portion of the Portuguese cavalry in a dismounted garrison role, and the steady reduction of the numbers of artillery batteries of that nation serving in the field, may also be attributed to the same cause.[43]

An additional problem stemming from the fact that horses were obtained by the individual regiments was that it was not always possible for them to obtain the most suitable animals. The 12th Light Dragoons was particularly lucky in this regard since the regiment's colonel from 1791 to 1815 was the veteran cavalry officer Lt. General Sir James Steuart, who saw his duties very much in the model of the eighteenth-century colonel-proprietor and vetted the quality of any augmentation to his regiment, be that the appointment of a new cornet or the purchase of a batch of remounts.[44] The month of July 1811, for example, found Steuart writing to Captain James Bridger, commanding the regimental depot, to see how well the latest remounts were doing;[45] a year later, he was taking an even more active role. A hope that Preston Fair would be a good source of remounts having proved false, Steuart turned his attentions from Lancashire to Scotland in response to an earlier arrangement with a Mr. Hugh Taylor to supply fifty horses at Dumfries. Steuart's standards, however, were exacting; not only would he "take no Horse which is not strong and active fit to carry 18 stone on the road," and between fourteen hands two inches and fifteen hands, he also specified a preference for "dark brown in point of colour—any blacks must have a dash of blood." Rather than entrust the task to anyone else, Steuart, then in his sixty-eighth year, informed Taylor that he would inspect and approve the horses personally.[46]

But if Steuart's attention to detail was thorough, it was also un-usual, and not all regiments were able to maintain the high stan-dards that he set for the 12th Light Dragoons. The fact that other regiments were obliged at times to make do with what they could get is emphasized by the bizarre situation that was revealed when the whole of Wellington's cavalry was inspected during the winter of 1812–13. It was then found not only that several of the heavy cavalry regiments were mounted on horses of altogether too light a description but also that certain of the light dragoon regiments were struggling with weighty mounts better suited to the heavy cavalry service. The 4th Dragoons was particularly noted as fall-ing into the former category, and the 11th Light Dragoons, the latter, whilst the unfortunate 13th Light Dragoons was left with a batch of remounts that touched the other extreme and were of a size barely even sufficient for light cavalry work. Other regi-ments had received replacement horses that were nearly as bad— in several cases insufficiently trained to the service—but, again by contrast, the 5th Dragoon Guards and 16th Light Dragoons had both received remounts of such a quality that they represented a marked improvement on the existing horses.[47] It was a decidedly mixed situation, and one of the few comforts that Wellington was able to take in York's having recalled four of his cavalry regiments during the following spring was that their departure enabled him to make provision to redress the situation by redistributing their horses. Not only were suitable mounts drafted off from the departing regiments, but a double transfer was instituted so that, for example, the 1st KGL Hussars received forty-three picked light cavalry horses from the departing 2nd KGL Hussars, which in turn enabled it to send an equivalent number of unsuitable heavier mounts to be redistributed amongst those regiments of dragoons and dragoon guards that were to remain in the peninsula.[48]

Even when assigned to regiments appropriate to their size and strength, the larger, more powerful, horses needed for heavy cavalry service posed other problems, not least that they required more in the way of forage to keep them in good condition. Wellington's discussion of the provision of remounts for the 1811 campaign highlights the problems inherent in maintaining the larger horses needed for an effectively mounted heavy cavalry arm, although, as ever with Wellington, the pragmatic desire not to deprive himself of a single experienced man if it could be avoided is also apparent: "As the appointments of the Heavy cavalry are so much more weighty than those of the Light Dragoons, and the larger horses of the former are with difficulty kept in condition, it would have been desirable to have a larger proportion of the Light Dragoons or Hussars with this Army; but as the officers, the men and their horses are now accustomed to the food they receive and to the climate, I do not recommend that the regiments should be changed, or that any additional regiments be sent out, excepting possibly the remaining two squadrons of the 3rd [sic, 2nd] Hussars, of which two Squadrons are already at Cadiz."[49] The dislike for heavy cavalry at that moment, and the desire that no further regiments be sent, is a reflection of the strategic situation at the time of writing rather than lasting prejudice against the heavy cavalry arm. As the theater of war shifted, the requirements for specific types of cavalry, and, indeed, more cavalry or less, changed. Thus, in 1810 and again during the fighting in the Pyrenees, Wellington had more cavalry than he could usefully employ—or, indeed, easily feed.

The year 1812 saw a new demand for heavy cavalry to operate with effect on the open plains of central Spain, and for the rest of the time light cavalry was of the greatest utility due to the almost constant need for a secure system of outposts. These preferences can be tracked by Wellington's comments at various junctures in the course of his correspondence with London.[50] Yet wastage amongst the light and heavy regiments is by no means as clear-cut as Wellington's 1811 letter might suggest. Similarly, as has been noted, preconceptions regarding the British soldier as a horseman generally, in particular when compared with his KGL counterpart, are also not as pronounced as might be assumed. All in all, when one moves from the big picture down to the regimental level, the situation is by no means straightforward.

A Vast Super-excellence?

In attempting to make some sense of the situation with regard to strategic consumption of horses at a regimental level, we first need to establish an overall average against which such an analysis can be made. Since the orthodoxy has it that the horses of the KGL were far better cared for than those of British units, figures for cavalry and artillery horses are given for British and German units separately in table 12, alongside an average for the whole. Figures for the horses of the Wagon Train, where mortality was inevitably higher due both to the nature of the work and the use of worn-out animals in this role, have also been presented separately so as not to skew any analysis. Whilst the average loss rate for the train horses is appreciably higher than it is for the other arms of service, there is little appreciable difference between losses for cavalry and artillery, or between British and German units. Indeed, for the cavalry, the difference is down to a matter of one-third of a percentage point; to all intents and purposes, there is no difference to be found between British and KGL regiments. For the artillery, the distinction is somewhat more pronounced, but it should be noted that the bulk of the KGL artillery in the peninsula was eventually assigned to garrison duty, with only a single battery present in the field after 1811.[51] Even then, the figure for both the British and German artillery is within one percent of the overall average for artillery horses.

Such results, however, fly directly in the face of much of our understanding of the relative merits of the British and KGL, and the

Table 12.   Comparative Equine Mortality in the Peninsula

	Average Horse Strength/ Month	Average Horse Deaths/ Month	Deaths as % of Strength	Months of Data
British Cavalry	3,906	134	3.32	70
German Cavalry	881	30	3.29	67
Total Cavalry	4,749	163	3.32	70
British Artillery	1,893	77	3.91	70
German Artillery	132	4	2.94	68
Total Artillery	2,079	82	3.79	70
Total British	5,799	212	3.53	70
Total German	1,029	33	3.11	70
Grand Total	6,828	244	3.45	70
Train	378	27	6.67	70
Grand Total + Train	7,206	271	3.62	70

Source: Data derived from Monthly Returns, TNA, WO17/2464, 2465, 2467–2476. Note that totals are over all figures and not cumulative totals.

validity of the assumption that KGL troops took better care of their animals is valid. This latter issue has, in particular, become an oft-repeated truism regarding the peninsular army, with Ian Fletcher commenting in his study of Wellington's cavalry that the Germans "are universally regarded as being far better at the job [of horse care]."[52] Fletcher's use of the present tense is in this case largely valid, for it is hard to find the matter discussed without a similar conclusion being drawn, even if other sources, unlike Fletcher who relies rather heavily on the partisan August Schaumann as his main authority, are not quite so insistent.[53] At the same time, if Schaumann is the most outspoken contemporary source for KGL superiority, he is by no means the only one. All this then begs the question, if the statistics show no appreciable distinction, why do contemporary accounts, and the histories that draw on them, per-petuate what would seem to be a complete fallacy?

The answer only begins to become apparent when cavalry losses are compared on a regiment-by-regiment basis in table 13, which ranks the twenty-six peninsular cavalry regiments in accordance with their average monthly figure for horse deaths. However, the results are slightly skewed by the presence of the Household and Hussar Brigades, which arrived only in time for the Vitoria campaign and thus spent very little time on service in the field. Understandably, these units have the best survival rates and fill the first seven lines of the table. Similarly, at the other end of the scale are the 23rd Light Dragoons and the 3rd KGL Hussars, neither of which served very long in the theater before being sent home. For the 23rd, this was because of extremely heavy losses of horses and men at Talavera, whilst in the case of the 3rd Hussars the data relates to the detachment left behind when the rest of the regiment marched with Moore, composed of those horses and men unfit for active service. Like the seven late arrivals, these units need to be set aside, accordingly leaving us with a central block of seventeen regiments that served in the theater for a reasonable length of time to form the basis for a further analysis.

When we look at this central block, one of the first things that becomes apparent is the marked divide between heavy and light cavalry. Almost without exception, the former constitute the top portion of the block, with better survival rates, and the latter form the lower. Only a single anomaly each way breaks the distinction, which reflects the differing roles played by these two subdivisions of the mounted arm. Whilst the tactical distinction between lights and heavies in the British service was not as pronounced as in many continental armies—particularly early in the war, when brigades often contained a mixture of both types—there was still a doctrinal reluctance to employ heavy cavalry in outpost work.[54] Similarly, the preference for light cavalry for reconnaissance and picket duties meant that there were times when these regiments, or some of them, were retained with the main army whilst the remainder of the cavalry remained in the rear, as with the defense of the Portuguese frontier or the final advance through southern France.

Finally, whilst one would expect lighter casualties for the heavies during campaigning to be canceled out by higher battlefield

Table 13.   Peninsular Equine Mortality Rates by Regiment

Regiment	Average Horse Strength/ Month	Average Horse Deaths/ Month	Deaths as % of Strength	Months of Data
Royal Horse Guards	318	2	0.63	18
15th Hussars	526	6	1.13	17
2nd Life Guards	280	4	1.41	18
10th Hussars	529	10	1.86	17
18th Hussars	498	11	2.16	19
1st Life Guards	250	6	2.34	18
7th Hussars	573	14	2.39	10
3rd Dragoon Guards	413	12	2.82	62
2nd KGL Dragoons	367	11	2.91	30
1st KGL Hussars	461	14	2.95	61
3rd Dragoons	410	13	3.07	34
1st Dragoons	461	15	3.15	58
4th Dragoons	409	14	3.31	62
1st KGL Dragoons	360	13	3.49	30
5th Dragoon Guards	381	14	3.54	34
13th Light Dragoons	409	15	3.54	51
2nd KGL Hussars	260	10	3.70	29
16th Light Dragoons	445	18	3.89	62
4th Dragoon Guards	353	15	4.08	20
12th Light Dragoons	393	17	4.15	36
14th Light Dragoons	461	20	4.16	66
11th Light Dragoons	362	16	4.23	24
20th Light Dragoons	221	10	4.33	10
9th Light Dragoons	374	17	4.35	20
3rd KGL Hussars	208	11	5.02	7
23rd Light Dragoons	314	48	13.26	6

Source: Data derived from Monthly Returns, TNA, WO17/2464, 2465, 2467–2476.

losses, the doctrinal tendency to use both arms for the mounted charge negates this expectation.[55] Even in the figures for July 1812, which one would expect to be heavily skewed by the losses of Le Marchant's Brigade at Salamanca and of Bock's Brigade at Garcia Hernandez, the five heavy regiments, representing half the total engaged in the Salamanca campaign, still only account for two-thirds of the cavalry's losses.[56] A similar situation can be seen with most peninsular battles, with the heavies generally taking more losses than the lights but not to an overwhelming degree. Only at Waterloo, where the two heavy brigades not only suffered during their own famous charge but also continued to play a role throughout the day, is the ratio hugely distorted toward the heavy arm.[57] Whilst good heavy cavalry mounts were harder to come by, the very fact that their selection was more exacting, along with a smaller requirement due to the greater prevalence of light regiments, may also have ensured a better quality of remount being sent to these units. There are certainly fewer such complaints to be seen, relative to those for the light dragoons, when one looks at the inspection reports on heavy cavalry units. A combination of better remounts and less attrition through strategic consumption can therefore be seen to have more than compensated for a higher battlefield loss rate to ensure that equine mortality in the heavy cavalry was lower than it was in the lights.

But what of the alleged superiority of the KGL as horsemen? Whilst the 2nd KGL Dragoons comes second in the ranking of the eight heavy regiments in the main block of data, its sister regiment comes in sixth. However, the two remaining KGL Hussar regiments both appear extremely high up in the listing of the light cavalry, taking first and third place respectively. It is perhaps unwise to draw too much from the high ranking of the 2nd KGL Hussars, since this regiment spent most of its time with the 2nd Cavalry Division in the southern theater and saw far less arduous service than those regiments serving with the main army in the north; the fact that the 13th Light Dragoons has the best survival record of any British light cavalry regiment may also be attributed to this cause. On the other hand, there is no denying the fact that the 1st KGL Hussars has an equine survival rate far better than any other light cavalry regiment and, indeed, better than many heavy regiments too. But whilst it is undoubtedly creditable, the high position of this one regiment in fact goes a long way to explaining the distortion of the record in favor of the KGL cavalry *as a whole* and stands as a warning against overreliance on the few core memoirs from which so many myths and half-truths originate.

When the memoir sources for the favorable view of the KGL are considered, two things swiftly become apparent. Firstly, these sources largely refer to this one regiment, and, secondly, they almost all stem from the early years of the war when the 1st KGL Hussars served alongside that hotbed of future memoirists, the Light Division. So far as the good reputation of the other KGL regiments is concerned, that of the dragoons is based largely on their heroics at Garcia Hernandez, whilst the 2nd KGL Hussars was in fact numbered amongst the four ineffective regiments sent home after the 1812 campaign. Furthermore, many of these complementary accounts refer only to the regiment in general terms, or as excellent outpost soldiers, rather than specifically noting their superiority as horse masters.[58] Still, if we add this general praise for one exceptional regiment to Schaumann's lauding of the KGL cavalrymen and denigration of their British counterparts, it is easy to see how the myth began to evolve.

However laudable his motives in singing the praises of the 1st KGL Hussars, to which he was attached between October 1810 and January 1812, Schaumann was first and foremost a consummate storyteller, and his exploits—military and romantic—are inevitably told with himself as the hero. Furthermore, although Schaumann relates other tales that display the British cavalry and its leaders in a less-than-flattering light, many of his most stinging indictments come from the time of his new posting, after some time spent on leave, to the 9th Light Dragoons. The 9th was not in good shape when Schaumann joined it, and, indeed, would shortly be ordered home as a result. Schaumann found little positive to say about the regiment, and was equally scathing about the newly arrived 18th Hussars to which he was then assigned. But the 18th, as we have seen, was another one of Wellington's problem regiments, and when Schaumann joined the regiment it was still adapting to the return to active service.[59] So, from serving with a regiment that was indisputably one of the best in the theater,

Schaumann went on to successively serve with two of the worst—hardly surprising, then, that his comparisons were not the most flattering.

All this, though, simply serves to demonstrate that Schaumann had distaste of what he perceived to be the aristocratic disdain of British cavalry officers toward a foreigner. It still does nothing to answer the specifics of his case against the British cavalrymen as horse masters, to the consequent credit of their KGL counterparts, made earlier in his memoirs. Even that complaint, though, is tainted with a clear bitterness that the omissions of British officers were giving him extra work that kept him from the arms of the "exquisite child called Joaquina Cavaleira," his current mistress. Schaumann relates how despite his diligent efforts, the horses of the 4th Dragoons, to which regiment he was then attached, remained in poor condition. It took a tip-off from the local Portuguese magistrate to reveal that the dragoon troopers were trading their corn in exchange for brandy, which led to new orders to ensure that feeding was always to be supervised by an officer. Nevertheless, although Schaumann claims that his intervention solved the problem, he remained critical of the British trooper, writing, "The English cavalry soldier looks upon his horse as a machine, as an incubus, which is the cause of all his exertions and punishments. He ill treats it. And even when forage lies within his reach, he will not, of his own accord, lift a finger to get it. The commissary must procure everything, and actually hold the food to his own and his horse's mouth. . . . How different things are in the German cavalry regiments of the legion!"[60] Yet even Schaumann, however partisan and however determined to tell a good tale, could not but concede that the 14th and 16th Light Dragoons must be omitted from this castigation of the British cavalry.

Schaumann's story took place in 1809, yet his characterization of the British trooper seems to have become attached to the whole era. However, thanks to the abstracts of Regimental Courts Martial included with most mid- and late-war inspection returns, it is possible to test the validity of this assumption. Using the records for the year 1812, when a large and mixed contingent of British and German cavalry was deployed to the peninsula, the regimental returns of courts-martial were checked, firstly for soldiers tried for stealing and selling their horses' corn and secondly for more general crimes relating to mistreatment of animals. There are certain flaws in this methodology, firstly that such records only tell us of those soldiers unfortunate enough to be caught in the act, and secondly that a significant number of sentences were awarded under the catchall term, "unsoldierlike conduct," which may cover a multitude of crimes including some of relevance here. There is also a problem in that records of the first half-yearly inspections are rather patchy, with the only case of a British soldier recorded as being tried for stealing forage being that of a private in the 4th Dragoons, who was acquitted; other crimes may be hidden by the fact that the relevant records are missing. Within the two KGL dragoon regiments, a total of seven men were brought to trial during the first half of the year, six being found guilty, although there were no cases in either regiment of hussars.[61]

For the second round of inspections, all sixteen cavalry regiments then with Wellington—which included four KGL units—were put through a full inspection, and here the situation is rather different. In the twelve British regiments, three men were tried for stealing corn, two from the 9th Light Dragoons of whom one was acquitted and one found guilty but "forgiven"; only one man, a private of the 14th Light Dragoons, was actually punished.[62] In the German regiments, by contrast, only one case of theft of forage was reported, although since the enterprising individual in question, a Private Schäfer of the 1st KGL Hussars, was implicated in the theft not only of two cartloads of long forage but also the carts themselves, getting 260 lashes for his pains, this may be considered as to some extent redressing the imbalance.[63] Schäfer's crime cannot be considered as amounting to cruelty as such, since he was not stealing directly from either his own horse or, indeed, any one specific animal, and with this in mind a comparison of more generic cases of cruelty to, or neglect of, animals is also instructive. In the same set of sixteen regiments, there are three relevant courts-martial amongst the British: two being for leaving a horse unattended, resulting in one case in the loss of the animal, and the last being the case of Private McKean's mule with which we began the chapter. A similar case represents the only direct cruelty trial from the KGL cavalry, and here, again, there was an element of wider insubordination involved, the case being that of Private Henry Bork of the 1st KGL Dragoons, sentenced to three hundred lashes on December 13 after being found guilty on two counts:

1. For making a false report in stating that his horse had died on the road on the 30th Sept.
2. For cutting and ill-using his horse with his sword on the same day.[64]

Again, the date places us in the Burgos retreat and the resulting collapse in order within even some of the best regiments.

Whilst a slightly higher incidence exists of theft and/or cruelty in British regiments, this is by no means as marked as might be expected from the traditional understanding of things. More telling, however, is the differing response to such cases. In contrast to their reception in the slackly run 9th Light Dragoons, outbreaks of theft in the KGL regiments resulted in draconian punishments—five hundred lashes per man in the 1st KGL Dragoons, as compared with only seven hundred ordered for desertion. Furthermore, whilst the deserter received only 515 lashes, the remainder being remitted, corn thieves received the full 500 awarded.[65] Theft of forage fell off considerably in the latter half of the year, both in general terms but particularly in the KGL heavy brigade, where it was evidently deemed not worth the risk after the punishments of the spring. No doubt this general decline in theft was in part due to the fact that the army was active and the contacts required to sell stolen fodder were not available. There may also be an additional issue here, relating to horse care more generally, in that the trooper on campaign had to take better care of his mount lest he risk death or capture as a result, hence fewer cases of stealing forage. This suggestion is borne out by the second half-yearly inspection of the 3rd Dragoons in which Brigadier General William Ponsonby attributed the unequal condition of the troop horses to the fact that men had every interest in looking after their own mounts, on which their safety depended, but none in caring for surplus horses belonging to the sick.[66]

Even from this later part of the war, there are a small number of accounts by memoirists whose views support Schaumann's assertion of the KGL's superiority in horse care. John Fox Burgoyne, then a lt. colonel with the Royal Engineers, noted the closer attachment of German troopers to their mounts, and consequent greater care paid to them, whilst George Gleig also made a direct comparison between British and German cavalry and came out in favor of the latter. What is particularly interesting with regard to Gleig is that his experience was later in the war, and was with the dragoons of the Legion rather than the usually cited hussars.[67] Yet even if one concedes a small element of superiority for the German cavalry when taken in comparison with the British, it still remains the case that the bulk of the support for the former as horse soldiers par excellence comes from the period in 1810 when the single regiment then in the peninsula formed the army's outposts along the Côa. During this time, the 1st KGL Hussars provided an extremely effective outpost service and, what is more, did so with the loss of extremely few horses. There are periods elsewhere in the course of the war where equine mortality figures drop to these levels, but these coincide with periods of inactivity, such as the time within the Lines of Torres Vedras or the lull in operations prior to Vitoria campaign, and not to active service. In other words, the superiority of this regiment's reputation, whilst owing something to its general achievements, has been inflated by an anomalously short period of operations during which the unit was operating with distinction, and unusually low equine mortality, before the eyes of some of the war's best-known future eyewitnesses. Had these eyewitnesses had experience of the cavalry as a whole, they might also have noted that all six regiments experienced a low equine mortality rate in these months.[68]

In this context, the comments of Troop Sergeant Major William Dawes of the 15th Hussars, writing with reference to the later years in the peninsula, become significant as an aid to reconciling matters. Whilst noting a degree of envy with regard to the sleek and well-fed appearance of both men and horses of the 1st KGL Hussars, Dawes makes it clear that he did not accept the arguments of those who credited them with "such a vast super-excellence over the English dragoon."[69] He believed that the KGL troopers were permitted considerably more freedom in how they obtained their forage, with no questions asked. Whilst Dawes attributed this to indiscipline, it is more likely that the officers and men of the 1st Hussars, with several years' experience of campaign service, had simply picked up far more tricks of the trade than might be expected to be known by the newly arrived 15th. In other words, the enviable condition of the 1st Hussars was not due to its being a KGL regiment, but to its being an experienced regiment.

Schaumann highlights the 14th and 16th Light Dragoons as deserving exemption from his criticisms of the British cavalry, but these were the two longest-serving British light cavalry regiments, with records almost as noteworthy as that of the 1st Hussars. It is therefore unsurprising that these two regiments have equine survival rates superior to most other British light cavalry regiments; nor is it to be wondered that the 3rd Dragoon Guards, one of the longest-serving regiments of heavies, has the best average of all for the main sample. Inevitably, it took time to pick up the experience necessary to ensure such a good performance, so wastage was higher until this process was complete. Alone of the peninsular cavalry regiments, the 1st KGL Hussars came to the theater with considerable prior experience of active service.[70] British regiments, by contrast, had to learn on the job, and, by and large, did so. As time went on, the numbers of horses lost by long-serving British regiments such as the 14th and 16th Light Dragoons, 1st and 4th Dragoons, and 3rd Dragoon Guards began to drop. Troopers gained more experience in caring for their mounts; officers and commissaries became more aware of the best ways of obtaining forage and of supervising its issue. At the same time, as the initial veteran strength of the 1st KGL Hussars became diluted, its equine mortality rates rose to meet those of the British regiments.

The suggestion, then, is that British regiments were largely able, over time, to acquire many of the same skills that the KGL cavalry of 1809 and 1810 already possessed, leading, over time, to an equalization of performance as judged by equine mortality, and, indeed by other factors such as performance of outpost duty.[71] There is ample evidence that late-arriving British cavalry units went through a similar learning curve to those joining earlier in the period, even if it was still incomplete at the war's end.

When the Household Brigade joined Wellington prior to the Vitoria campaign, its regiments discarded combs and brushes, as representing unnecessary equipment now that they were no longer required for ceremonial duties. The result was that the coats of their horses rapidly became matted and clogged, blocking pores and leading to sickness. Combing and brushing were swiftly recognized as having more than just an aesthetic benefit, and accordingly were reinstated.[72] The learning curve extended to the new arrivals of the Hussar Brigade as well, with Lt. George Woodberry of the 18th Hussars remarking on the great amount of unnecessary impedimenta brought on campaign by the officers of both formations.[73] The same officer was also initially concerned that "We are actually oblig'd to send near two leagues for green forage. The whole produce of the fields near have already been consumed by the 15th and us; those very fields of corn, barley &c was [the locals'] principal support for the summer."[74] After a year's active service, however, the men of the18th had adapted to the more practical elements of supplying themselves and their mounts on campaign, with Captain Arthur Kennedy referring to his men, with affection, as "Drogheda's Cossacks" for their prowess in obtaining food and clothing from the surrounding countryside as they advanced toward Toulouse.[75] The hussars had thus begun to acquire the same skills and experience that Dawes had identified in the KGL cavalry, but had evidently picked up the bad habits along with the good; as we have already seen with regard to the survival strategies of the men themselves, experience by no means equated to perfection.

Conclusion

A SYSTEM REASSESSED

There is a tendency amongst the historiography of the Napoleonic Wars to treat the British Army as unique, and to see the British military experience of the Napoleonic Wars as something apart from the norms on the continent. Yet the similarities are more marked than is often allowed, in particular when comparison is made with those powers that were able to resist Napoleonic France without ever being entirely overrun or subjugated. Britain, like Austria and Russia, entered the conflict with an established system that was refined throughout the conflict without ever being the subject of revolutionary change. Differing geographical circumstances naturally dictated differences in scale and priority, but the similarities remain marked. Even the upsurge of nationalism seen amongst the continental allies is mirrored in Britain's volunteer movement,[1] and the abortive deployment of Militia on active service in 1814 adds an additional level of commonality with the continental powers.

The similarities should not be overstressed. The British Army never underwent such a large-scale process of overhaul as the forces of Prussia; nor did its manpower difficulties reach the sort of levels faced by France after 1812. Britain's avoidance of full-fledged conscription also distinguishes the British and continental experiences of warfare. Geographical isolation kept the home islands largely remote from the horrors of war but necessitated different strategic imperatives as reflected in the priority accorded the Royal Navy and the preference for paying subsidies rather than committing troops to the continent. However, the French Revolutionary Wars had indicated the drawbacks of British

286

policy, and by the period of this study the relative lack of support for the British Army had begun to represent a definite strategic handicap.[2] The Royal Navy remained essential, but the concentration of so much military manpower into the Militia and Volunteers now represented a diversion of resources away from the real struggle. The British Army, which after 1808 became the arm that bore the brunt of active service, was left with too much competition for finite resources.[3] Conversely, the distribution of resources between its army, navy, and subsidies gave Britain a wider range of strategic options than was available to many continental powers. When the right balance was struck between the three, as with the Peninsular War, the result was success, but the extent of Britain's commitments between 1808 and 1815 meant that there were also times when the British Army was left with an unequal share of the burden.

Two things enabled the British Army to maintain itself in the field long enough for the war to be won: the flexibility of the regimental system and the esprit de corps that it helped foster. When the British Army is compared with its foreign counterparts, the advantages inherent here become apparent, and this helps explain how an army that was ostensibly out of step with the times could in fact sustain itself in the field. This sets up a series of further comparisons, for there is no denying that continental armies could cultivate esprit de corps that was just as strong as in British regiments, nor that manpower organization in many of these large conscript armies led to far greater efficiency than that of the British system. Somehow, however, and against all reasonable expectations, that system continued to function and enabled the British Army, by the skin of its teeth, to meet the nation's strategic demands. If the recipe for military success entails a successful balancing act between identity on the one hand and efficiency on the other, this then begs the question of what factors enabled Britain to continue to maintain that balance.

Manpower and Identity

For the bulk of the Napoleonic Wars, the small size of the British Army enabled it to field a far more cohesive force than the bulk of the continental powers. Even when units were sent on service a brigade at a time, the units forming these brigades had no permanent organizational or command connection, and so could easily be transferred as necessary. Furthermore, as we have seen, newly arrived brigades could be fitted into the divisional system in order that any lingering imbalance could be absorbed at that level. Were a typical continental army to receive a similar reinforcement, this would in likelihood be composed of multi-battalion regiments, which could not easily be broken up and redistributed without creating problems of command and control. More likely still would be the dispatch of a whole new division or even army corps to a theater, leading to marked disparity in troop quality at a very high organizational echelon. The strength returns for Masséna's Armée de Portugal indicate a marked distinction between the effectiveness ratio of VIII Corps, comprising inexperienced troops including many provisional units, and the more seasoned troops of II and VI Corps. Even within VI Corps, the survival rate is better for the two veteran divisions that had served as such since 1805 than the corps's third division, created out of odds and ends and only sent to the front in 1810.[4] In the British service, fewer units, and no fixed requirement to keep battalions in any particular command grouping, made manpower management at a lower level far more practical. This represents a key strength of the British system.

To extend the comparison at a regimental level, multi-battalion regiments on the continental model also meant that, in the French service at least, new manpower tended to be formed into new battalions and sent to the front as such.[5] Many French regiments condensed their original three battalions into two after Masséna's 1811 retreat from Portugal, sending home the third battalion cadre, but received in its stead a new fourth battalion composed of conscripts.[6] A French regiment was that army's primary organization for combat as well as administrative purposes, which presented tactical problems for a French colonel that his British counterpart never experienced. Whereas the former, serving under Masséna in 1811, was left with three battalions of uneven quality, for the latter even a large draft could be broken down and distributed amongst ten companies, facilitating easy integration and substantially increasing total manpower without unduly affecting unit quality. Other continental armies formed replacement battalions or *battalions de marche*, which could be broken up once they reached the front and manpower allocated where needed; this enabled manpower to be distributed to create numerically balanced units, but frequently meant men being assigned to a regiment with which they had no prior connection. Not only would the unit contributing men to a replacement or *marche* battalion frequently skimp on what it supplied to the men so assigned, but such a reassignment also destroyed any feelings of unit identity or esprit de corps that the recruits in question might have gained through their training.[7] In contrast, the British Army had learned of the iniquities of drafting men off to other regiments, and thus recruits going overseas could be assured that they would ultimately find themselves in a battalion of their own regiment. Only when the army was being rapidly—and prematurely—reduced during late 1814, and many second battalions broken up, were men from disbanding units reassigned to other regiments.[8]

Regimental tradition, in established units or those with a good record of recent service, was a strong unifying factor, and may explain why units with such strong traditions were able to survive spells of service under the command of less-than-stellar officers. The legacy of a good commander could help a unit survive service under a bad one; a series of bad commanders led to disaster, as with the case of the 2nd. As formations higher up the organizational chain also began to develop identities of their own, those that flourished to the greatest degree were those where effective and/or charismatic leadership was involved. This can be particularly seen in the peninsular Third Division under Picton, the Light Division under Craufurd, and, to some extent, the whole peninsular army under Wellington.

Yet whilst the importance of leadership remained constant during the period, shifts can be identified in its nature. Early models of charismatic personality-based leadership give way to a quieter, more professional, style. Men like Inglis, Gough, and Cadogan serve to blur the boundary, as do the likes of Harry Smith and Kincaid at a lower level, and old-school models of leadership were still apparent to 1815. York was keen to instill a professional ethos in his officers, and so too was Wellington, whose stinging rebukes were at times required to remind erring individuals of their duties.[9] At a unit level this dedication to duty can also be linked to the fact that many of the officers who were commanding battalions and regiments by the end of the period were those for whom the British Army represented a career for life, in which advancement could not be obtained through interest but only as a result of diligent service. If such an officer possessed charismatic skills of personal leadership, then so much the better, but professional competence ensured that such men were respected. The backgrounds and motives of such men are apparent from Graves's biography of Thomas Pearson of the 23rd, or Urban's life of George Scovell,[10] but similar cases can be identified amongst the officers encountered during the course of this study, such as Hamilton of the 30th, Harris of the 73rd, and Morrison of the 89th. That many such officers, like Pearson or Hamilton, served for years in the same regiment, rather than jumping from corps to corps as they purchased their way up the ladder, helped tie them to their units. Donald Graves speculates that Pearson's greatest ambition—eventually achieved—was command of the 23rd, a hypothesis that fits with the more regimentally focused career ambitions encountered later in the nineteenth century.[11]

Of whatever nature, effective leadership was not only militarily valuable but also appreciated by those being led. Strong evidence of the latter can be seen in the number of occasions when a poor combat performance was directly linked by those involved to the absence of an established leader. Perhaps the simplest and most evocative example comes from Cooper's account of Albuera: "Had he come sooner we should have had more confidence of victory. This may appear from the brief dialogue which took place between Horsefall and myself, when marching to attack the dark columns on the hill. Turning to me, Horsefall dryly said, 'Whore's ar *Arthur*?' meaning Wellington. I said, 'I don't know, I don't see him.' He rejoined, 'Aw wish he wore here.' So did I."[12] At the divisional level, the absence of Leith at Villamuriel during the retreat from Burgos was partly to blame for the disorder of the Fifth Division,[13] whilst the Light Division did not take kindly to the disastrous Erskine temporarily replacing Craufurd when the latter went on leave in early 1811.[14] We have already encountered instances where the absence of a regimental or battalion commander led to problems, but it is particularly telling in terms of unit cohesion that the most serious problems arose when command passed to officers little-known to their men, as with erstwhile staff officer Major Kelly's unfortunate stint in command of the 2/73rd. Bonds of shared experience not only bound units together at the lowest level but also facilitated effective command relationships extending to the regimental level and beyond.

Although these bonds were tight, it is not to be thought that they never began to fail. However, the distinction must be made between a temporary collapse and a total breakdown; after all, as Wellington said in relation to combat, "He thought nothing of men running, provided they came back again."[15] In a similar fashion, men did fall out and straggle, but, by and large, they came back again and on occasion made immense efforts to do so. If anything, this emphasizes the strengths—moral and physical—of the regimental system. Moral strength because the regiment embodied the group and personal loyalties that helped its men define their place in what must frequently have been a confusing and terrifying world, and physical strength because those bonds could only exist in a body of men that had the organizational framework created through its being part of a larger functioning whole. Edward Coss encapsulates the social benefits of the system by defining the regiment as "a secondary social group, [which] worked to give men a sense of worth and bind them to each other."[16] This emphasis on the group can be extended further to explain how those elements of the British Army that served in subregimental bodies were able to function in a similar fashion; an artillery battery, or train detachment, could replicate, on a smaller scale, the loyalties and identities associated with the infantry battalion or cavalry regiment. Having made his point, however, Coss's overprioritization of the physical element at the expense of the moral then leads him to belittle his analysis. His belief that "The value of the regimental system was muted, if not negated, by the inability of the army to look after the basic needs of the men"[17] misses the point, because any rank-and-file reaction to those failings, which Coss in any case overstates, took on a far more abstract form, as

discussed in chapter 6. In all but the most extreme cases, the regiment represented a group of fellow sufferers, and bonds could
even be reinforced by shared hardship.

In meeting the nation's military manpower needs, Britain's
regimental system in force during the Napoleonic Wars appears
outdated and wasteful. Far from following the continental example and mobilizing the nation for war through conscription,
recruitment remained decentralized. Although efforts were made
to impose greater order, the available manpower remained in 140-
odd regimental pools, of varying size, rather than a single central
source. Failure to implement conscription, and the attractions
of alternative modes of service through the Militia, prevented a
steady supply of manpower, and, whilst quotas could be set, this
did not guarantee that they would be met.[18] Some units struggled
to maintain their strengths, whilst others had men to spare. Some
units swelled to field three or four battalions, whilst others failed
to raise even a second. Whatever the other advantages of the system, efficient it certainly was not.

Thankfully for the British Army, although the manpower numbers demanded by new strategic commitments grew, the manpower requirements to replace strategic consumption fell, at least
in relative terms. This decrease was due to the benefits obtained
through greater experience of campaigning and thus the development of the skills and hardiness required to survive it. Based
on the available data for the Peninsular War, the general trend in
sickness and mortality was downward as Wellington's army grew in
experience. This took place both on an individual level, as men
became more personally acclimatized to the rigors of active service, and institutionally through the institution of new practices
such as McGrigor's overhaul of the hospital system, and in turn
directly links to, and reinforces the value of, the increasing recognition of the importance of logistical factors in analysis of the
success of the British Army during the Peninsular War.[19] It is not to
be suggested that older debates concerning the tactics of line and
column, or the particular secrets of Wellington's undoubted ge-
nius as a commander, are invalid, but without the means of keeping the troops in the field, their ability to utilize any particular
nuance of minor tactics simply becomes irrelevant. The fact that

Britain by the second decade of the nineteenth century was able to maintain armies successfully in the heart of the Iberian Peninsula or the depths of Upper Canada, when only two decades previously it had failed to do so even in Flanders, is concrete proof of the effectiveness of the military reforms instituted after the nadir of the 1790s.[20] If the systems of manpower management remained somewhat lacking at times, this should not take anything away from the great advances in staff, logistics, and command during the first years of the nineteenth century. The year 1808 found Britain with a system completing the process of reform; 1815 saw that same system tested to the extreme but still—just—holding together. What was more, these reforms coincided with Britain's rediscovery, discussed above, of direct continental intervention in conjunction with allies as the only successful strategy by which a decisive settlement could be obtained. Together, this allowed the British Army an operational effectiveness and significance far in excess of its numbers.

Traditionally, Britain had compensated for its manpower shortfalls by hiring foreign auxiliaries. Indirectly, through subsidies, this practice continued during the Napoleonic era and made a strong contribution to allied victory.[21] But the direct use of foreign manpower was harder to justify, and if foreigners were to serve in nominally British units it would be harder still to assuage the fears of those who associated foreign troops with repression. Line units in which significant enlistment of foreigners took place— the 60th, 97th, and 102nd—were never long in Britain, with the exception of the 97th after 1811 by which time it had had its foreigners diluted with a large influx of Irishmen.[22] The all-Canadian 104th spent its entire service on the other side of the Atlantic.[23] Equally, Wellington's Spanish recruits, and the Germans enlisted in 1813 and 1814, remained overseas where their presence could be quietly ignored. Potentially more damaging was the active employment of some of the more dubious foreigners, with American propaganda justifiably making much of the excesses committed by the Independent Foreigners during the attack on Hampton.[24] Part of the problem, as Wellington indicated, was that Britain provided no incentives for Frenchmen, or troops of Napoleon's vassal states, to desert to the British.[25] The bulk of those who deserted to the British acquired a one-way ticket to the fever islands, with the result that the only men available were such desperate characters as those who found their way into the Independent Foreigners and 7/60th. This lack of a sustained or effective policy meant that the better foreign units in British service, from the KGL on down, deteriorated due to lack of suitable manpower. As a result, such collections of bad characters as the Independent Foreigners had to be employed in combat rather than their envisaged penal garrison role, to the detriment of the efficiency, and reputation, of the commands to which they were assigned.

Yet the very fact that domestic recruits were not a finite resource, and that foreign recruits were a supplement rather than an alternative, meant that the system had to allow for flexibility if it were to function. The outwardly rigid two-battalion regimental system for the infantry was rarely followed to the letter, not least because the number of battalions per regiment was never standardized. This flexibility allowed junior battalions to provide vital services. Deploying second battalions certainly caused problems—either through the low strength of the units concerned, the diversion of manpower away from the first battalions, or both—but without this measure the British forces in the first half of the Peninsular War would have been slim indeed. For the cavalry, the existence of four or more squadrons within the regiment allowed for greater flexibility, particularly since the composite troops could be shuffled about in a way that infantry companies could not. It was also possible to shuffle the regiments and battalions themselves in order to place units in the station most suited to their state of readiness. However, such measures could only be used to good effect when there was a sufficient supply of campaign-ready units. Once all the effective battalions were on active service, nothing further could be gained by moving units around.

Even then, however, benefits could still be obtained by employing an element of care in the assignment of units to higher formations. The divisional system as refined under Wellington was geared to creating a balanced and homogenous army, and by 1813 this had been achieved for the forces in the peninsula. This system attempted to ensure that units of poor or uncertain quality were supported by more seasoned troops, but could also be manipulated to facilitate the management of manpower. At the simplest level, the ways in which Wellington's system simplified manpower management can be seen in the tendency to place battalions of the same regiment in the same division or brigade, to allow easy amalgamation if necessary. At a more sophisticated level, formations could be created outside of the norm either to concentrate specialist troops for a particular purpose, as with the Light Division, or to create second-line formations through which new or poor quality units could be eased into active service. Lastly, on those occasions where a balanced divisional system had not yet been fully established, divisions of different capabilities could be assigned tasks appropriate to their state of readiness, as with the second-line roles assigned to the Seventh Division in 1811 or the Fourth Division in 1815.[26] This divisional imbalance could even be deliberately maintained, as in 1812 when the First Cavalry Division got the pick of Wellington's mounted troops whilst the Second, in the quieter southern theater, contained many of the less effective regiments.[27] Only in North America did space and a lack of manpower prevent the evolution of such a sophisticated system of field command; elsewhere, as at the regimental echelon, the key to successful manpower management at a higher level was flexibility.

The injection of flexibility at various hierarchical levels enabled the regimental system to function notwithstanding its drawbacks, but the basic concept of the two-battalion infantry regiment was by no means a bad one in the first place. It could certainly have been better had the time existed to implement a uniform system of multi-battalion regiments; such units could then have absorbed some of the home service duties of the Militia, as Sir Harry Calvert proposed at the time.[28] As it was, an initial lack of strategic direction meant that the British Army was condemned to manage with the best of a bad lot and try and make the most of a system that was never complete but by no means irredeemably flawed. It is telling, too, that when Wellington was asked to advise on the best methods by which the regimental systems of the Spanish and Portuguese armies under his command might be managed, his response in both cases was a two-battalion model not far removed from that employed by Britain. In his proposals, both battalions would serve together if possible, but if numbers fell off the second battalion would revert to a noncombat depot role to support the first battalion and keep it in the field.[29] Similar levels of flexibility already existed in those elements of the British infantry that existed outside of the regular line. The Foot Guards, KGL infantry, and 95th Rifles were all able to use more sophisticated systems, prioritizing subunits above and below battalion level to maximize the effective use of available manpower. One might ask why such systems were not extended to the line, but with the level of hostilities that Britain faced during the period it was never possible to countenance such a degree of upset. When the British Army finally did go over to a universal system of paired battalions, it was decades later and, crucially, in the middle of a long European peace.[30]

Legacies

The British military system was placed under many of the same pressures as those of other European powers, but different strategic choices meant that the British Army was able to respond in ways that were unique to it. The continuation of the regimental system facilitated what must be seen as the primary advantage enjoyed by the British Army over its adversaries and its allies: that is, a strong and successfully maintained sense of unit identity. Peter Dietz rightly terms the period from 1793 to circa 1840 the "golden age of the regiments,"[31] for not only did the majority of regiments enter it already possessed of strong traditions, but their wartime service enhanced this. From new nicknames won through hard service, such as the 50th's appellation as the "Dirty Half-Hundred" after Vimeiro, or new distinctions to be treasured such as the collar flash adopted, defended, and finally permitted to the 23rd Fusiliers, campaigning between 1808 and 1815 added much to regimental traditions.[32] Older traditions could also be reaffirmed and reinforced; staying with the 23rd, there is little difference between the Saint David's Day celebration described by Browne in 1808 and accounts of the same festivities a quarter of a century before.[33] Apart from the increased number of nicknames and traditions apparent by 1815, Sylvia Frey's argument in favor of regimental identity during the American War of Independence can readily be applied to the Napoleonic era British Army.[34]

This self-identity found reflection in the great crop of regimental histories published throughout the nineteenth century, the majority of which seem not to have let fact get in the way of preserving regimental conceits that were still of practical use.[35] It is unlikely in the extreme that Wellington at Waterloo really called for the absent 43rd to reinforce his line, but a hundred years later the fact that the story was "current at the time and never called into question" was sufficient evidence to ensure its inclusion in the history of that regiment's successor unit, and the assumed indication of "the perfection of the regiment's work [and] confidence of their chief" to be presented as a strong focus for pride.[36] Such was the importance of the Napoleonic period in terms of regimental identity that reraised units having only a tangential connection with the regiments of 1815 sought to appropriate their histories. As late as the Second World War, the new 23rd Hussars promptly laid claim to the somewhat dubious laurels of Wellington's erstwhile 23rd Light Dragoons and thereby sought to ground its new identity within old traditions.[37]

Just as regimental conceits, traditions, and identity remained strong after 1815, so too was much retained of the newer identities created through the divisional system. Not only did this organization become the norm in all the major campaigns of the nineteenth century from the Crimea onward, but a deliberate attempt to secure a continuity can be seen in the way in which these divisional identities have been perpetuated and appropriated down the years. The Army of the East during the Crimean War had its Light Division in emulation of Craufurd's command, even though by then nearly the whole of the British Army was equipped with rifles, whilst the First Division of the same force was again composed of the "Gentlemen's Sons" with its brigade of Foot Guards. One new feature in that war, however, was that the Highland Brigade ultimately became the Highland Division, extending the idea of massing the army's self-identified elites.[38] This practice of giving divisions a title as well as a number would develop further in the wars of the twentieth century, with the creation of numbered divisions that also had territorial or status affiliations:

Highland, Light, North Midland, London, and so forth.[39] The attempt to seek continuity with historical namesake formations has continued from the peninsula to the present day, with the monument of the Second Division at York Minster listing battle honors from Talavera and Vitoria through the Alma and Sevastopol to Kohima and Imphal, even though the only constant feature over a century and a half was the formation's number.

Lastly, in terms of identity, chapter 5 suggested that the pattern of regimental and divisional identification could be extended upward to postulate identification with the force as a whole, at least for men serving in the peninsula. This sense of theater-army identity may be directly compared with that generated in North America between 1754 and 1763, and again between 1775 and 1783, identified in the works of Brumwell, Urban, and Frey.[40] The specific nature of such identities, along with the added complication that that of the peninsular army was built on a partnership between British and Portuguese regiments,[41] makes further extension of the concept problematic. Linda Colley postulates military service as leading to an increased sense of national identity, but the bulk of her evidence relates to the Volunteers and Militia, making only some dubious and assertive links to the regular army.[42] What Colley identifies is not the extension of soldiers' identities through the regiment and the British Army to the nation as a whole, but rather the appropriation of military identities by outside individuals or groups. This is particularly evident in the use of Wellington as a patriotic and uniting figure, but the erecting of monuments to the Duke, for all manner of motives, implies only that the British nation had recognized "Wellington's Army," and not that its rank and file necessarily saw themselves as part of a united Britain.[43] This is all well and good, but not very helpful in getting into the heads of the men in the ranks, where British identity existed largely by default, in terms of self-definition against enemies and allies.[44] Within a regimental "family" whose typical demographic had more than its share of Irish Catholics, the Francophobic Protestant Britain postulated by Colley could only have been an alien concept.[45] On the other hand, acceptance of English, Scots, and Irish comrades as part of the same social group, as vocalized in the lyrics of "The British Bayoneteers," suggests an acceptance of

Union and unity, supporting Colley's contention that war served as a motor for unity by breaking down parochial attitudes.[46]

In Britain's earlier wars, the creation of loyalties and identities that were theater-specific rather than directed toward the British Army as a whole prevented many of the tactical and operational lessons of these conflicts from being applied more widely, and so too were many of the lessons of 1808 to 1815 dismissed, ignored, or forgotten. Ironically, whereas in 1763 and 1783 lessons were discounted as representing a colonial anomaly, after 1815 many of the lessons of the previous two decades seemed out of place in an army fighting small wars around the British Empire. The discontinuing of standing brigades and divisions prior to the 1850s, and the distribution of the army by battalions and regiments, is a case in point and was resisted by those who had experience of the effectiveness of such methods before 1815.[47]

This trend of forgotten lessons is particularly apparent when the issues of manpower are considered. Even at the time, Wellington felt that his temporary manpower solutions could be expanded upon and incorporated into the established system. During the controversy over provisional battalions, he submitted a proposal to Bathurst for forwarding to York, proposing the universal adoption of similar methods. In the proposed system, whenever a unit on active service fell below 350 effective rank and file, it would send company cadres home to recruit up to strength and then rejoin; in the meantime, the cut-down battalion remaining in the field would form part of a provisional battalion.[48] Wellington's proposals had merit, and would surely have been less damaging if widely established than Bathurst's counterproposal, which involved sending understrength battalions home but drafting off to other units any volunteers who wished to remain in the theater, as was already done with units leaving the Indies.[49] But it would be another decade before a juggling of companies within units similar to Wellington's proposal, albeit without the creation of provisional battalions from the residue, became accepted practice, and even then this was implemented primarily as part of a policy to produce more, smaller, battalions for colonial service.[50] This system, and those that followed it, retained the onus of recruitment and manpower administration securely at the regimental level.

Because ultimate victory seemed to endorse existing practices, little thought was given to the narrow margin by which that victory had been obtained. By 1813, the amount of flexibility in the early part of the period, made available through manipulation of the system, had come to represent a double-edged sword. On the one hand, this flexibility enabled Britain's military to function successfully throughout most of the period of this study. On the other hand, this success hid the limitations of what was, for all its positives, in many ways an outdated system. Only when events toward the end of the Napoleonic Wars pushed that system as far as it could go did its failings become apparent, but because there was no major disaster an element of complacency entered the equation. In truth, the effects of the manpower crisis of 1813 and 1814 were such that the British Army had still not fully regained its former levels of effectiveness even by the time of Waterloo. The forces sent to the Netherlands in 1813, and retained there after the peace, had a function that was primarily political rather than military, but their lack of quality and numbers precluded their obtaining Britain any great advantage during the closing months of hostilities or giving it any political leverage during the peace settlement. Indeed, with the bulk of its regiments either worn down after long service or shifted to North America, Britain was forced to enlist French support during the Vienna negotiations in order to give its position some level of military credibility, with Liverpool insisting, "It would be quite impossible to embark this country in a war at present."[51]

Just as the European war had precluded serious military endeavors in North America before 1814, so did the shifting of focus across the Atlantic now prevent Britain from using its military resources on the continent. The peace settlement at Ghent thankfully allowed for a renewed European commitment during the Hundred Days, but by no means one on a par with the forces that Britain had fielded two years previously.[52] Not only was the army that fought at Quatre Bras and Waterloo organizationally incomplete, but it contained elements—British as well as foreign—whose dubious performance escaped notice only because of the eventual triumph.[53] The Waterloo campaign saw an unusually high level of officer casualties, right up to the senior level with one British divisional commander and three brigadiers killed at Waterloo, as compared to two and eight respectively for the whole Peninsular War.[54] Other than the exploits of the heavy cavalry, the epics of the fighting in 1815 were essentially defensive, which, in conjunction with the high level of officer casualties incurred through the need to lead by personal example, suggests a definite awareness of relative inferiority. Yet this situation is only to be expected when the strain under which the system had been placed over the previous eight years is taken into account, and the depth of the crisis in which it had fallen little more than a year before the final victory. Wellington's "nearest run thing"[55] can be applied to more than just a single day in June 1815.

The military system with which the British Army fought the Napoleonic Wars was that of the eighteenth century. When Fortescue sums up the workings of the system during the campaigns of Marlborough, many of the points that are made with regard to manpower could be taken word for word and inserted into this analysis.[56] But it is easy to seek commonality, and far harder to identify those areas where change was taking place. A system that had given problems throughout the eighteenth century was, by the end of the Napoleonic Wars, rapidly approaching the end of its useful life. That the system had survived for so long, and that its vestiges in terms of regimental identity linger to this day, represents a fair tribute to its motivational benefits, and in the Napoleonic Wars these benefits did, by and large, enable it to overcome the organizational drawbacks that became more and more apparent as time went on. But the fact remains that by the end of this period the individual soldier was becoming far more aware of his place in the system and his place in the world. The British Army was forced to recognize the rank-and-file soldier as a man rather than a cog in a machine or as a figure on a return, a recognition that is attested to by the institution of a Waterloo Medal for all ranks and, much later, by the belated award of the Military General Service Medal in 1847.[57]

Perhaps the greatest legacy of the period 1808–15 was that it served to imprint the regimental system yet more firmly on the British Army, thus doing much to shape the nature of that force for the remainder of the nineteenth century and beyond.

Organizational flexibility during the Napoleonic Wars enabled the British Army to fight above its weight, but only to a point and never to the degree that it could ever expect to compete on equal terms with the continental military heavyweights. In the context of broad British strategy, this military inequality was not a detriment because British gold, ploughed into subsidies for its allies, meant that the British Army was almost always operating within a coalition framework. That the British Army performed as well on campaign as it did between 1808 and 1815 is nevertheless tribute not to the system itself, but to the men who made it work. But by making an outdated system function, they gave it a new lease of life and institutionalized it to a degree that would bedevil their successors for much of the nineteenth century. For Britain and its allies, it was essential that the British military system held together long enough to contribute to the final victory, but this success was not necessarily beneficial for the future development of the British Army.

Appendix 1

DETAILS OF DATA SAMPLE

Data was sampled from a total of eleven theaters of war, utilizing a total of 345 individual monthly returns. The selection includes all theaters where significant campaigning took place, but also incorporates adjacent secondary and garrison stations. In part, this inclusion allows comparison between inactive and active stations in similar climatic conditions, but it also enables track to be kept of units being rotated between major and minor theaters. Data for Europe is taken from September 1808, the first month for which data from the peninsula is available, until July 1815 when hostilities finally ceased. Data for North America has been taken from January 1812 until July 1815, allowing the transition to and from hostilities to be tracked. The full data is now available online, courtesy of the Napoleon Series website.

Peninsula (main army): September 1808 to June 1814 (70 months). Data from TNA, WO17/2464–2465, 2467–2476.

Sir John Moore's army in Spain: November and December 1808 (2 months). Data from TNA, WO17/2464.[1]

Cadiz: April 1810 to July 1814 (52 months). Data from TNA, WO17/1486–1488.

Gibraltar (includes detachments at Ceuta and Tarifa): September 1808 to July 1815 (83 months). Data from TNA, WO17/1796–1801.

East Coast of Spain: April 1813 to April 1814 (13 months). Data from TNA, WO17/2478.

Walcheren: August to November 1809 (4 months). Data from TNA, WO17/2479.

303

Canada: January 1812 to July 1815 (43 months). Data from
TNA, WO17/1516–1519.

Nova Scotia (includes Newfoundland and Bermuda):
January 1812 to July 1815 (43 months). Data from TNA,
WO17/2241–2243.

American Coast (including forces on passage): June 1814 to
February 1815 (9 months). Data from TNA, WO17/1218.

Germany: July to December 1813 (6 months). Data from TNA,
WO17/1773.[2]

Flanders (includes march on Paris): December 1813 to July
1815 (20 months). Data from TNA, WO17/1773, 1760.

Within this sample, in addition to detachments of artillery, engineers, and train troops serving in all stations, data was taken for
the following selection of units. For each of these, data for the
whole period has been collected into a unit-specific spreadsheet.

British infantry: 133 battalions, plus 6 sub-battalion
detachments.

British provisional units: ten battalions, in some cases
composed of units also listed individually under the
previous heading.

British cavalry: twenty-six regiments, plus one subregimental
detachment.

British garrison units: six battalions, plus three sub-battalion
detachments.

Foreign infantry: eighteen battalions (of which one is
provisional), five sub-battalion detachments, and two
independent companies.

Foreign cavalry: six regiments, one subregimental detachment,
and one independent troop.

Foreign garrison units: one battalion and one independent
company.

Canadian provincial units: six battalions, plus three
sub-battalion detachments.

Royal Marine units: two battalions and one artillery company.

Appendix 2

MEN RETURNED AS SICK AFTER
THE CORUNNA CAMPAIGN

Battalion	Feb. 25	March 25	April 25	May 25	June 25
1/1st Foot Guards	No Data	84	77	43	36
3/1st Foot Guards	No Data	245	115	69	56
3/1st Foot	22	119	105	47	0
2nd Foot	No Data	17	52	76	33
1/4th Foot	107	88	36	17	6
1/5th Foot	102	68	36	21	37
1/6th Foot	31	40	11	35	45
1/9th Foot	46	454	318	325	315
2/14th Foot	55	159	64	41	47
20th Foot	79	176	148	149	111
2/23rd Fusiliers	109	106	38	27	20
1/26th Foot	187	148	71	34	11
1/28th Foot	267	255	197	159	146
1/32nd Foot	109	141	17	12	11
1/36th Foot	247	201	109	100	88
1/38th Foot	294	225	137	119	No Data
1/42nd Highland	174	118	54	28	16
1/43rd Light	322	201	127	No Data	No Data
2/43rd Light	No Data	384	106	133	133
1/50th Foot	No Data	103	152	129	119
51st Foot	284	166	80	29	26

continued

Battalion	Feb. 25	March 25	April 25	May 25	June 25
1/52nd Light	419	297	67	0	0
2/52nd Light	423	404	319	179	159
2/59th Foot	176	195	67	65	32
1/71st Highland	115	172	106	68	75
76th Foot	No Data	No Data	No Data	No Data	No Data
1/79th Highland	39	34	14	8	0
2/81st Foot	208	190	53	41	33
1/82nd Foot	414	342	50	64	28
1/91st Highland	429	104	51	24	18
1/92nd Highland	265	229	76	59	44
1/95th Rifles	370	227	101	0	No Data
2/95th Rifles	98	110	103	62	0
1st KGL Light	No Data	No Data	No Data	No Data	No Data
2nd KGL Light	No Data	No Data	No Data	No Data	No Data
Totals	5,391	5,802	3,057	2,163	1,645

Note: Figures are taken from monthly Battalion Returns in TNA WO17/64–219, 806–808, and refer to those sick in hospitals.

Appendix 3

SAMPLE OF RETURNED DESERTERS
IN THE PENINSULA

1810	British Deserted	British Returned	Foreign Deserted	Foreign Returned	Total Deserted	Total Returned
Jan.	35	6	5	7	40	13
Feb.	26	18	4	0	30	18
Mar.	16	10	46	4	62	14
Apr.	13	20	1	7	14	27
May	13	6	3	1	16	7
June	17	14	4	3	21	17
July	12	8	2	1	14	9
Aug.	21	9	3	2	24	11
Sept.	12	6	9	2	21	8
Oct.	56	18	22	1	78	19
Nov.	59	5	25	1	84	6
Dec.	16	7	2	0	18	7
Total	296	127	126	29	422	156
		= 43%		= 23%		= 37%

continued

1812	British Deserted	British Returned	Foreign Deserted	Foreign Returned	Total Deserted	Total Returned
Jan.	22	10	14	5	36	15
Feb.	28	28	2	6	30	34
Mar.	17	15	4	2	21	17
Apr.	27	8	62	8	89	16
May	21	11	20	4	41	15
June	29	17	23	8	52	25
July	57	11	32	1	89	12
Aug.	36	11	11	1	47	12
Sept.	42	10	6	4	48	14
Oct.	59	7	147	7	206	14
Nov.	167	0	134	0	301	0
Dec.	41	0	3	0	44	0
Total	546	128	458	46	1,004	174
		= 23%		= 10%		= 17%

Note: Data derived from Monthly Returns in TNA, WO17/2465, 2470–2471.
"British" is taken to include all units that were part of the regular line.
Although it is appreciated that the 5/60th and 97th did contain substantial
numbers of foreign nationals, and that other "British" units did also recruit
foreigners, the nature of the data prevents this sort of distinction being made.
"Foreign" deserters are therefore those belonging to KGL and émigré units.

Notes

Introduction

1. Sherer, *Recollections*, 5.
2. Reproduced in Austin, *1812: The March on Moscow*, 398–399.
3. Bowden, *Napoleon's Grande Armée*.

1. The British Army and Its Campaigns

1. See Linch, "Recruitment."
2. A more detailed account of the system, with case studies based on Moore's activities in 1808, can be found in Muir and Esdaile, "Strategic Planning," 1–90; see also Haythornthwaite, *Armies of Wellington*, 12–21, 267–268.
3. Muir, *Britain and the Defeat of Napoleon*, 44–45.
4. Gregory, *No Ordinary General.*
5. Partridge and Oliver, *Handbook*, 5–9; Linch, "Recruitment," 246–247.
6. Muir and Esdaile, "Strategic Planning," 8.
7. Petty, "Wellington's General Orders," 145–146.
8. There are thus two "monthly" returns for June 1809, but for the purposes of statistical analysis, these have been consolidated using the strengths from June 25 and combining the losses of both returns to cover the period from May 1 to June 24, 1809.
9. Adjutant General's Office, *General Regulations*, 266; see also ibid., 267–271 for detailed instructions as to how the forms were to be completed and what additional information was to be included.
10. See, for example, GO of October 9, 1810, in Gurwood, *General Orders* 2:179–181. Regarding Cadiz, see notes to the Monthly Returns for Portugal of April 25 and August 25, 1810, TNA, WO17/2465.
11. Adjutant General's Office, *General Regulations*, 261; italics in the original.
12. Adjutant General's Office, *General Regulations*, 260, 262.
13. Green, *Soldier's Life*, 17.
14. Sherer, *Recollections*, 5.
15. Adjutant General's Office, *General Regulations*, 279–288.
16. See, for example, Campbell to Calvert, July 3, 1811, TNA, WO27/102, part 2; Campbell to Calvert, September 15, 1812, TNA, WO27/111, part 1.

17. Haythornthwaite, *Armies of Wellington;* Partridge and Oliver, *Handbook,* 1–90.

18. Holmes, *Redcoat,* 106.

19. Haythornthwaite, *Armies of Wellington,* 274.

20. Haythornthwaite, *Armies of Wellington* 76–77; Fortescue, *British Army,* vol. 4, bk. 2, 931–934.

21. Oman, *Wellington's Army,* 178–179.

22. Haythornthwaite, *Armies of Wellington,* 274.

23. MacArthur, "British Army Establishments: Part 1," 154–159.

24. Glover, *Peninsular Preparation,* 232.

25. Sherer, *Recollections,* 180–181.

26. Adjutant General's Office, *General Regulations,* 72–78; see also Wellington to Bentinck, July 20, 1813, in Gurwood, *Dispatches* 10:552–555.

27. Grattan, *Adventures,* vi.

28. Inspection Report on 2/89th by Major General Burgoyne, June 29, 1811, TNA, WO27/102. For examples in Britain, see Inspection Report on 2/9th by Major General Manningham, May 4, 1808, TNA, WO27/92; and Inspection Report on 2/86th by Major General Anson, May 19, 1814, TNA, WO27/127, part 2.

29. Linch, "Recruitment," 42–55.

30. Returns of 2/91st, TNA, WO17/212, recording the dispatch of men to the first battalion.

31. Battalion Returns, TNA, WO17/146, 157. But see also "Troops Embarked from Cork under Lieut. General Sir Arthur Wellesley KB," TNA, WO17/2464, which gives slightly different figures and would imply that, at least on paper, these men were entered first into the respective second battalions.

32. Glover, *Peninsular Preparation,* 228–229; Fletcher, *"A Desperate Business,"* 20.

33. McGuigan, "Origin of Wellington's Peninsular Army," 40–41.

34. Cannon, *Historical Record of the Ninth,* 43–44.

35. Holmes, *Redcoat,* 52.

36. Dempsey, *Albuera,* 238.

37. See "Return of Deserters from the French Army enlisted in Portugal for the Undermentioned Corps to the 4th of July 1810," TNA, WO1/246, 155.

38. Wellington to Bathurst, November 9, 1813, in Gurwood, *Dispatches* 11:272–273.

39. Partridge and Oliver, *Handbook,* 93.

40. Glover, *Peninsular Preparation,* 81–93; Kiley, *Artillery of the Napoleonic Wars,* 157–180; Leslie, *Services of the Royal Regiment of Artillery.*

41. Monthly Return of August 25, 1809, TNA, WO17/2464.

42. Glover, *Peninsular Preparation,* 103–109; Haythornthwaite, *Armies of Wellington,* 114–117. On the Staff Corps of Cavalry, see Urban, *Man Who Broke Napoleon's Codes,* 256–257, 265–266.

43. Haythornthwaite, *Armies of Wellington,* 119–122.

44. Figures from TNA, WO17/2814.

45. Fortescue, *British Army* 6:41–47.

46. Glover, *Peninsular Preparation,* 184–185.

47. Gregory, *Sicily.* See also Fortescue, *British Army* 5:253–255; and Hall, *British Strategy,* 86–87.

48. Hopton, *Maida*; Fortescue, *British Army* 5:338–355; 6:5–27.

49. Muir, *Britain and the Defeat of Napoleon*, 37.

50. Fortescue, *British Army* 6:102–104, 113–115, 185–199, 122–137; Muir, *Britain and the Defeat of Napoleon*, 43–47; Oman, *Peninsular War* 1:220–300; Weller, *Wellington in the Peninsula*, 29–58.

51. Oman, *Peninsular War* 1:263–290, 625–628; Fortescue, *British Army* 6:235–250; Muir, *Britain and the Defeat of Napoleon*, 51–53.

52. Oman, *Peninsular War* 1:123–205, 334–375, 630–645; Muir, *Britain and the Defeat of Napoleon*, 60–62; Chandler, *Campaigns of Napoleon*, 612–625.

53. Summerville, *March of Death*; Oman, *Peninsular War* 1:486–602. On the exploits of Paget's cavalry, see also Fletcher, *Galloping at Everything*, 90–100.

54. *Times* (London), January 9 and 10, 1809. The confusion was between Marshal Lefebvre, Duke of Danzig, and Général de Division Charles Lefebvre-Desnouettes, who was captured at Benavente.

55. *Times* (London), January 28, 1809. See also Fortescue, *British Army* 7:28–31; and Muir, *Britain and the Defeat of Napoleon*, 87.

56. "Recapulation of Provisions and Stores, Lost, Taken, or Destroyed during the late Campaign in Spain; as far as can be ascertained from Documents in the office of John Erskine Esq. Commissary-in-Chief," *House of Commons Parliamentary Papers 1809* 11:252.

57. See Monthly Returns in TNA, WO17/2464.

58. Memorandum of March 7, 1809, in Gurwood, *Dispatches* 3:181–183.

59. Bond, *Grand Expedition*; Burnham, "The British Expeditionary Force."

60. Monthly Returns in TNA, WO17/2464, 2479.

61. See service histories in Partridge and Oliver, *Handbook*, 55–85; McGuigan, "Origin of Wellington's Peninsular Army," 61–70; and Oman, *Wellington's Army*, 343–346.

62. The Oporto campaign lacks a dedicated study, but see Glover, *Wellington as Military Commander*, 115–123; and Oman, *Peninsular War* 2:286–366.

63. Field, *Talavera*; Oman, *Peninsular War* 2:433–608; 3:105–110.

64. Costa, "Army Size"; Oman, *Peninsular War* 3:171–182.

65. Glover, *Wellington as Military Commander*, 80–83; Robertson, *Commanding Presence*, 139–158; Oman, *Peninsular War* 3:153–196.

66. Robertson, *Commanding Presence*, 159–177; Oman, *Peninsular War* 3:231–281, 341–481.

67. Monthly Return for March 25, 1811, TNA, WO17/2467 shows 48,026 British rank and file, of whom 35,983 were present for duty.

68. Oman, *Peninsular War* 4:23–63, 247–292.

69. Buttery, *Wellington against Massena*, 122–139; Oman *Peninsular War* 4:131–205. On the supply problems, see, for example, "Arrangements for 20th March [1811]," as communicated by the quartermaster general, in Gurwood, *Dispatches* 4:682.

70. Wellington to Wellesley-Pole, July 2, 1811, in Wellesley, *Supplementary Despatches* 7:176.

71. Glover, *Wellington as Military Commander*, 142–151; Oman, *Peninsular War* 4:288–362.

72. Dempsey, *Albuera*; comparative casualty figures are from Smith, *Napoleonic Wars Data Book*, 360–363.

73. Oman, *Peninsular War* 4:404–457, 542–582; 5:157–186.

74. See Returns in TNA, WO17/1486 (1810–1811) and WO17/1487 (1812). For more detail, see Bamford, "British Forces at Cadiz."

75. Oman, *Peninsular War* 4:91–130.

76. Barker, "Debacle of the Peninsular War."

77. The detachment of troops to Tarifa from Gibraltar, and later Cadiz, may be tracked through returns in TNA, WO17/1796–1800 and WO17/1486–1488 respectively. On the siege, see Oman, *Peninsular War* 5:111–139; and Musteen, "Nelson's Refuge and Wellington's Rock," 184, 200–216.

78. Oman, *Peninsular War* 4:593–607.

79. Oman, *Peninsular War* 5:215–279, 594–595; Glover, *Wellington as Military Commander*, 162–171.

80. Wellington to Liverpool, May 26, 1812, in Gurwood, *Dispatches* 9:172–178. The letter as a whole contains a detailed appreciation of the strategic situation after the fall of Badajoz, and the means by which the writer proposed to take advantage of it.

81. Muir, *Salamanca*; see also Oman, *Peninsular War* 5:297–518.

82. Oman, *Peninsular War* 6:1–180; Glover, *Wellington as Military Commander*, 171–177.

83. Calculation of losses comes from Glover, *Wellington as Military Commander*, 104–106. For Wellington's memorandum, addressed to "Officers Commanding Divisions and Brigades," see Gurwood, *Dispatches* 9:574–577.

84. See Monthly Returns in TNA, WO17/1516, 2241.

85. McGuigan, "British North America."

86. Latimer, *1812*, 47–48.

87. Elting, *Amateurs*, 19–58; Latimer, *1812*, 35–83.

88. See Anglo-Portuguese Morning States for May 1813, and Spanish returns for May–July, reproduced in Oman, *Peninsular War* 6:750–753.

89. Oman, *Peninsular War* 6:314–450, 754–762; Weller, *Wellington in the Peninsula*, 247–269.

90. Robertson, *Wellington Invades France*, 56–98; Oman, *Peninsular War* 6:523–556, 587–740.

91. Robertson, *Wellington Invades France*, 47–55, 99–111; Oman, *Peninsular War* 6:557–586; 7:8–36.

92. These campaigns warrant a study in their own right. In the absence of such, see Oman, *Peninsular War* 5:565–575, 609–610; 6:275–298, 488–522; and 7:63–109, 406–420, 429–432.

93. See Wellington to Bathurst, November 11, 1813, in Gurwood, *Dispatches* 11:275–277, proposing that the force be broken up; Wellington to Clinton, March 4, 1814, in Gurwood, *Dispatches* 11:544–546, with arrangements for its dispersal; and Weekly State for April 8, 1814, TNA WO17/2478, for details of the troops forwarded to Wellington.

94. Fortescue, *British Army* 10:60–65.

95. Monthly Return for November 25, 1813, TNA, WO17/2474.

96. Graves, *Field of Glory*; see also Elting, *Amateurs*, 136–155; and Latimer, *1812*, 195–216. On reinforcements, see Grodzinski, "Much to be Desired"; and McGuigan, "British North America."

97. Elting, *Amateurs*, 103–115; Latimer, *1812*, 174–192.

98. The task of this force is outlined in Bathurst to Gibbs, June 30, 1813, in Wellesley, *Supplementary Despatches* 8:47–48. See also Beamish, *History of the KGL* 2:140–154, 171–218; and Fortescue, *British Army* 9:387–388.

99. Monthly Return for February 25, 1814, TNA, WO17/1773.

100. This is another campaign meriting more study, but see Fortescue, *British Army* 10:1–11, 33–54; and Great Britain War Department, *British Minor Expeditions*, 80–88.

101. Figures from TNA, WO17/2814.

102. For example, Wellington to Bathurst, November 9, 1813, and March 11, 1814, in Gurwood, *Dispatches* 11:272–273 and 11:571–572.

103. Robertson, *Wellington Invades France*, 123–186; Oman, *Peninsular War* 7:110–340.

104. Robertson, *Wellington Invades France*, 187–249; Oman, *Peninsular War* 7:341–405, 433–512.

105. Wellington to Bathurst, April 7, 1814, in Gurwood, *Dispatches* 11:626–628.

106. Strengths from Monthly Returns, TNA, WO17/1518–1519, 2241–2243. Sherbrooke's operations are related in Latimer, *1812*, 345–348.

107. Graves, *Where Right and Glory Lead*. See also Elting, *Amateurs*, 175–197, 244–252; and Latimer, *1812*, 277–300.

108. Bathurst to Wellington, January 28, 1814, in Wellesley, *Supplementary Despatches* 8:547; York to Bathurst, April 13, 1814, TNA, WO1/657, 635; Wellington to Hill, May 5, 1814, in Gurwood, *Dispatches* 12:2; Wellington to Bathurst, May 5 1814, in Gurwood, *Dispatches* 12:2–3.

109. Elting, *Amateurs*, 252–263; Latimer, *1812*, 345–368.

110. Elting, *Amateurs*, 198–243; Latimer, *1812*, 301–344; strengths from TNA, WO17/1218.

111. Elting, *Amateurs*, 282–309; Latimer, *1812*, 369–388; strengths from TNA, WO17/1218.

112. Latimer, *1812*, 360–368, 389–392.

113. Beamish, *History of the KGL* 2:315–319; Lt. General Baron Linsingen to Duke of Cambridge, July 18, 1814, TNA, WO1/659, 153–155.

114. Fortescue, *British Army* 10:227–228.

115. One could provide a library's worth of references on this campaign. For the most relevant or provoking accounts, see Barbero, *The Battle*; Fletcher, *"A Desperate Business"*; Haythornthwaite, *Waterloo Armies*; Hofschröer, *1815*; and Robinson, *Quatre Bras*.

116. Figures from "Grand Total of the British, Dutch, Hanoverian, and Brunswick Troops Under the Command of the Duke of Wellington," in Wellesley, *Supplementary Despatches* 10:750.

117. Partridge and Oliver, *Handbook*, 89–90.

2. Regimental Identity and Leadership

1. Inspection Report of April 27, 1814, TNA, WO27/127, part 2.

2. Morris, *Recollections*, 146–149.

3. Haythornthwaite, *Armies of Wellington*, 271–274.

4. Holmes, *Redcoat*, 26, 403. For a discussion of the "Die hard!" legend, see Dempsey, *Albuera*, 291–295.

5. Cooper, *Rough Notes*, 68.

6. Cookson, *Armed Nation*, 8, 24–27.

7. Haythornthwaite, *Armies of Wellington*, 272–273.

8. Fusilier Regiments: 7th (Royal Fusiliers), 21st (Royal North British Fusiliers), 23rd (Royal Welch Fusiliers). Highland Regiments: 42nd (Royal Highland), 71st (Glasgow Highland), 72nd, 73rd, 74th, 75th, 78th (Ross-shire Buffs), 79th (Cameron Highlanders), 91st (Argyllshire), 92nd (Gordon Highlanders), 93rd (Sutherland Highlanders). See Partridge and Oliver, *Handbook*, 39–46; and Haythornthwaite, *Armies of Wellington*, 76.

9. Mackerlie, *Scottish Regiments*; Linch, "Recruitment," 178, 187–189.

10. Fassiefern, *Memoir of Colonel John Cameron*.

11. Glover, *Peninsular Preparation*, 128–129.

12. Mackesy, *British Victory in Egypt*, 36, 84, 90.

13. On the 20th, see Partridge and Oliver, *Handbook*, 22–23; and Oman, *Peninsular War* 1:645; on the 2/78th, see Reid, *Highland Warriors*, 180.

14. Urban, *Rifles*, 31–34; Haythornthwaite, *Armies of Wellington*, 93–96; Glover, *Peninsular Preparation*, 122–134.

15. Oman, *Wellington's Army*, 212. See also Urban, *Rifles*, which makes it clear that the 95th was also not free from such individuals.

16. Cook and Burnham, "Nicknames."

17. Quoted in Field, *Talavera*, 82. The "old Buffs" were, of course, the 3rd Foot.

18. Cook and Burnham, "Nicknames." A "Havercake" was a West Riding oatcake, allegedly offered to potential recruits, "Jaggers" was a corruption of "Jägers," and "Pot Hooks" refers to the similarity between a hook and a figure seven.

19. Ibid. Note, therefore, that the 11th's nickname predates their Victorian adoption of red legwear.

20. A distinction that has produced a not inconsiderable body of historiography; see, for example, Cookson, *Armed Nation*, as opposed to Colley, *Britons*.

21. Grattan, *Adventures*, 39–40.

22. Gleig, *Subaltern*, 61–67.

23. Haythornthwaite, *Die Hard*, 14–15, 35.

24. Quoted in Dempsey, *Albuera*, 291.

25. For example, Inspection Return of May 1, 1813, TNA, WO27/117.

26. Partridge and Oliver, *Handbook*, 44.

27. Le Couteur, *Merry Hearts*, 93–104.

28. Haythornthwaite, *Armies of Wellington*, 73.

29. Oatts, *Proud Heritage* 1:97–98, 101–103, 124.

30. Inspection Report of January 16, 1813, TNA, WO27/111, part 1.

31. McGuigan, "Wellington's Generals," 179.

32. Donaldson, *Eventful Life*, 129–131; emphasis in the original.

33. Coss, *King's Shilling*, 139–141.

34. GO of October 14, 1811, in Gurwood, *General Orders* 3:214–215.

35. For example, GO of September 30, 1810, in Gurwood, *General Orders* 2:170–172. See also Oman, *Wellington's Army*, 206–207.

36. GO of October 2, 1811, in Gurwood, *General Orders* 3:198–200.

37. Aitchison to father, October 7,1809, in Aitchison, *Letters*, 64–65.

38. Haythornthwaite, *Armies of Wellington*, 80. On the 1/3rd, see Dempsey, *Albuera*, 296–299.

39. Aitchison to father, July 31, 1809, in Aitchison, *Letters*, 55–57.

40. Grattan, *Adventures*, 136–137.

41. Calculations are based on data in Oman, *Peninsular War* 1:646–648.

42. Gordon, *Cavalry Officer*, 191.

43. Summerville, *March of Death*, 149–163.

44. Monthly Return for November 1812, TNA, WO17/2470.

45. This has arguably been a fault typical of elite units throughout history. See MacKenzie, *Revolutionary Armies*, 153–154.

46. "Journal by Sergeant Archibald Johnston, of the 2nd (or R.N.B.) Regiment of Dragoons," in Glover, *Waterloo Archive* 1:50. See also Fletcher, *Galloping at Everything*, 235–266.

47. Oman, *Peninsular War* 7:486–487, 559. Regarding concepts of precedence within the Highland corps, and the nature of the 42nd's attempts to secure and retain its preeminent status, see Reid, *Highland Warriors*, 33–54.

48. Battalion Return for May 1809, TNA, WO17/217; see also Cope, *Rifle Brigade*, 47.

49. Linch, "Recruitment," 200–201; the latter anecdote comes from Napier, *Military Life*, 21.

50. Cooper, *Rough Notes*, 3; *Memoirs of a Sergeant*, 21; Wheeler, *Letters*, 17.

51. The 4th Dragoon Guards, 6th Dragoons, 8th Hussars, and 18th, 27th, 86th–88th, and 99th–101st Regiments of Foot all bore Irish titles, whilst the 2nd Dragoons and 1st, 21st, 26th, 70th, 90th, and 94th Regiments of Foot, in addition to the Highland corps already noted, complete the list of Scots units. See Haythornthwaite, *Armies of Wellington*, 271–274.

52. Linch, "Recruitment," 180; Haythornthwaite, *Armies of Wellington*, 272.

53. Bois, "The Inniskillings at Waterloo."

54. Holmes, *Redcoat*, 54–55; Graves, *Dragon Rampant*.

55. Browne, *War Journal*, 73; Grattan, *Adventures*, 320–321.

56. Sherer, *Recollections*, 224; emphasis added.

57. Adjutant General's Office, *General Regulations*, 279.

58. Adjutant General's Office, *General Regulations*, 279–281.

59. Hay, *Reminiscences*, 44.

60. Ibid., 91–92, 156, 163.

61. Inspection Report of December 24, 1812, TNA, WO27/111, part 1.

62. James, *General Courts Martial*, 456–477. See also Burnham and McGuigan, "The Impeccable Timing of Sir George Brown"; Haythornthwaite, *Armies of Wellington*, 36; and Holmes, *Redcoat*, 173–174. For the improved condition of the unit under its new officers, see Inspection Report by Major General Mackenzie of May 31, 1813, TNA, WO27/117; and Gleig, *Subaltern*.

63. Gurney, *Trial of Colonel Quentin*; see also Haythornthwaite, *Armies of Wellington*, 36; and Oman, *Wellington's Army*, 248–249.

64. This trend is discussed with examples from the history of the 30th Foot—which seems to have enjoyed a largely consistent run of good commanders—in Divall, *Inside the Regiment*, 18–24.

65. Wheeler, *Letters*, 19, 21, 43–44, 55–56, 66.

66. On Inglis, see Woolright, *Fifty-Seventh*, 370–371. On Donellan, see Garvey, *Northamptonshire Regiment*, 133, citing an account in *The Naval and Military Sketch Book* of 1844; and Field, *Talavera*, 112–113. On Beckwith, see Urban, *Rifles*, 11–15, 27–31; and Surtees, *Rifle Brigade*, 53.

67. Ward, *Old Soldier*, 104.

68. Graves, "'Every Horror.'"

69. See Harris, *Recollections*, 201–202; and Urban, *Rifles*, 27–30.

70. Grattan, *Adventures*, 16–20; Myatt, *Peninsular General*, 78–80.

71. Inspection Report by Major General Drummond, May 9, 1809, TNA, WO27/94.

72. Adjutant General's Office, *General Regulations*, 125–126, 296; Hough, *Military Law*, 797–799.

73. Elting, *Swords*, 429–431, 508; Harris, *Recollections*, 280–282.

74. Lawrence, *Autobiography*, 48–50.

75. Harris, *Recollections*, 204–205; although there is the sneaking suspicion that these may be the views of his ex-officer ghostwriter.

76. Hoghton to Graham, July 6, 1810, TNA, WO27/99.

77. Hoghton to Graham, July 6, 1810, TNA, WO27/98.

78. Cope, *Rifle Brigade*, 66.

79. Inspection Report by Lt. General Sir John Doyle, May 4, 1808, in TNA, WO27/92; emphasis in the original. In partial qualification, it should be noted that Sir John was both Colonel of the 87th and uncle to Charles Doyle.

80. Hoghton to Graham, July 6, 1810, TNA, WO27/99.

81. Circular letter dated 25, March 1812, quoted in Marshall, *Military Miscellany*, 184–185.

82. On the politico-legal element, see Steiner, "Separating the Soldier from the Citizen."

83. Cook and Burnham, "Nicknames."

84. Smythies, *40th Regiment*, 134.

85. *General Courts Martial*, 495.

86. Ibid., 495–496.

87. Ibid., 496.

88. Hay to Wellington, March 2, 1813, in James, *General Courts Martial*, 496–497.

89. Smythies, *40th Regiment*, 137–138.

90. Hough, *Military Law*, 797–798.

91. Ibid., 798. I am much obliged to Ron McGuigan for information on Archdall's second career.

92. *Gazette* (London), March 15, 1808, 386.

93. Inspection Report of May 12, 1808, TNA, WO27/92.

94. Oman, *Peninsular War* 1:646.

95. Inspection Return of November 26, 1810, TNA, WO27/100.

96. Inspection Report of November 26, 1810, TNA, WO27/100.

97. Inspection Report of May 12, 1808, TNA, WO27/92.

98. Wellington to Liverpool, May 15, 1811, in Gurwood, *Dispatches* 7:565–567.

99. *Gazette* (London), May 7, 1811, 828.

100. Inspection Report of May 2, 1812, TNA, WO27/106, part 2.

101. Ibid. Although here referred to as a major, Kingsbury was a brevet lt. colonel.

102. Ibid.

103. Ibid.

104. Clinton to Manners, May 12, 1812, TNA, WO27/106, part 1.

105. Obituary in *The Gentleman's Magazine* 5 (January 1836), 88–90.

106. Cannon, *Historical Record of the Third Light Dragoons*, 51.

107. "Remarks" to Monthly Returns August–October 1812, TNA, WO17/2470; Brown, "British Regiments."

108. Bingham to mother, December 12, 1812, in Bingham, *Wellington's Lieutenant*, 168–169.

109. Inspection Report of May 12, 1808, TNA, WO27/92.

110. Details of the case, and background, are from Oman, *Peninsular War* 7:270, 280. Sir Nathaniel is not to be confused with Warren Peacocke, the distinguished and effective commandant of Lisbon.

111. Wellington to Torrens, March 8, 1814, cited in Millar, "Further Notes"; see also York to Wellington, March 23, 1814, in James, *General Courts Martial*, 611–612.

112. Elting, *Swords*, 179; Muir, *Experience of Battle*, 177.

113. Inspection Report of May 11, 1813, TNA, WO27/116.

114. Wheeler, *Letters*, 66.

115. Millar, "Duncan Macdonald" and "Further Notes."

116. Napier, *Military Life*, 264–268.

117. Inspection Report of October 17, 1814, TNA, WO27/130, part 1.

118. Napier, *Military Life*, 265–266: Denis Pack, by then a major general, had commanded the battalion before Cadogan; Victorian delicacy in the original replaces Peacocke's name with an ellipsis.

119. Barker, "Debacle of the Peninsular War"; Oman, *Peninsular War* 3:335–337.

120. Inspection Report by Major General Burgoyne, December 9, 1810, TNA, WO27/101.

121. Inspection Report of June 29, 1811, TNA, WO27/102, part 2.

122. Campbell to Calvert, July 3, 1811, TNA, WO27/102, part 2.

123. Graves, *Where Right and Glory Lead*, 180.

124. Inspection Report of October 29, 1811, TNA, WO27/104, part 2.

125. Graves, *Field of Glory*, 134–136, 262–264.

126. Graves, *Where Right and Glory Lead*, 180–181.

127. Graves, *Where Right and Glory Lead*, 67.

128. Cooper, *Rough Notes*, 64.

3. The Regimental System in Practice

1. Inspection Report of May 14, 1808, TNA, WO27/92, part 2.

2. Monthly Returns, TNA, WO17/2470–2476.

3. As, for example, in Oman, *Wellington's Army*, 178–181, where the "rules" as the author understood them are laid down.

4. Partridge and Oliver, *Handbook*, 60–81.

5. Partridge and Oliver, *Handbook*, 71.

6. Battalion Return of October 1, 1808, TNA, WO17/154.

7. Monthly Returns, TNA, WO17/2464.

8. "Troops Embarked from Cork Under the Command of Lieut. General Sir Arthur Wellesley KB," TNA, WO17/2464.

9. Monthly Return, TNA, WO17/2464.

10. Lawrence, *Autobiography*, 47.

11. Oman, *Wellington's Army*, 344.

12. Linch, "Recruitment," 204.

13. Willis, "George Lake."

14. Casualty figures from Oman, *Peninsular War* 1:240; embarkation strength from McGuigan, "Origin of Wellington's Peninsular Army," 41; September figures from Monthly Return, TNA, WO17/2464.

15. Sherer, *Recollections*, 61.

16. Wellington to Castlereagh, September 12, 1809, in Gurwood, *Dispatches* 5:146–147.

17. Oman, *Wellington's Army*, 353–354. Figures from Monthly Returns, TNA, WO17/2467–2468.

18. GO of October 3, 1811, *General Orders* 3:201–202.

19. Inspection Report of May 14, 1813, TNA, WO27/116.

20. Partridge and Oliver, *Handbook*, 60–81.

21. Latimer, *1812*, 299; Le Couteur, *Merry Hearts*, 196.

22. Monthly Returns, TNA, WO17/1516–1519.

23. Oman, *Wellington's Army*, 338, 345, 348.

24. Cassidy, *Marching with Wellington*, xvii–xix, 1–6; Bois, "The Inniskillings at Waterloo"; Oman, *Wellington's Army*, 334.

25. York to Wellington, January 10, 1814, in Wellesley, *Supplementary Despatches* 3:495–498.

26. Latimer, *1812*, 186–192, 225.

27. Monthly Returns, TNA, WO17/1517–1518.

28. Fortescue, *British Army* 7:190; Jourdain and Fraser, *Connaught Rangers* 1:57. On the issue of inexperience, an Inspection Return of June 14, 1810 (TNA, WO27/98) shows 594 of 672 rank and file with two years' service or less.

29. Oman, *Wellington's Army*, 346–349.

30. Ibid., 349.

31. Monthly Return, TNA, WO17/2467.

32. GO of July 10, 1811, in Gurwood, *Dispatches* 5:146.

33. Jourdain and Fraser, *Connaught Rangers* 1:59.

34. GO of November 17, 1810, in Gurwood, *General Orders* 2:207–208.

35. Oman, *Wellington's Army*, 189, 354.

36. Monthly Returns, TNA, WO17/1760.

37. McGuigan, "Anglo-Allied Army."

38. MacArthur, "British Army Establishments: Part 2," 331–337.

39. Oman, *Wellington's Army*, 191.

40. Fletcher, *Galloping at Everything*, 26.

41. MacArthur, "British Army Establishments: Part 2," 337–340.

42. Oman, *Wellington's Army*, 192. Oman's misconception may stem from the fact that, in his day, the Household Brigade frequently *did* deploy only a single combined regiment when required for active service. See White-Spunner, *Horse Guards*, 391, 413–415.

43. Oman, *Wellington's Army*, 365–367, 371; Fletcher, *Galloping at Everything*, 238.

44. This was the case with the 12th Light Dragoons; see Steuart to Captain James Bridger, July 12, 1811, in Bound letters of Colonel Sir James Stewart [Steuart].

45. "Detachments proposed to be sent to the Force under the Command of the Marquess of Wellington by the 1st February next," January 2, 1814, TNA, WO1/657, 13–14.

46. Wellington to Cotton, September 17, 1811, in Gurwood, *Dispatches* 5:279; see also Oman, *Wellington's Army*, 193.

47. Wellington to York, December 26, 1812, in Gurwood, *Dispatches* 10:5. See also Gerges, "Command and Control," 342–347; and Oman, *Peninsular War* 6:232.

48. Tomkinson, *Diary*, 2; Monthly Return for May 1809, TNA, WO17/2464.

49. Monthly Return for October 1811, TNA, WO17/2468.

50. Tomkinson, *Diary*, 273; Monthly Return for May 1815, TNA, WO17/1760.

51. Wellington to Cotton, October 7, 1811, in Gurwood, *Dispatches* 8:326.

52. Tomkinson, *Diary*, 7, 22–23, 98, 225, 228–229.

53. Tomkinson, *Diary*, 120–121.

54. Guy, *Oeconomy and Discipline*, 118; Houlding, *Fit For Service*, 4–23; "Tables of Relief," TNA, WO379/1.

55. Wellington to Liverpool, January 19, 1811, TNA, WO1/248, 71.

56. Wellington to Peacocke, January 24, 1810, in Gurwood, *Dispatches* 5:440–442.

57. Oman, *Wellington's Army*, 354, 361, 368.

58. GO of September 24, 1809, in Gurwood, *General Orders* 1:184–186.

59. Oman, *Wellington's Army*, 345–346, 348, 354, 360; Wellington to Peacocke, August 30, 1810, in Gurwood, *Dispatches* 6:370.

60. Monthly Return for August 1809, TNA, WO17/2464.

61. Monthly Returns, TNA, WO17/2464, 2465.

62. "Memorandum of Instruction," July 20, 1813, in Gurwood, *Dispatches* 10:548–549; Wellington to Hope, November 5, 1813, in Gurwood, *Dispatches* 11:255. See also Bathurst to Wellington, July 5, 1813, in Wellesley, *Supplementary Despatches* 8:60, which notes that the 2/84th was originally intended to relieve the 1/37th at Gibraltar.

63. "Memorandum of Instruction," July 20, 1813, in Gurwood, *Dispatches* 10:548–549; Oman, *Wellington's Army*, 368.

64. Wellington to Hope, October 18, 1813, in Gurwood, *Dispatches* 11:203. Hope seems slightly to have missed the point, since Wellington had to reiterate himself two days later; see ibid., 208.

65. Wellington to Hope, March 2, 1814, in Gurwood, *Dispatches* 11:541–542.

66. Gleig, *Subaltern*.

67. McGuigan, "Anglo-Allied Army."

68. Monthly Returns, TNA, WO17/2472–2476.

69. Beamish, *History of the KGL* 2:157.

70. Monthly Return for November 1813, TNA, WO17/2241.

71. Monthly Returns, TNA, WO17/2241–2243.

72. Dunn, "Royal Veterans."

73. Wellington to Frere, February 5, 1810, in Gurwood, *Dispatches* 5:488–489; Wellington to William Stewart, February 27, 1810, in Gurwood, *Dispatches* 5:523–525; notes to Monthly Returns for June and July 1810, TNA, WO17/2465.

74. See, for example, Wellington to Campbell and Wellington to Bathurst, both December 26, 1812, in Gurwood, *Dispatches* 10:4 and 10:7.

75. Monthly Returns, TNA, WO17/2464.

76. Wellington to Cradock, May 27, 1809, in Gurwood, *Dispatches* 4:338; Oman, *Wellington's Army*, 345.

77. McGuigan, "Origin of Wellington's Peninsular Army," 48; Oman, *Wellington's Army*, 346.

78. Muir, *Britain and the Defeat of Napoleon*, 115–116; Oman, *Peninsular War* 3:128–152, 315–326.

79. Hall, *Wellington's Navy*, 114. See also "Journal of the Proceedings of the British Army serving at Cadiz and the I. of Leon," entry of September 23, 1810, TNA, WO28/339.

80. Oman, *Peninsular War* 3:145–148; 5:537–539.

81. Report accompanying inspection return of June 12 1810, TNA, WO27/98.

82. Divall, *Redcoats against Napoleon*, 38–50.

83. The composition of the Cadiz Garrison may be tracked by reference to the "Journal of the Proceedings of the British Army," TNA, WO28/339–341, and correspondence, TNA, WO1/247, 252, 264–266, which are together summarized in Bamford, "British Forces at Cadiz."

84. Graham to Liverpool, September 13, 1810, TNA, WO1/247, 583.

85. "Journal of the Proceedings of the British Army," September 23, 1810, TNA, WO28/340.

86. With respect to instructions from home see, for example, Wellington to Liverpool, September 12, 1810, TNA, WO1/246, 529–530.

87. Wellington to Cooke, September 9, 1812, TNA, WO1/265. See also Oman, *Peninsular War* 5:539–541; 6:95.

88. Wellington to Donkin, June 23, 1809, in Gurwood, *Dispatches* 4:431–432; Donkin to Gough, September 15, 1809, in Cunliffe, *Royal Irish Fusiliers*, 38. See also Oman, *Peninsular War* 2:649–651.

89. Wellington to William Stewart, February 7, 1810, in Gurwood, *Dispatches* 5:489–490.

90. Oman, *Wellington's Army*, 363, 369; Monthly Return, TNA, WO17/2476.

91. Stanhope, *Letters and Journals*, 106.

92. Monthly Returns, TNA, WO17/2470–2472.

93. Musteen, "Nelson's Refuge and Wellington's Rock," 218.

94. Reports accompanying Monthly Returns for July 1813 through May 1814, TNA, WO17/1799–1800.

95. Inspection Report by Major General Widdrington, December 10, 1811, TNA, WO27/111, part 2.

96. Battalion Return for April 1813, TNA, WO17/259; Monthly Return for March 1813, TNA, WO17/1799.

4. The Limits of the System

1. York to Bathurst, March 14, 1814, TNA, WO1/657, 455, quoted in full on pages 170–171.

2. Muir, *Britain and the Defeat of Napoleon*, 89–90, 101–106. The extent to which Dundas kept out of strategic decisions is reinforced by the fact that Muir mentions him on only four occasions.

3. Cameron to Craddock, January 16, 1809, quoted Burnham, "Filling the Ranks," 202.

4. Burnham, "Filling the Ranks," 203.

5. Ibid.

6. Wellesley to Castlereagh, 12, May 1809, in Gurwood, *Dispatches* 4:297–301.

7. Stewart to Castlereagh, June 15, 1809, quoted Field, *Talavera*, 14.

8. GO of September 14, 1809, in Gurwood, *General Orders* 1:168–170.

9. GO of September 22, 1809, in Gurwood, *General Orders* 1:179–181.

10. Dundas to Castlereagh, July 7, 1809, TNA, WO1/641, 233–234.

11. Monthly Returns, TNA, WO17/2464–2465, 2467. From November 1809 the detachments from each battalion are listed separately, implying the existence of two companies.

12. McGuigan, "Origin of Wellington's Peninsular Army," 68; Oman, *Wellington's Army*, 344.

13. Return for the 9th's depot at Canterbury, August 25, 1809, TNA, WO17/110, lists the arrival of 126 men joined from "Sick Absent in Spain and Portugal." See TNA, WO17/2464 for strengths whilst in Spain with 2/9th.

14. Dundas to Castlereagh, May 10, 1809, submitting "Return of all the Battalions of the Line now in England, with the effective strength of each," TNA, WO1/641, 49–50

15. TNA, WO1/641, 77; this led to the dispatch of seven battalions, detailed in Dundas to Castlereagh, May 27, 1809, TNA, WO1/641, 109.

16. Dundas to Castlereagh, May 13, 1809, TNA, WO1/641, 59–60.

17. TNA, WO1/641, 171–180.

18. Dispatch of July 3, 1809, TNA, WO1/641, 221–224; signatory and recipient illegible.

19. "List of the several corps, General and Staff Officers, comprising a large Division of His Majesty's Army, to be employed upon a Particular service," July 7, 1809, *House of Commons Parliamentary Papers 1809* 15:21–24. See also Bond, *Grand Expedition*, 168; and Burnham, "British Expeditionary Force."

20. Monthly Return, TNA, WO17/2479.

21. Battalion Returns, TNA, WO17/234.

22. Partridge and Oliver, *Handbook*, 62–79.

23. MacVeigh, *78th Highlanders*, 127.

24. On the inadequacy of British strategy for 1809, and the overkill inherent in the numbers of troops sent to Walcheren, see Muir, *Britain and the Defeat of Napoleon*, 87–92, 101–104.

25. Muir, *Britain and the Defeat of Napoleon*, 110–113.

26. Aubrey-Fletcher, *Foot Guards*, 184.

27. The initial proposal was for the flank battalions to have only the companies from the second battalions, but in practice the flank companies of the First Guards Brigade were also added; see Monthly Return for August 1809, TNA, WO17/2479.

28. Hamilton, *Grenadier Guards* 2:410.

29. "Disposition of His Majesty's Land Forces; and the Foreign Troops in British Pay—1776," TNA, WO379/1; Fortescue, *British Army* 3:178; Great Britain War Department, *British Minor Expeditions*, 26.

30. Dilkes to Graham, 16, June 1810, TNA, WO27/98.

31. Monthly Returns for April and May 1811, TNA, WO17/1486, recording transfer of 210 men from 2/1st Foot Guards to 3/1st Foot Guards.

32. Oman, *Peninsular War* 4:112–113.

33. Compare Oman, *Peninsular War* 4:612 with Monthly Returns, TNA, WO17/1486.

34. Oman, *Wellington's Army*, 346–349; Bamford, "British Forces at Cadiz."

35. Urban, *Rifles*, 53, 185–186; Oman, *Wellington's Army*, 357–372.

36. GO of June 26, 1811, in Gurwood, *General Orders* 3:115–117.

37. Battalion Returns, TNA, WO17/126, and Monthly Return for August 1809, TNA, WO17/2479; see also Graves, *Dragon Rampant*, 103–107.

38. Strengths respectively from "State of the Force about to proceed on Foreign Service," Horse Guards, June 22, 1810, *House of Commons Parliamentary Papers 1809* 15:16–17; "List of the several Corps," 21–23; "Return of the Regiments now under orders for Foreign Service, showing the probable number of Rank and File, which will embark with each corps, leaving behind such men as are at present unfit for duty," Adjutant General's Office, July 15, 1809, in Great Britain War Department, *British Minor Expeditions*, 76–77.

39. Cannon, *Historical Record of the Eighth*, 81–83.

40. Inspection Report by Major General Sir John Murray, October 25, 1810, TNA, WO27/100.

41. Battalion Returns, TNA, WO17/206; Monthly Return for March 1811, TNA, WO17/2467; see also Barrett, *85th*, 59.

42. Oatts, *Proud Heritage* 1:93.

43. Latimer, *1812*, 215, 277–279.

44. York to Wellington, January 20, 1813, in Wellesley, *Supplementary Despatches* 7:528–529.

45. Fletcher, *Bloody Albuera*, 100; Holmes, *Redcoat*, 106.

46. Wellington to Liverpool, June 6, 1811, and Wellington to York, June 27, 1811, in Gurwood, *Dispatches* 5:75–76 and 5:120–121; and GOs of June 6 and 26, 1811, in Gurwood, *Dispatches* 5:74 and 5:118.

47. Inspection Report of May 12, 1813, TNA, WO17/117.

48. Wellington to Liverpool, May 7, 1811, in Gurwood, *Dispatches* 7:521–525.

49. Ibid.

50. "Confidential Half-Yearly Report of the Provisional Battalion composed of the 2nd Battalions 31st and 66th Regiments, Villafranca, May 15, 1812," accompanying Inspection Return for 2/31st, TNA, WO27/106, part 1.

51. l'Estrange, *Recollections*, 165.

52. GO of December 6, 1812, in Gurwood, *General Orders* 4:229–235.

53. Oman, *Wellington's Army*, 345–364; Holmes, *Redcoat*, 403.

54. Burnham, "Filling the Ranks," 216–219.

55. Monthly Return, TNA, WO17/2470.

56. "List of Courts Martial of the 2nd Provisional Battalion," filed with 1813 first half-yearly inspection returns for the 2/53rd, TNA, WO17/117.

57. See Muir, "Order of Battle," 150–157, in respect of the deployment of Beresford's corps at Toulouse.

58. Wellington to Bathurst, August 11, 1813, in Gurwood, *Dispatches* 6:674–675.

59. Monthly Returns, TNA, WO17/2470.

60. Fifty-six wounded and six dead; see Oman, *Peninsular War* 7:554.

61. See, generally, Oman, *Peninsular War* 1:225, and on this particular issue, see ibid., 6:223–224, 231–235.

62. Muir, *Britain and the Defeat of Napoleon*, 276.

63. Oman, *Peninsular War* 6:232–233; this rendering is a fraction of the length of the original, and even those passages presented as direct quotations are only loosely based on the text in Wellesley, *Supplementary Despatches*, which is nevertheless cited by Oman as their origin.

64. York to Wellington, January 13, 1813, in Wellesley, *Supplementary Despatches* 7:524–525.

65. Ibid.

66. Ibid.

67. Ibid.

68. Ibid.

69. York to Wellington, January 10, 1814, in Wellesley, *Supplementary Despatches* 8:495–496.

70. Enclosure accompanying York to Wellington, January 10, 1814, in Wellesley, *Supplementary Despatches* 8:497–498.

71. See, for example, correspondence with the War Office, TNA, WO1/657–658.

72. Inspection Return and Report, October 21, 1813, TNA, WO27/122, part 1; Monthly Return of July 1813, TNA, WO17/1773.

73. Report of October 10, 1813, accompanying return, TNA, WO27/122, part 1.

74. Report of October 12, 1813, accompanying return, TNA, WO27/122, part 1.

75. Beamish, *History of the KGL* 2:173; Smith, *Leipzig*, 210–211.

76. Wallmoden to Graham, January 18, 1814, TNA, WO1/199, 590.

77. Beamish, *History of the KGL* 2:143, 171–172, 213–214; returns, TNA, WO17/270.

78. Linch, "Recruitment," 250–251, not including the incorporation of the 102nd through 104th into the line, or the two battalions of foreigners added to the 60th.

79. York to Bathurst, January 17, 1814, TNA, WO1/658, 89–92.

80. Bathurst to Graham, nos. 1 to 4, December 4, 1813, TNA, WO1/199, 29–49.

81. Memorandum dated November 21, 1813, TNA, WO6/16, 18–19.

82. Hamilton, *Grenadier Guards* 2:484–485.

83. Battalion Returns, TNA, WO17/262–263, 267–268.

84. Battalion Returns, TNA, WO17/262–263.

85. Cope, *Rifle Brigade*, 175–179.

86. Inspection Report by Major General Wilder, November 4, 1813, TNA, WO27/121, part 1.

87. Inspection Report by Major General Mackenzie, October 8, 1813, TNA, WO122, part 1.

88. Report of October 1, 1813, TNA, WO27/121, part 2; Report of October 27, 1813, TNA, WO27/122, part 1.

89. Graham to Bathurst, December 27, 1813, TNA, WO1/199, 277–280.

90. Torrens to Bunbury, 19, December 1813, TNA, WO1/198, 39.

91. Inspection Reports, TNA, WO27/121–122; Divall, *Redcoats against Napoleon*, 124–129.

92. Monthly Return, TNA, WO17/1773.

93. Enclosure accompanying Graham to Bathurst, 15, April 1814, TNA, WO1/200, 557; Inspection Report for 2/21st, 21, May 1814, TNA, WO27/126, part 2.

94. "Statement of Forces in Holland and Under Orders," January 4, 1814, TNA, WO1/657; Inspection Report for 33rd, April 25, 1814, TNA, WO27/126, part 2.

95. York to Bathurst, October 18, 1813, TNA, WO1/657; Monthly Return for January 1814, TNA, WO17/1488.

96. Graham to Liverpool, January 3, 1811, no. 2, TNA, WO1/252, 5–7.

97. Abstract of Courts Martial for 7/60th, TNA, WO27/127.

98. Calvert to Doyle, November 23, 1813, TNA, WO3/60, 225.

99. Inspection Report by Major General Darroch, TNA, WO27/127.

100. Inspection Reports by Major General Darroch on 7/60th, October 31st, 1814, and Major General Widdrington on 8/60th, January 19, 1815, TNA, WO27/130, part 1.

101. Histman and Sorby, "Independent Foreigners," 16, quoting Beckwith to Admiral Sir John Borlase Warren, July 5, 1813.

102. Graves, "'Every Horror'"; Histman and Sorby, "Independent Foreigners," 14–15.

103. Inspection Report by Major General Horsforth, May 4, 1814, TNA, WO27/127, part 2.

104. "Remarks" to Monthly Return, TNA, WO17/2470.

105. Burnham, "Filling the Ranks," 210–215.

106. Inspection Report by Gibbs, October 12, 1813, TNA, WO27/122, part 1.

107. Return of troops at Breda, February 28, 1814, TNA, WO1/200, 37.

108. The battalion had 580 total rank and file in February and 611 in March, with 29 deaths and 1 desertion to be factored in; Monthly Returns, TNA, WO17/1773. Regarding the draft from Britain, see "Statement of Forces in Holland," January 4, 1814, TNA, WO1/658, 25.

109. Report by Lt. General Ferguson on 33rd and 54th Regiments, April 25, 1814, accompanying Inspection Returns, TNA, WO27/126, part 2.

110. Castlereagh to Clancarty, December 8, 1813, TNA, WO1/198, 217–219.

111. Graham to Bathurst, December 19 and 26, 1813, TNA, WO1/199, 163–169 and 239–244 respectively.

112. Graham to Bathurst, January 3, 1814, TNA, WO/199, 347–350.

113. Inspection Returns, TNA, WO27/126, 127.

114. Inspection Report by Major General Cooke, May 21, 1814, TNA, WO27/126, part 2.

115. Inspection Report by Brigadier General Halkett, May 17, 1814, TNA, WO27/127, part 1.

116. Inspection Report by Major General Anson, May 19, 1814, TNA, WO27/127, part 2.

117. Bathurst to Wellington, September 11, 1813, in Wellesley, *Supplementary Despatches* 8:249; see also Fortescue, *British Army* 9:416–420.

118. Haythornthwaite, *Armies of Wellington*, 173.

119. They were still slated to go to Wellington as of March 10, and were again reassigned before the end of the month. See Bathurst to Wellington, March 10, and 29, 1814, in Wellesley, *Supplementary Despatches* 8:635–636 and 8:702–703; and Bunbury to Graham, April 1, 1814, TNA, WO6/16, 81–82.

120. Burnham, "Filling the Ranks," 221. See also Haythornthwaite, *Armies of Wellington*, 176, and, regarding the detachment of an NCO from the 95th to drill Wynn's Rifle Companies, Calvert to Rosslyn, February 17, 1814, TNA, WO3/60, 414.

121. Torrens to Wellington, January 26, 1814, in Wellesley, *Supplementary Despatches* 8:544.

122. "Embarkation Return of the Provisional Battalions of Militia under the Command of Major General Bayley at Pouillac on the 22, and 23, days of May 1814," TNA, WO17/2476.

123. Burnham, "Filling the Ranks," 221–223.

124. Wellington to Bathurst, April 7, 1814, in Gurwood, *Dispatches* 11:626–628.

125. York to Bathurst, March 14, 1814, TNA, WO1/657, 455.

126. Circular to General Officers commanding relevant districts, March 15, 1814, TNA, WO3/60, 475.

127. Calvert to Green, March 15, 1814, TNA, WO3/60, 483.

128. Notes to Battalion Return for March 1814, TNA, WO17/277.

129. Notes to Battalion Return for March 1814, TNA, WO17/274.

130. Darling to Acland, March 26, 1814, TNA, WO3/61, 4–5.

131. Calvert to Chatham, March 18, 1814, TNA, WO3/60, 485; Calvert to Richmond, March 18 and 23, 1814, TNA, WO3/60, 485 and 494; Calvert to Don, March 31, 1814, TNA, WO3/61, 16.

132. Bathurst to Graham, March 17, 1814, TNA, WO6/16, 69–71.

133. Calvert to Don, April 12, 1814, TNA, WO3/61, 48; Calvert to Chatham, April 14, 1814, TNA, WO3/61, 57.

134. However, see Oman, *Peninsular War* 6:233.

135. Hall, *British Strategy*, 195.

136. Oman, *Peninsular War* 6:558–561; Muir, *Britain and the Defeat of Napoleon*, 197; Hall, *British Strategy*, 47, 94.

5. Beyond the Regiment

1. Cole to Wellington, May 28, 1815, in Wellesley, *Supplementary Despatches* 10:389–390.

2. Graham to Bunbury, January 15, 1814, TNA, WO1/199, 459–464.

3. On eighteenth-century doctrines of organization by unit seniority, see also Muir, "Order of Battle," 98–101.

4. Ross, "Combat Division," 84–94; Boycott-Brown, *Road to Rivoli*, 47–61.

5. Mackesy, *Statesmen at War*, 238, 258, 320; Great Britain War Department, *British Minor Expeditions*, 34–35, 43–44.

6. Smith, *Napoleonic Wars Data Book*, 195–196.

7. Weller, *Wellington in India*, 140–142, 202, 211, 301–303.

8. "Memorandum of Troops arrived in the Weser on and previous to the 1, January 1806," TNA, WO1/186, 441; see also Fortescue, *British Army* 5:285–298.

9. Great Britain War Department, *British Minor Expeditions*, 55.

10. Fortescue, *British Army* 5:306–307, 339; 6:7; Hopton, *Maida*, 96–99.

11. Fletcher, *Waters of Oblivion*, 73, 133–134.

12. McGuigan, "Origin of Wellington's Peninsular Army," 41–42. This excellent study details the British order of battle in the Peninsula to the end of the Oporto campaign, from which point the record is picked up by C. T. Atkinson's Appendix 2, "Divisional and Brigade Organisation and Changes 1809–1814," in Oman, *Wellington's Army*, 343–373.

This chapter makes extensive use of McGuigan and Atkinson to provide details of organizational changes, supplemented where necessary by other sources in order to fill the very small number of gaps in their data.

13. Oman, *Peninsular War* 1:220–262. See also Hadaway, "Rolica: A Most Important Affair."

14. McGuigan, "Origin of Wellington's Peninsular Army," 43.

15. Ibid., 41–42, 44, 64–68.

16. Wellesley remained chief secretary of Ireland until his reappointment to Portugal in 1809. Longford, *Wellington: Years of the Sword*, 220.

17. Chart filed with returns, TNA, WO17/2464. Although undated, this chart was drawn up before the arrival of the 1/42nd, which is a penciled addition, but after the arrival of Beresford since he appears in the original replacing Barnard Bowes. This places it around September 1, 1808.

18. McGuigan, "Origin of Wellington's Peninsular Army," 48–50.

19. Oman, *Peninsular War* 1:287–290.

20. Ibid., 287.

21. The 51st and 1/71st, also with Moore, were not re-designated as light infantry until *after* their return from Corunna.

22. McGuigan, "Origin of Wellington's Peninsular Army," 55–58.

23. Great Britain War Department, *British Minor Expeditions*, 57–79; Burnham, "British Expeditionary Force." Not all accounts give Paget's command its divisional number, possibly because it was soon broken up.

24. Bond, *Grand Expedition*, 167–168.

25. Return of September 7, 1809, in Great Britain War Department, *British Minor Expeditions*, 78; see also Monthly Return, September 25, 1809, TNA, WO17/2466.

26. McGuigan, "Origin of Wellington's Peninsular Army," 40–44.

27. Haythornthwaite, *Armies of Wellington*, 156–157.

28. Memorandum of March 19, 1808, WO17/2464. See also McGuigan, "Origin of Wellington's Peninsular Army," 65.

29. AGO of May 8, 1809, in Gurwood, *General Orders* 1:24–25.

30. Wellesley to Castlereagh, May 12, 1809, in Gurwood, *Dispatches* 4:297–301.

31. GO of May 4, 1809, in Gurwood, *General Orders* 1:9–14; McGuigan, "Origin of Wellington's Peninsular Army," 69.

32. GO of June 18, 1809, in Gurwood, *General Orders* 1:70–72.

33. Strictly speaking, Hill's appointment was also temporary, but was made permanent five days later, and Major General Tilson posted to command Hill's old brigade. See AGO of June 23, 1809, in Gurwood, *General Orders* 1:80.

34. Oman, *Wellington's Army*, 345–346. But for a qualification of the status of Cole, and later Picton with the Third Division, see Muir, "Order of Battle," 134–135.

35. Wellesley to Castlereagh, June 26, 1809, in Gurwood, *Dispatches* 4:438–439. See also Wellesley to Cotton, June 23, 1809, in Gurwood, *Dispatches* 4:432–433; and Dundas to Castlereagh, August 4, 1809, TNA, WO1/641, 281.

36. Oman, *Wellington's Army*, 344. In qualification of Oman's stance on seniority of Brigades within a Division, see also Graves, "'We Have an Immediate Opening.'"

37. Oman, *Peninsular War* 2:435–438.

38. Fletcher, *Galloping at Everything*, 102.

39. Coss, *King's Shilling*, 130.

40. McGuigan, "Wellington's Generals," 195; Fletcher, *Galloping at Everything*, 109.

41. By 1814 there were five British and four Portuguese battalions, two of the latter being from a line regiment; Oman, *Wellington's Army*, 349–373.

42. GO of February 22, 1810, in Gurwood, *General Orders* 2:24–26; see also Myatt, *Peninsular General*, 76–78.

43. Muir, "Order of Battle," 138.

44. McGuigan, "Wellington's Generals," 194, 196; Centeno, "General Officers in the Portuguese Army"; Moon, *Wellington's Two-Front War*, 191.

45. Wellington to Beresford, April 1, 1811, *Dispatches* 8:424–425. It is to be inferred that Hoghton's men would have joined the Portuguese Division, which would then presumably have become the Eighth Division.

46. Oman, *Wellington's Army*, 353–354.

47. Wellington to Beresford, July 17, 1811, in Gurwood, *Dispatches* 5:162; the opinion is reiterated in Wellington to Bathurst, May 5, 1815, in Gurwood, *Dispatches* 12:354. With reference to later plans for the Portuguese contingent to be reorganized into a national corps, see also Moon, *Wellington's Two-Front War*, 190–192.

48. Nichols, *Wellington's Mongrel Regiment*, 70.

49. Ibid., 20.

50. Oman, *Peninsular War* 4:650.

51. Oman, *Wellington's Army*, 349, 357.

52. Ibid., 361, 364.

53. Muir, "Order of Battle," 138.

54. Oman, *Wellington's Army*, 252.

55. States, TNA, WO17/2475, with explanation in Gerges, "Command and Control," 394–395; Oman, *Wellington's Army*, 372–373. This idea had first been mooted over a year previously; see Wellington to Torrens, December 2, 1812, in Wellesley, *Supplementary Despatches* 7:485–486.

56. Gerges, "Command and Control," 324, 353–355, 378–381, 401–403; Fletcher, *Galloping at Everything*, 210–219.

57. Oman, *Wellington's Army*, 170–171, 357–358.

58. Ibid., 170–171, 347–350.

59. Ibid., 345, 357. This practice is exemplified by the assembly of Lightburne's and Catlin Craufurd's Brigades, for which see also Dundas to Castlereagh, May 27, 1809, TNA, WO1/641.

60. Thoumine, *Scientific Soldier*, 152–153.

61. Alten to officers of the 18th Hussars, September 22, 1813, quoted in Hunt, *Charging against Napoleon*, 145.

62. On the necessity for time for a division to retrain tactically with a new sequence of regiments and brigades, see Muir, "Order of Battle," 149.

63. Reproduced in Oman, *Peninsular War* 4:750–752.

64. Muir, *Salamanca*, 161–158, 178–183.

65. Oman, *Wellington's Army*, 355–372.

66. Wellington to Fane, December 8, 1813, in Gurwood, *Dispatches* 11:353–354; Wellington to Torrens, April 21, 1815, in Gurwood, *Dispatches* 12:317–318.

67. Stewart was brave to a fault but lacking in common sense; see Wellington to Torrens, December 6, 1812, in Wellesley, *Supplementary Despatches* 7:494–495.

68. Tilson had acquired a new surname since 1809, but no additional military aptitude.

69. Glover, *Wellington as Military Commander*, 179–181.

70. GO of April 11, 1815, in Wellesley, *Supplementary Despatches* 10:62–63.

71. Wellington to Torrens, April 21, 1815, in Gurwood, *Dispatches* 12:317–318.

72. Clayton, *British Officer*, 115, 122.

73. For the original KGL organization, see GO of October 12, 1814, TNA, WO30/122.

74. Wellington to Torrens, May 2, 1815, in Gurwood, *Dispatches* 12:340–341.

75. "Reminiscences of Captain Carl Jacobi, Luneberg Field Battalion, 1st Hanoverian Brigade," in Glover, *Waterloo Archive* 2:121–123.

76. Tomkinson, *Diary*, 133.

77. Holmes, *Acts of War*, 307 (however, see also 327–328 for an exception). Coss (*King's Shilling*, 191) ignores the division entirely in suggesting a scale of identity levels running from squad to company, battalion, brigade, army—this may reflect modern soldiering but does not fit the realities of the Napoleonic era.

78. Wellington to Bathurst, August 1, 1813, in Gurwood, *Dispatches* 10:576–589. Oman (*Wellington's Army*, 172) falsely expresses the opinion that their advance at Albuera was the source of the nickname.

79. Tomkinson, *Diary*, 133.

80. Cook and Burnham, "Nicknames."

81. Quoted in Nichols, *Wellington's Mongrel Regiment*, 131.

82. Tomkinson, *Diary*, 286.

83. Kincaid, *Adventures in the Rifle Brigade*, 55.

84. Grattan, *Adventures*, 238–254.

85. Robertson, *Commanding Presence*, 237.

86. l'Estrange, *Recollections*, 124

87. Colley, *Britons*, 282–319; Cookson, "Regimental Worlds," 23–42.

88. McGuigan, "Wellington's Generals," 185–196.

89. Urban, *Rifles*, 156–157; Coss, *King's Shilling*, 128.

90. Wellington to Leith, December 21, 1813, in Gurwood, *Dispatches* 11:383; Wellington is quoting Leith's own description back to him.

91. Bingham to mother, letters of March 13, April 4, November 10, and December 12, 1812, in Bingham, *Wellington's Lieutenant*, 105–106, 110–112, 164–165, and 168–169.

92. Stewart to Wellington, March 6, 1813 (including enclosure of previous correspondence with Lt. Col. Guise, 1st Foot Guards); and York to Wellington, April 7, 1813, in Wellesley, *Supplementary Despatches* 7:574–576 and 7:600–601.

93. Crowe, *Eloquent Soldier*, 145.

94. Costello, *Adventures of a Soldier*, 120.

95. Donaldson, *Eventful Life*, 198–199.

96. Cole to Wellington, May 28, 1815, in Wellesley, *Supplementary Despatches* 10:389–390.

97. Wellington to Clinton, June 15, 1815, in Gurwood, *Dispatches* 8:469–470.

98. Wellington to Cole, June 2, 1815, in Gurwood, *Dispatches* 8:435–436.

99. Smith, *Autobiography*, 62.

100. Grattan, *Adventures*, 196–206.

101. Oman, *Peninsular War* 6:557–586; 7:6–36, 529–530.

102. Watson, *When Soldiers Quit*, 67–88, 157–167.

103. Oman *Peninsular War* 7:15–17.

104. Wellington to Graham, August 25, 1813, in Gurwood, *Dispatches* 11:31–33.

105. Wellington to Graham, August 28, 1813, in Gurwood, *Dispatches* 11:50.

106. Watson, *When Soldiers Quit*, 84.

107. See Coss, *King's Shilling*, 211–234. On the issue of command breakdown, Oman (*Peninsular War* 7:33, 529–530) records officer casualties in excess of 50 percent. See also Watson, *When Soldiers Quit*, 86.

108. Campbell, *British Army*, 256–258.

6. Strategic Consumption

1. McGrigor, "Medical History."

2. To be exact, there were 8,889 deaths through combat and 24,930 through sickness; see Holmes, *Redcoat*, 249.

3. Howard, *Wellington's Doctors*. See also Blanco, "British Military Medicine," and chapter-length treatments in Haythornthwaite, *Armies of Wellington*, 132–144; and Holmes, *Redcoat*, 249–262.

4. For details of data sources for this sample, see appendix 1.

5. McGrigor, "Medical History," 389, 394, 467–468.

6. Ibid., 397; Monthly Returns in TNA, WO17/1796–1801.

7. Elting, *Amateurs*, 18–19.

8. Latimer, *1812*, 309.

9. McGrigor, "Medical History," 471.

10. The suggestion of indifference by officers is made in Coss, *King's Shilling*, 123–125, based largely on the omission of such matters from correspondence and memoirs.

11. Return prepared by the Adjutant General's office, February 1810, reproduced in Great Britain War Department, *British Minor Expeditions*, 80.

12. Howard, *Wellington's Doctors*, 198–201.

13. The role of the Walcheren scandal is placed in this wider context in Crowe, "Walcheren Expedition."

14. Summerville, *March of Death*, 81–82; Harris, *Recollections*, 134–135.

15. Note by Major George McGregor, accompanying battalion return, TNA, WO17/176. Of the fit men, 57 were recruits who had not taken part in the Peninsula campaign, leaving only 176 fit from those who had served under Moore.

16. Battalion Returns in TNA, WO17/148 and WO17/101.

17. Battalion Return of June 25, 1809, TNA, WO17/131.

18. Report by Robert Craufurd, April 30, 1809, accompanying inspection return for 1/43rd, TNA, WO27/94; see also Urban, *Rifles*, 2, 13, 35.

19. It should be stressed that this is not a new argument—McGrigor made the connection as far back as 1815 ("Medical History," 401)—but its significance has not generally been appreciated.

20. Harris, *Recollections*, 256.

21. Ibid., 261–265.

22. Green, *Soldier's Life*, 39.

23. Ibid., 40–43.

24. Harris, *Recollections*, 265–268.

25. "Journal of the Proceedings of the British Army," TNA, WO28/339, entry for April 19, 1810.

26. Harris, *Recollections*, 268–280.

27. McGuffie, "Walcheren Expedition," 201.

28. The hospitalized numbered 11,296 out of 33,373; see return reproduced in Great Britain War Department, *British Minor Expeditions*, 80.

29. Inspection Report of May 3, 1811, by Major General Robinson, TNA, WO27/102, part 2; Inspection Report of May 13, 1813, by Major General Burgoyne, TNA, WO27/116.

30. Inspection Report of August 18, 1812, TNA, WO27/106, part 2.

31. Inspection Report of February 16, 1813, TNA, WO27/111, part 1.

32. Oman, *Peninsular War* 5:501.

33. McGrigor to Fitzroy Somerset, April 11, 1813, TNA, WO27/111, part 1.

34. McGrigor, "Medical History," 383–385.

35. Schaumann, *On the Road*, 37.

36. Elting, *Swords*, 469–470.

37. Howard, *Wellington's Doctors*, 97–98.

38. Battalion Returns, TNA, WO17/136 (29th); WO17/141 (1/32nd); WO17/164 (1/50th); WO17/203 (1/82nd).

39. Monthly Return for October 1808, TNA, WO17/2464.

40. Lawrence, *Autobiography*, 18–20, 39–41, 42–43.

41. Levinge, *Forty-Third Regiment*, 106. Note, though, that the 2/43rd was nevertheless considered fit to march with Moore, suggesting that the problem was not deep-seated.

42. Oman, *Peninsular War* 1:646–648; see also Harris, *Recollections*, 104–106.

43. GO of August 9, 1809, in Gurwood, *General Orders* 1:122–123; see also Fortescue, *British Army* 7:196–197; and Oman, *Peninsular War* 2:484–486.

44. Comparative figures from contemporary digest, TNA, WO17/2814.

45. Monthly Returns October 1809 to January 1810, TNA, WO17/2464–2465. See also Gordon to Aberdeen, November 16, 1809, in Gordon, *Wellington's Right Hand*, 68–69.

46. For a full monthly breakdown of losses by units, see Bamford "Guadiana Fever."

47. Howard, *Wellington's Doctors*, 166–170.

48. Mackinnon, "Campaign in Portugal and Spain," 40–42.

49. Howard, *Wellington's Doctors*, 95–99.

50. Cooper, *Rough Notes*, 33–34.

51. Simmons, *British Rifle Man*, 35–40.

52. Monthly Return, TNA, WO17/2464. On the shortage of medical personal, see Wellington to Liverpool, November 14, 1809, in Gurwood, *Dispatches* 5:280–282.

53. Coss, *King's Shilling*, 97–104, 272–287.

54. McGrigor, "Medical History," 469–470.

55. Ibid., 470.

56. Graham to Bathurst, December 31, 1813, TNA, WO1/198, 339–342. McGrigor, who asserted the value of a warm breakfast in "Medical History" (470), would certainly have approved.

57. Brett-James, *Life in Wellington's Army*, 25.

58. The assertion made by Coss (*King's Shilling*, 105–107) that plundering stemmed directly from the nutritional deficiency of the standard ration is, therefore, too simplistic and, like the rations themselves, is best taken with a pinch of salt.

59. The nature of these reforms is discussed in Howard, *Wellington's Doctors*, 122–128, 137–138; and McGrigor, "Medical History," 473–476. On the establishment of general hospitals during 1813 and 1814, see McGrigor, 392.

60. McGrigor, "Medical History," 388–389.

61. GO of March 1, 1813, in Wellesley, *Supplementary Despatches* 7:562–564.

62. Wellington to Hill, October 11, 1810, in Gurwood, *Dispatches* 6:477–478; Wellington to Bathurst, February 13, 1814, in Gurwood, *Dispatches* 11:517.

63. l'Estrange, *Recollections*, 129–130.

64. McGrigor, "Medical History," 394–395, 471–472.

65. Elting, *Swords*, 467.

66. GO of October 23, 1810, in Gurwood, *General Orders* 2:190–193.

67. On the battle of wits between malingering soldiers and a long-suffering medical staff, see Haythornthwaite, *Armies of Wellington*, 135; and Howard, *Wellington's Doctors*, 206–208.

68. Howard, *Wellington's Doctors*, 88–90. For an officer's perspective (that of Lt. Charles Crowe), see Crowe, *Eloquent Soldier*, 205–230; for the view from the ranks, see Cooper, *Rough Notes*, 35.

69. GO of November 29, 1813, in Wellesley, *Supplementary Despatches* 8:404.

70. "Return of the number of men belonging to Corps in the Peninsula who could not be found in the General Hospitals, and who have been struck off as Dead," March 25, 1814, TNA, WO17/2475.

71. McGrigor, "Medical History," 397.

72. Linch, "Recruitment," 202–233.

73. Officers, by contrast, were more commonly granted leave and frequently overstayed it, with some of the most flagrant offenders being cashiered as a result. See, for example, GO of May 29, 1810, copying Horse Guards circular of April 23, relating to the case of Lt. Alexander Winckstern, 7th KGL Line Battalion, in Gurwood, *General Orders* 2:78–80.

74. Urban, *Fusiliers*, 292–293.

75. Hall, *British Strategy*, 7, notes only fifty-two Britons executed for desertion in the peninsula. See also Oman, *Wellington's Army*, 243.

76. GO of August 30, 1811, in Gurwood, *General Orders* 3:169–172; a largely identical case, relating to Private George Lever of the 51st, can be found in AGO of July 11, 1811, in Gurwood, *General Orders* 3:143–145.

77. Battalion Returns for 5/60th TNA, WO17/179.

78. Gough to father, June 23, 1809, quoted in Rait, *Gough* 1:31–36.

79. Monthly Return of June 1, 1809, TNA, WO17/2464.

80. GO of June 26, 1809, in Gurwood, *General Orders* 1:86-89; see also Oman, *Wellington's Army*, 257.

81. Summerville, *March of Death*, 149–163.

82. Notes to Monthly Return of February 25, 1811, TNA, WO17/2467.

83. Inspection Report, June 19, 1810, TNA, WO27/99.

84. Note to Monthly Return of November 25, 1813, TNA, WO17/1773.

85. Note to Monthly Return of June 25, 1810, TNA, WO17/2465.

86. Haythornthwaite, *Armies of Wellington*, 209–210; for Wellington's views on the Almeida case, see Wellington to Charles Stuart, August 31, 1810, in Gurwood, *Dispatches* 6:371—even then, he considered such recourse acceptable only for the rank and file and not for officers.

87. Wellington to Liverpool, November 10, 1810, TNA, WO1/246, 353–355. An appeal to the men under his command, couched in very similar terms, was made in GO of November 10, 1810, in Gurwood, *General Orders* 2:203–204.

88. GO of September 2, 1811, in Gurwood, *General Orders* 3:176–180.

89. Latimer, *1812*, 38–39, 45–46.

90. Notes to Monthly Return of October 25, 1810, TNA, WO17/2465.

91. Nichols, *Wellington's Mongrel Regiment*, 71.

92. Gray, "'Prisoners, Wanderers and Deserters.'"

93. Monthly Return of September 25, 1814, TNA, WO17/1518.

94. Inspection Report of October 8, 1813, TNA, WO27/124.

95. Monthly Returns, TNA, WO17/2476.

96. Grattan, *Adventures*, 333–334; see also Oman, *Peninsular War* 7:513–514.

97. Brumwell, *Redcoats*, 43–47, 155, 127–136.

98. Cooper, *Rough Notes*, 121–141.

99. Urban, *Rifles*, 137–141, 161–163.

100. Linch, "Recruitment," 204.

101. Haythornthwaite, *Armies of Wellington*, 50–51.

102. Linch, "Recruitment," 218–221.

103. See, for example, Lawrence, *Autobiography*, 16–20; and Surtees, *Rifle Brigade*, 6–7. The tendency is also discussed more generally in Urban, *Rifles*, 4–9.

7. Beasts of Burden

1. Return of Regimental Courts Martial for 5th Dragoon Guards, TNA, WO27/111, part 1.

2. That only a dismounted trooper could be employed to tend the camp kettle mule is made clear by the GO of June 2, 1811, in Gurwood, *General Orders* 3:103–104.

3. Memorandum dated November 24, 1813, in TNA, WO1/414, 51–54.

4. Ibid. Note that the measurement was from the ground to the withers.

5. Tennant, "Wellington's Mules." On the issue of forage mules for the cavalry, see circular from Wellington to cavalry brigade commanders, August 31, 1811, in Gurwood, *Dispatches* 8:251.

6. "Journal of the Proceedings of the British Army," March 11, 1810, TNA, WO28/339.

7. Henegan, *Seven Years* 1:175–176.

8. Hunt, *Charging against Napoleon*, 267.

9. Brett-James, *Life in Wellington's Army*, 25; Robertson, *Commanding Presence*, plate three. Regarding the use of artillery oxen, see "Memorandum for Lieut. Col. Framingham RA," August 27, 1811, in Gurwood, *Dispatches* 8:227.

10. Henegan, *Seven Years* 1:35.

11. Monthly Return for April 1814, TNA, WO17/2476

12. Cited in Fletcher, *Galloping at Everything*, 20.

13. Monthly Return for June 1812, TNA, WO17/2469.

14. Wellington to Liverpool, December 7, 1810, TNA, WO1/246, 481–485.

15. Fletcher, *Galloping at Everything*, 13; Brett-James, *Life in Wellington's Army*, 179. Note that in Graham's case this figure includes horses for servants and baggage in addition to the general's own mounts.

16. Gleig, *Subaltern*, 2.

17. Aitchison to father, May 2, 1812, in Aitchison, *Letters*, 151–152.

18. Hay, *Reminiscences*, 26–31.

19. Brett-James, "Diary of Captain Neil Douglas," 103.

20. December figure from Monthly Return for Moore's army, TNA, WO17/2646.

21. Hunt, *Charging against Napoleon*, 36.

22. Monthly Return TNA, WO17/2646.

23. Inspection Return and Report by Major General Victor Alten, February 12, 1813, TNA, WO27/111, part 1.

24. Fletcher, *Galloping at Everything*, 15–16; see also Wellesley to Castlereagh, June 22, 1809, in Gurwood, *Dispatches* 4:427–428.

25. Wellington to Bathurst, April 14, 1814, in Gurwood, *Dispatches* 10:295–297.

26. Wellington to Liverpool, December 7, 1810, TNA, WO1/246, 481–485.

27. GO of May 5, 1809, in Gurwood, *General Orders* 1:14–16.

28. Oman, *Peninsular War* 1:225.

29. Cradock to Castlereagh, March 17, 1809, quoted in Napier, *War in the Peninsula* 2:488.

30. Maitland to Bathurst, August 30, 1812, TNA, WO1/316, 5.

31. Taylor to Bathurst, December 4, 1813, TNA, WO1/414, 59–69; ibid., December 16, 1813, TNA, WO1/414, 189–198.

32. Robertson, *Commanding Presence*, 60.

33. Figures are from Woodman, *Victory of Seapower*, 158.

34. Diary of Lt. George Woodberry, February 2, 1813, in Hunt, *Charging against Napoleon*, 61–62.

35. Wellington to Vice Admiral Berkeley, September 16, 1811, in Gurwood, *Dispatches* 8:282–283.

36. Tomkinson, *Diary*, 1.

37. Ibid., 2

38. GO of November 17, 1810, in Gurwood, *General Orders* 2:207–208.

39. See, for example, Wellington to Liverpool, September 22, 1810, in Gurwood, *Dispatches* 6:405–406.

40. Brett-James, *Life in Wellington's Army*, 180–181.

41. A point emphasized in Fletcher, *Galloping at Everything*, 14–16.

42. Wellington to Bathurst, April 14, 1813, in Gurwood, *Dispatches* 10:295–297; ibid., May 25, 1813, in Gurwood, *Dispatches* 10:400.

43. Artillery appendices are in Oman, *Peninsular War* 2:654–655; 3:558–560; 4:650–652; and 5:619–622; also Partridge and Oliver, *Handbook*, 146–147.

44. Stewart, *XII Royal Lancers*, 58–63.

45. Steuart to Bridger, July 12,1811, Bound letters of Colonel Sir James Stewart.

46. Steuart to Taylor, January 17, 1812, Bound letters of Colonel Sir James Stewart.

47. Summarized from Inspection Returns and Reports in WO27/111, part 1.

48. GO of April 25, 1813 in Gurwood, *General Orders*, Alphabetical Volume, 60–61; see also Wellington to York, April 27, 1813, in Gurwood, *Dispatches* 10:332–333.

49. Wellington to Liverpool, December 7, 1810, TNA, WO1/246, 481–485.

50. See, for example, Wellington to Payne, January 2, 1810 in Gurwood, *Dispatches* 5:402, and, in contrast, Wellington to Liverpool, August 27, 1811, in

Gurwood, *Dispatches* 8:230–233. These points were discussed at some length on the Napoleon Series forum in 2007, and I am obliged in particular to Ron McGuigan, David Buttery, and Anthony Gray for their input on this topic. The topographical element is discussed in Fletcher, *Galloping at Everything*, 67–77; and Oman, *Wellington's Army*, 94–97.

51. See artillery appendices in Oman, *Peninsular War* 2:654–655; 3:558–560; 4:650–652; and 5:619–622.

52. Fletcher, *Galloping at Everything*, 20.

53. For example, Robertson, *Commanding Presence*, 147–148.

54. Fletcher, *Galloping at Everything*, 44–45.

55. Ibid., 26–30; Glover, *Peninsular Preparation*, 134–137.

56. Monthly Return for July 1812, TNA, WO17/2470.

57. Monthly Return for June 1815, TNA, WO17/1760.

58. See for example, Smith, *Autobiography*, 24–27.

59. Schaumann, *On the Road*, 360–366.

60. Ibid., 218–219.

61. Returns of Regimental Courts Martial accompanying 1812 first half-yearly Inspection Reports, TNA, WO27/106, part 1.

62. Returns of Regimental Courts Martial for 9th and 14th Light Dragoons, TNA, WO27/111, part 1.

63. Return of Regimental Courts Martial for 1st KGL Hussars, TNA, WO27/111, part 1.

64. Return of Regimental Courts Martial for 1st KGL Dragoons, TNA, WO27/111, part 1.

65. Returns of Regimental Courts Martial for the 1st and 2nd KGL Dragoons, TNA, WO27/106, part 1, and those for the 9th Light Dragoons, TNA, WO27/111, part 1.

66. Inspection Report of January 29, 1813, TNA, WO27/106, part 1.

67. Burgoyne quoted in Brett-James, *Life in Wellington's Army*, 188–189; Gleig, *Subaltern*, 138.

68. Monthly Returns, TNA, WO17/2465; see also Oman, *Wellington's Army*, 111.

69. "A Chelsea Pensioner," *Jottings from my Sabretache* (London, 1847), 189–90, quoted in Fletcher, *Galloping at Everything*, 22. Fletcher attributes the account to Sergeant William Tale; for the Dawes attribution, see Griffith and Philips, *Corunna to Waterloo*, 10.

70. Partridge and Oliver, *Handbook*, 58.

71. The latter point is made in Haythornthwaite, *Armies of Wellington*, 107.

72. Fletcher, *Galloping at Everything*, 17–18.

73. Hunt, *Charging against Napoleon*, 76.

74. May 1813 diary entry, quoted in Hunt, *Charging against Napoleon*, 80.

75. Kennedy to Elizabeth Kennedy, April 1813, quoted in Hunt, *Charging against Napoleon*, 205.

Conclusion

1. Emsley, "The Volunteer Movement," in Guy, *Road to Waterloo*; Colley, *Britons*, 283–319.

2. See Mackesy, *Statesmen at War*, for analysis of this with reference to the 1799 campaign.

3. Hall, *British Strategy*, 1–15.

4. Oman, *Peninsular War* 3:540–543.

5. Elting, *Swords*, 328. For a critique of these methods as applied in the peninsula, see also Alexander, "French Replacement Methods."

6. Oman, *Peninsular War* 4:625–626.

7. Elting, *Swords*, 328.

8. Memo from Adjutant General's Office, October 19, 1814, TNA, WO1/659, 457. Note that this applies only to units disbanded whilst overseas.

9. Clayton, *British Officer*, 67–68.

10. Graves, *Fix Bayonets*, 7–29; Urban, *Man Who Broke Napoleon's Codes*, 7–8, 25–29.

11. Graves, *Fix Bayonets*, 410; Clayton, *British Officer*, 102.

12. Cooper, *Rough Notes*, 63.

13. Divall, *Redcoats against Napoleon*, 113–114.

14. Kincaid, *Adventures in the Rifle Brigade*, 50.

15. Sir Walter Scott, quoted in Haythornthwaite, *Die Hard*, 125.

16. Coss, "King's Shilling," 140.

17. Ibid.

18. Linch, "Recruitment," 164, 239–242.

19. See in particular Robertson, *Commanding Presence*. Recent attempts to analyze Britain's success in the field include Muir, "Wellington and the Peninsular War," 1–38; Fletcher, "Wellington, Architect of Victory," 149–162; and Arnold, "A Reappraisal of Column Versus Line in the Peninsular War."

20. Evelyn, "'I Learned What One Ought Not to Do,'" in Guy, *Road to Waterloo*, 16–22, which stresses the distinction between tactical ability and failings of manpower and supply.

21. Muir, *Britain and the Defeat of Napoleon*, 380.

22. Inspection Return of May 14, 1808, TNA, WO27/92, part 2.

23. Le Couteur, *Merry Hearts*, 10–19.

24. Histman and Sorby, "Independent Foreigners," 11.

25. Wellington to Bathurst, August 18, 1813, in Gurwood, *Dispatches* 11:11–12.

26. Oman, *Wellington's Army*, 171; Hofschröer, *1815* 2:63.

27. Oman, *Wellington's Army*, 352–353.

28. Fortescue, *British Army* 10:185–186.

29. On the Spanish organization, see Wellington to the Conde de la Bisbal, February 27, 1813, in Gurwood, *Dispatches* 10:325–326; with reference to similar proposals for the Portuguese service, see Partridge and Oliver, *Handbook*, 139.

30. Clayton, *British Officer*, 98–100; Fortescue, *British Army* 13:561–562.

31. Dietz, *Last of the Regiments*, 40–59.

32. Cook and Burnham, "Nicknames"; Graves, *Fix Bayonets*, 137–138, 414–415.

33. Compare with the description of the 1775 festivities in Boston in Urban, *Fusiliers*, 7–13.

34. Frey, *British Soldier in America*, 117–125.

35. Gilbert, "Tale of Two Regiments," 275.

36. Newbolt, *Oxfordshire and Buckinghamshire Light Infantry*, 136.

37. Keegan, *Six Armies in Normandy*, 194.

38. Fortescue, *British Army* 13:50, 226.

39. Holmes, *Tommy*, 639–640.

40. Brumwell, *Redcoats*; Frey, *British Soldier in America*; Urban, *Fusiliers*.

41. Oman, *Peninsular War* 7:513–514.

42. Colley, *Britons*, 177–191, 364–368.

43. Cookson, "Duke of Wellington Statues," 23–40; Colley, *Britons*, 257–258.

44. Cookson, "Regimental Worlds," 23–24, 37–39; however, this admittedly chimes with Colley's broader construct of national identity defined against an outside "other" presented in *Britons*.

45. Haythornthwaite, "'Carrying On the War as It Should Be,'" 115–130; Kincaid, *Adventures in the Rifle Brigade*, 26–27.

46. Colley, *Britons*, 313–314; lyrics quoted in Graves, *Field of Glory*, 251.

47. Strachan, *Wellington's Legacy*, 159–175.

48. Burnham, "Filling the Ranks," 219.

49. Glover, *Peninsular Preparation*, 184–185.

50. Strachan, *Wellington's Legacy*, 211–216.

51. Zamoyski, *Rites of Peace*, 366, 372–373.

52. Fortescue, *British Army* 10:232–242.

53. In particular the near-collapse of Halket's Fifth Brigade at Quatre Bras and again at Waterloo—see Hofschröer, *1815* 1:303; 2:137; and Divall, *Redcoats against Napoleon*, 148–150, 174–175.

54. Oman, *Peninsular War*, 343–373; McGuigan, "Anglo-Allied Army."

55. Fortescue, *British Army* 10:419.

56. Ibid., 1:554–580.

57. Smurthwaite, "Glory is Priceless" in Guy, *Road to Waterloo*, 164–183.

Appendix 1

1. No January return, and includes data for Baird's corps for December only. Includes some units that returned to Portugal, as well as those that went to Corunna. Surviving January Battalion Returns for 1/32nd and 1/52nd have been incorporated into data for these units.

2. See also individual data for KGL units serving under Wallmoden, for which additional data has been obtained from battalion returns.

$\mathscr{Bibliography}$

Archival Sources

Brotherton Library, University of Leeds.
  House of Commons Parliamentary Papers 1809.

The National Archives, Kew.
  WO1—War Office and Predecessors. Secretary-at-War, Secretary of State for War, and Commander-in-Chief. In-letters and Miscellaneous Papers, 1732–1868.
  WO6—War Department and Successors. Secretary of State for War and Secretary of State for War and the Colonies. Out-letters, 1793–1859.
  WO12—Commissary General of Musters Office and successors. General Muster Books and Pay Lists, 1732–1878.
  WO17—Office of the Commander in Chief. Monthly Returns to the Adjutant General, 1754–1866.
  WO27—Office of the Commander-in-Chief and War Office. Adjutant General and Army Council. Inspection Returns, 1750–1912.
  WO28—War Office. Records of Military Headquarters, 1746–1926.
  WO379—Office of the Commander-in-Chief and War Office. Adjutant General's Office. Disposition and Movement of Regiment. Returns and Papers (Regimental Records), 1737–1967.

9th/12th Royal Lancers Regimental Museum, Derby.
  Bound letters of Colonel Sir James Stewart [Steuart] dating from 1806–18, 12th Light Dragoons. 912L:2088/44.

Books, Articles, and Theses

Adjutant General's Office. *General Regulations and Orders for the Army.* London: W. Clowes, 1811.
Aitchison, John. *An Ensign in the Peninsular War: The Letters of John Aitchison.* Edited by W. F. K. Thompson. London: Michael Joseph, 1981.
Alexander, Don W. "French Military Problems in Counterinsurgency Warfare in Northeastern Spain, 1808–1813." *Military Affairs* 40, no. 3 (October 1976): 117–122.

————. "French Replacement Methods during the Peninsular War, 1808–1814." *Military Affairs* 44, no. 4 (December 1980): 192–197.

Aubrey-Fletcher, Major H. L. *A History of the Foot Guards to 1856.* London: Constable, 1927.

Austin, Paul Britten. *1812: The March on Moscow.* London: Greenhill, 1993.

Barbero, Alessandro. *The Battle: A New History of Waterloo.* Translated by John Cullen. London: Atlantic, 2005.

Barker, Thomas M. "A Debacle of the Peninsular War: The British-Led Amphibious Assault against Fort Fuengirola 14–15 October 1810." *Journal of Military History* 64, no.1 (January 2000): 9–52.

Barrett, C. R. B. *The 85th King's Light Infantry.* London: Spottiswoode, 1913.

Beamish, North Ludlow. *History of the King's German Legion.* 2 vols. London: Thomas and William Boone, 1832–1837.

Bingham, George Ridout. *Wellington's Lieutenant, Napoleon's Gaoler: The Peninsula Letters and St. Helena Diaries of Sir George Ridout Bingham.* Edited by Gareth Glover. Barnsley: Pen and Sword, 2004.

Blanco, Richard L. "The Development of British Military Medicine, 1793–1814." *Military Affairs* 38, no. 1 (February 1974): 4–10.

Bond, Gordon C. *The Grand Expedition: The British Invasion of Holland in 1809.* Athens: University of Georgia Press, 1979.

Bowden, Scott. *Napoleon's Grande Armée of 1813.* Chicago: Emperor's Press, 1990.

Boycott-Brown, Martin. *The Road to Rivoli: Napoleon's First Campaign.* London: Cassell, 2001.

Brereton, J. M., and A. C. S. Savory. *The History of the Duke of Wellington's Regiment (West Riding) 1702–1991.* Halifax: privately printed, 1993.

Brett-James, Antony, ed. "The Diary of Captain Neil Douglas, 79th Foot, 1809–1810." *Journal of the Society for Army Historical Research* 41 (1963): 101–107.

————. *Life in Wellington's Army.* Hampstead: Tom Donovan, 1994.

Browne, Thomas Henry. *The Napoleonic War Journal of Captain Thomas Henry Browne, 1807–1816.* Edited by Roger Norman Buckley. London: Bodley Head for the Army Records Society, 1987.

Brumwell, Stephen. *Redcoats: The British Soldier and War in the Americas, 1755–1763.* Cambridge: Cambridge University Press, 2002.

Burnham, Robert. "Filling the Ranks: How Wellington Kept His Units up to Strength." In Muir et al., *Inside Wellington's Peninsular Army,* 201–225.

Burnham, Robert, and Ron McGuigan. "The Impeccable Timing of Sir George Brown." *War of 1812 Magazine,* Issue 10 (October 2008). http://www.napoleon-series.org/military/Warof1812/2008/Issue10/c_ElegantExtracts.html.

Buttery, David. *Wellington against Massena: The Third Invasion of Portugal, 1810–1811.* Barnsley: Pen and Sword, 2007.

Campbell, Lt. Colonel James. *A British Army, As It Was, Is, and Ought To Be.* London: T. and W. Boone, 1840.

Cannon, Richard. *Historical Record of the Eighth, or the King's Regiment of Foot Containing an Account of the Formation of the Regiment in 1685 and of Its Subsequent Services to 1844.* London: Parker, Furnivall, and Parker, 1844.

————. *Historical Record of the Ninth, or the East Norfolk Regiment of Foot Containing an Account of the Formation of the Regiment in 1685 and of Its Subsequent Services to 1847.* London: Parker, Furnivall, and Parker, 1848.

————. *Historical Record of the Third, or the King's Own Regiment of Light Dragoons, Containing an Account of the Formation of the Regiment in 1685 and of Its Subsequent Services to 1846.* London: Parker, Furnivall, and Parker, 1847.

Cassidy, Martin. *Marching with Wellington: With the Inniskillings in the Napoleonic Wars.* Barnsley: Leo Cooper, 2003.

Chandler, David. *The Campaigns of Napoleon.* London: Weidenfeld and Nicolson, 1966.

Clayton, Anthony. *The British Officer: Leading the Army from 1660 to the Present.* Harlow: Longman, 2006.

Colley, Linda. *Britons: Forging the Nation, 1707–1837.* New Haven: Yale University Press, 1992.

Cookson, J. E. *The British Armed Nation, 1793–1815.* Oxford: Clarendon Press, 1997.

————. "The Edinburgh and Glasgow Duke of Wellington Statues: Early Nineteenth-Century Unionist Nationalism as a Tory Project." *Scottish Historical Review* 83, no.1 (April 2004): 23–40.

————. "Regimental Worlds: Interpreting the Experience of British Soldiers during the Napoleonic Wars." In Forrest, Hagemann, and Rendall, *Soldiers, Citizens and Civilians,* 23–42.

Cooper, John Spencer. *Rough Notes of Seven Campaigns in Portugal, Spain, France and America During the Years 1809–1815.* Staplehurst, UK: Spellmount, 1996.

Cope, Sir William H., Bart. *The History of the Rifle Brigade (The Prince Consort's Own) Formerly the 95th.* London: Chatto and Windus, 1877.

Coss, Edward J. *All for the King's Shilling: The British Soldier Under Wellington, 1808–1814.* Norman: University of Oklahoma Press, 2010.

Costa, Fernando Dores. "Army Size, Military Recruitment and Financing in Portugal during the Period of the Peninsula War—1808–1811." *E-Journal of Portuguese History* 6, no. 2 (Winter 2008).

Costello, Edward. *The Adventures of a Soldier, or Memoirs of Edward Costello KFS.* London: Henry Colburn, 1841.

Crowe, Charles. *An Eloquent Soldier: The Peninsular War Journals of Lieutenant Charles Crowe of the Inniskillings, 1812–1814.* Edited by Gareth Glover. Barnsley: Frontline, 2011.

Crowe, Kate Elizabeth. "The Walcheren Expedition and the New Army Medical Board: A Reconsideration." *English Historical Review* 88, no. 349 (October 1973): 770–785.

Cunliffe, Marcus. *The Royal Irish Fusiliers, 1793–1950.* Oxford: Oxford University Press, 1952.

Dempsey, Guy. *Albuera 1811: The Bloodiest Battle of the Peninsular War.* Barnsley: Frontline, 2008.

Dietz, Peter. *The Last of the Regiments: Their Rise and Fall.* London: Brassey's, 1990.

Divall, Carole. *Inside the Regiment: The Officers and Men of the 30th Regiment during the Revolutionary and Napoleonic Wars.* Barnsley: Pen and Sword, 2011.

————. *Redcoats against Napoleon: The 30th Regiment during the Revolutionary and Napoleonic Wars.* Barnsley: Pen and Sword, 2009.

Donaldson, Joseph. *Recollections of an Eventful Life, Chiefly Passed in the Army.* Glasgow: W. R. M'Phun, 1824.

Eliot, William Granville. *A Treatise on the Defence of Portugal*. London: T. Egerton, 1811.

Elting, John R. *Amateurs, To Arms! A Military History of the War of 1812*. Chapel Hill, N.C.: Algonquin Books, 1991.

———. *Swords Around a Throne: Napoleon's Grande Armée*. London: Weidenfeld and Nicolson, 1989.

Emsley, Clive. "The Volunteer Movement." In Guy, *The Road to Waterloo*, 40–47.

Evelyn, G. J. "'I Learned What One Ought Not to Do': The British Army in Flanders and Holland, 1793–95." In Guy, *The Road to Waterloo*, 16–22.

Fassiefern, John Cameron. *Memoir of Colonel John Cameron Fassiefern, K.T.S., Lieutenant-Colonel of the Gordon Highlanders, or 92nd Regiment of Foot*. Edited by Archibald Clerk. Glasgow: T. Murray and Son, 1858.

Field, Andrew W. *Talavera: Wellington's First Victory in Spain*. Barnsley: Pen and Sword, 2006.

Fletcher, Ian. *Bloody Albuera: The 1811 Campaign in the Peninsula*. Marlborough: Crowood, 2000.

———. *"A Desperate Business": Wellington, the British Army and the Waterloo Campaign*. Staplehurst: Spellmount, 2001.

———. *Galloping at Everything: The British Cavalry in the Peninsular War and at Waterloo, 1808–15: A Reappraisal*. Staplehurst: Spellmount, 1999.

———, ed. *The Peninsular War: Aspects of the Struggle for the Iberian Peninsula*. Staplehurst: Spellmount, 1998.

———, ed. *Voices from the Peninsula: Eyewitness Accounts by Soldiers of Wellington's Army, 1808–1814*. London: Greenhill, 2001.

———. *The Waters of Oblivion: The British Invasion of the Rio de la Plata, 1806–07*. Stroud: Spellmount, 2006.

———. "Wellington, Architect of Victory." In Fletcher, *The Peninsular War*, 149–162.

———. *Wellington's Foot Guards*. London: Osprey, 1994.

Forrest, Alan, Karen Hagemann, and Jane Rendall, eds. *Soldiers, Citizens and Civilians: Experiences and Perceptions of the Revolutionary and Napoleonic Wars, 1790–1820*. Basingstoke: Palgrave Macmillan, 2009.

Fortescue, J. W. *A History of the British Army*. 13 vols. London: Macmillan, 1899–1930.

Frey, Sylvia R. *The British Soldier in America: A Social History of Military Life in the Revolutionary Period*. Austin: University of Texas Press, 1981.

Garvey, Lt. Colonel Russell. *A History of the Northamptonshire Regiment, 1742–1934*. Aldershot: Regimental History Committee, 1935.

Gerges, Mark T. "Command and Control in the Peninsula: The Role of the British Cavalry 1808–1814." PhD diss., Florida State University, 2005.

Gilbert, Arthur N. "A Tale of Two Regiments: Manpower and Effectiveness in British Military Units During the Napoleonic Wars." *Armed Forces and Society* 9, no. 2 (Winter 1983): 275–292.

Gleig, George Robert. *The Subaltern: A Chronicle of the Peninsular War*. Barnsley: Pen and Sword, 2001.

Glover, Gareth, ed. *The Waterloo Archive*. 4 vols. Barnsley: Frontline, 2010–2011.

Glover, Michael. *Wellington as Military Commander*. London: Batsford, 1968.

Glover, Richard. *Peninsular Preparation: The Reform of the British Army, 1795–1809*. Cambridge: Cambridge University Press, 1963.

Gordon, Alexander. *A Cavalry Officer in the Corunna Campaign, 1808–1809: The Journal of Captain Gordon of the 15th Hussars.* Edited by H. C. Wylly. London: J. Murray, 1913.

———. *At Wellington's Right Hand. The Letters of Lieutenant-Colonel Sir Alexander Gordon, 1808–1815.* Edited by Rory Muir. Stroud: Sutton, 2003.

Grattan, William. *Adventures with the Connaught Rangers, 1809–1814.* London: Greenhill, 2003.

Graves, Donald E. *Dragon Rampant: The Royal Welch Fusiliers at War, 1793–1815.* Barnsley: Frontline, 2010.

———. *Field of Glory: The Battle of Crysler's Farm, 1813.* Toronto: Robin Brass, 1999.

———. *Fix Bayonets! A Royal Welch Fusilier at War, 1796–1815.* Stroud: Spellmount, 2007.

———. *Where Right and Glory Lead! The Battle of Lundy's Lane, 1814.* Toronto: Robin Brass, 2003.

Gray, D. S. "'Prisoners, Wanderers and Deserters' Recruiting for the King's German Legion 1803–15." *Journal of the Society for Army Historical Research* 53 (1975): 148–158.

Great Britain War Department. *British Minor Expeditions 1746–1814, Compiled in the Intelligence Branch of the Quartermaster General's Department.* London: HMSO, 1884.

Green, John. *The Vicissitudes of a Soldier's Life.* Louth: J. and J. Jackson, 1827.

Gregory, Desmond. *No Ordinary General: Lt. General Henry Bunbury (1778–1860): The Best Soldier Historian.* Madison, N.J.: Fairleigh Dickinson University Press, 1999.

———. *Sicily: The Insecure Base: A History of the British Occupation of Sicily, 1806–1815.* London: Associated University Press, 1988.

Griffith, Edwin, and Frederick Philips. *From Corunna to Waterloo: The Letters and Journals of Two Napoleonic Hussars: Major Edwin Griffith and Captain Frederick Philips 15th (King's) Hussars, 1801–1816.* Edited by Gareth Glover. London: Greenhill, 2007.

Gurney, W. B. *The Trial of Colonel Quentin of the Tenth or Prince of Wales's Own Regiment of Hussars.* London: Gale, Curtis, and Fenner, 1814.

Gurwood, Colonel John, ed. *The Dispatches of Field Marshal the Duke of Wellington, during His Various Campaigns in India, Denmark, Portugal, Spain, the Low Countries, and France.* 12 vols. London: John Murray, 1844.

———. *The General Orders of Field Marshal the Duke of Wellington KG.* 5 vols. London: T. Egerton, 1838.

Guy, Alan J. *Oeconomy and Discipline: Officership and Administration in the British Army, 1714–63.* Manchester: Manchester University Press, 1985.

———, ed. *The Road to Waterloo: The British Army and the Struggle against Revolutionary and Napoleonic France.* London: National Army Museum, 1990.

Hall, Christopher D. *British Strategy in the Napoleonic War, 1803–1815.* Manchester: Manchester University Press, 1992.

———. *Wellington's Navy: Sea Power and the Peninsular War 1807–1814.* London: Chatham, 2004.

Hamilton, Lt. General Sir F. W. *The Origins and History of the First or Grenadier Guards.* 3 vols. London: J. Murray, 1874.

Harris, Benjamin. *The Recollections of Rifleman Harris.* Edited by Henry Curling. London: Clayton, 1848.

Hay, Captain William. *Reminiscences 1808–1815, under Wellington.* London: Simpkin, Marshall, 1901.

Haythornthwaite, Philip J. *The Armies of Wellington.* London: Arms and Armour, 1994.

———. "'Carrying On the War as It Should Be.'" In Fletcher, *The Peninsular War,* 115–130.

———. *Die Hard! Dramatic Actions from the Napoleonic Wars.* London: Arms and Armour, 1996.

———. *The Waterloo Armies: Men, Organization and Tactics.* Barnsley: Pen and Sword, 2007.

Henegan, Sir Richard D. *Seven Years Campaigning in the Peninsula and the Netherlands, 1808–1815.* 2 vols. Stroud: Nonsuch, 2005.

Histman, J. Mackay, and Alice Sorby. "Independent Foreigners or Canadian Chasseurs." *Military Affairs* 25, no.1 (Spring 1961): 11–17.

Hofschröer, Peter. *1815 The Waterloo Campaign.* 2 vols. London: Greenhill, 1999.

Holmes, Richard. *Acts of War: The Behaviour of Men in Battle.* London: Weidenfeld and Nicolson, 2003.

———. *Redcoat: The British Soldier in the Age of Horse and Musket.* London: HarperCollins, 2001.

———. *Tommy: The British Soldier on the Western Front, 1914–1918.* London: HarperCollins, 2004.

———. *Wellington: The Iron Duke.* London: HarperCollins, 2002.

Hopton, Richard. *The Battle of Maida, 1806: Fifteen Minutes of Glory.* Barnsley: Leo Cooper, 2002.

Hough, Lt. Colonel William. *Precedents in Military Law.* London: W. H. Allen, 1855.

Houlding, J. A. *Fit For Service: The Training of the British Army, 1715–1795.* Oxford: Clarendon Press, 1981.

Howard, Martin. *Wellington's Doctors: The British Army Medical Services in the Napoleonic Wars.* Staplehurst: Spellmount, 2002.

Hunt, Eric, ed. *Charging Against Napoleon: Diaries and Letters of Three Hussars.* Barnsley: Pen and Sword, 2001.

James, Charles, ed. *A Collection of the Charges, Opinions, and Sentences of General Courts Martial.* London, 1820.

Jourdain, Lt. Colonel H. F. N., and Edward Fraser. *The Connaught Rangers.* 3 vols. London: Royal United Service Institution, 1924–1928.

Keegan, John. *Six Armies in Normandy. From D-Day to the Liberation of Paris.* London: Pimlico, 1992.

Kiley, Kevin F. *Artillery of the Napoleonic Wars, 1792–1815.* London: Greenhill, 2004.

Kincaid, Sir John. *Adventures in the Rifle Brigade in the Peninsula, France and the Netherlands, from 1809 to 1815.* London: P. Davies, 1929.

Latimer, Jon. *1812: War with America.* Cambridge, Mass.: Belknap Press of Harvard University Press, 2007.

Lawrence, William. *The Autobiography of Sergeant William Lawrence, A Hero of the Peninsular and Waterloo Campaigns.* Edited by George Nugent Bankes. London: Sampson Low, 1886.

Le Couteur, John. *Merry Hearts Make Light Days: The War of 1812 Journal of Lieutenant John Le Couteur, 104th Foot.* Edited by Donald E. Graves. Ottawa: Carleton University Press, 1994.

Leslie, Major John H. *The Services of the Royal Regiment of Artillery in the Peninsular War, 1808–1814.* London: H. Rees, 1908.

l'Estrange, George. *Recollections of Sir George l'Estrange.* London: S. Low, Marston, Low, and Searle, 1874.

Levinge, Sir Richard George Augustus. *Historical Records of the Forty-Third Regiment, Monmouthshire Light Infantry.* London: Clowes, 1868.

Linch, Kevin B. "The Recruitment of the British Army 1807–1815." PhD thesis, University of Leeds, 2001.

Longford, Elizabeth. *Wellington: The Years of the Sword.* London: Weidenfeld and Nicolson, 1969.

MacArthur, Roderick. "British Army Establishments during the Napoleonic Wars: Part 1—Background and Infantry." *Journal of the Society for Army Historical Research* 87 (2009): 150–172.

———. "British Army Establishments during the Napoleonic Wars: Part 2—Cavalry, Artillery, and Supporting Units." *Journal of the Society for Army Historical Research* 87 (2009): 331–356.

MacKenzie, S. P. *Revolutionary Armies in the Modern Era: A Revisionist Approach.* London: Routledge, 1997.

Mackerlie, Peter Handyside. *An Account of the Scottish Regiments, With the Statistics of Each, From 1808 to March 1861.* Edinburgh: William P. Nimmo, 1862.

Mackesy, Piers. *British Victory in Egypt, 1801: The End of Napoleon's Conquest.* London: Routledge, 1995.

———. *Statesmen at War: The Strategy of Overthrow, 1798–1799.* London: Longman, 1974.

———. *The War in the Mediterranean, 1803–1810.* London: Longmans, Green, 1957.

Mackinnon, Major General Henry. "A Journal of the Campaign in Portugal and Spain from the Year 1809 to 1812." Reprinted in *Two Peninsular War Journals.* Cambridge: Ken Trotman, 1999.

MacVeigh, James. *The Historical Records of the 78th Highlanders or Ross-shire Buffs.* Dumfries: J. Maxwell, 1887.

Marshall, Henry. *Military Miscellany; Comprehending a History of the Recruiting of the Army, Military Punishments, &c., &c.* London: John Murray, 1846.

McGrigor, James. "Sketch of the Medical History of the British Armies in the Peninsula of Spain and Portugal During the late Campaigns." *Transactions of the Medico-Chirurgical Society* 6 (1815): 381–489.

McGuffie, T. H. "The Walcheren Expedition and the Walcheren Fever." *English Historical Review* 26, no. 243 (April 1947): 191–202.

McGuigan, Ron. "The Origin of Wellington's Peninsular Army, June 1808–April 1809." In Muir et al., *Inside Wellington's Peninsular Army*, 39–70.

———. "Wellington's Generals in Portugal, Spain and France: 1809–1814." In Muir et al., *Inside Wellington's Peninsular Army*, 172–200.

Memoirs of a Sergeant. Stroud: Nonsuch, 2005.

Moon, Joshua. *Wellington's Two-Front War: The Peninsular Campaigns, at Home and Abroad, 1808–1814.* Norman: University of Oklahoma Press, 2011.

Morris, Thomas. *Recollections of Military Service, in 1813, 1814, & 1815, through Germany, Holland, and France, including some Details of the Battles of Quatre Bras and Waterloo.* London: James Madden, 1845.

Muir, Howie. "Order of Battle: Customary Battle-Array in Wellington's Peninsular Army." In Muir et al., *Inside Wellington's Peninsular Army*, 84–171.

Muir, Rory. *Britain and the Defeat of Napoleon, 1807–1815.* London: Yale University Press, 1996.

———. *Salamanca 1812.* London: Yale University Press, 2001.

———. *Tactics and the Experience of Battle in the Age of Napoleon.* London: Yale University Press, 1998.

———. "Wellington and the Peninsular War." In Muir et al., *Inside Wellington's Peninsular Army*, 1–38.

Muir, Rory, Robert Burnham, Howie Muir, and Ron McGuigan. *Inside Wellington's Peninsular Army, 1808–1814.* Barnsley: Pen and Sword, 2006.

Muir, Rory, and Charles Esdaile. "Strategic Planning in a Time of Small Government: The Wars against Revolutionary and Napoleonic France, 1793–1815." In Woolgar, *Wellington Studies*, 1–90.

Musteen, Jason R. "Becoming Nelson's Refuge and Wellington's Rock: The Ascendancy of Gibraltar during the Age of Napoleon." PhD diss., Florida State University, 2005.

Myatt, Frederick. *Peninsular General: Sir Thomas Picton, 1758–1815.* London: David and Charles, 1980.

Napier, George. *Passages in the Early Military Life of General Sir George T. Napier, K.C.B.* London: J. Murray, 1884.

Napier, William. *History of the War in the Peninsula and in the South of France, from the Year 1807 to the Year 1814.* 3rd ed. 6 vols. London: D. Christie, 1835.

Newbolt, Sir Henry. *The Story of the Oxfordshire and Buckinghamshire Light Infantry (The old 43rd and 52nd Regiments).* London: C. Scibner, 1915.

Nichols, Alistair. *Wellington's Mongrel Regiment: A History of the Chasseurs Britanniques Regiment, 1801–1814.* London: Spellmount, 2005.

Oatts, Lt. Colonel Lewis Balfour. *Proud Heritage: The Story of the Highland Light Infantry.* 4 vols. London: Nelson, 1952–1963.

Oman, Sir Charles. *A History of the Peninsular War.* 7 vols. Oxford: Clarendon, 1902–1930.

———. *Wellington's Army, 1809–1914.* London: Arnold, 1912.

Partridge, Richard, and Michael Oliver. *Napoleonic Army Handbook.* Vol. 1, *The British Army and Her Allies.* London: Constable, 1999.

Petty, S. "Wellington's General Orders." In Woolgar, *Wellington Studies*, 141–146.

Rait, Robert S. *The Life and Campaigns of Hugh First Viscount Gough, Field Marshal.* 2 vols. Westminster: Constable, 1903.

Reid, Stuart. *Wellington's Highland Warriors: From the Black Watch Mutiny to the Battle of Waterloo, 1743–1815.* Barnsley: Frontline, 2010.

Robertson, Ian C. *A Commanding Presence: Wellington in the Peninsula, 1808–1814: Logistics, Strategy, Survival.* Stroud: Spellmount, 2008.

———. *Wellington Invades France: The Final Phase of the Peninsular War, 1813–1814.* London: Greenhill, 2003.

Robinson, Mike. *The Battle of Quatre Bras, 1815.* Stroud: History Press, 2009.

Ross, Steven T. "The Development of the Combat Division in Eighteenth-Century French Armies." *French Historical Studies* 4 no.1 (Spring 1965): 84–94.

Schaumann, August Ludolf Friedrich. *On the Road with Wellington: The Diary of a War Commissary.* London: Greenhill, 1999.

Sherer, Moyle. *Recollections of the Peninsula.* Staplehurst: Spellmount, 1996.

Simmons, Major George. *A British Rifle Man: Journals and Correspondence during the Peninsular War and the Campaign of Waterloo.* Edited by William Verner. London: A. and C. Black, 1899.

Smith, Digby. *1813, Leipzig: Napoleon and the Battle of the Nations.* London: Greenhill, 2001.

———. *The Greenhill Napoleonic Wars Data Book: Actions and Losses in Personnel, Colours, Standards and Artillery, 1792–1815.* London: Greenhill, 1998.

Smith, Sir Harry. *The Autobiography of Sir Harry Smith.* Edited by G. C. Moore Smith. London: Constable, 1999.

Smurthwaite, Lesley. "Glory is Priceless! Awards to the British Army during the French Revolutionary and Napoleonic Wars." In Guy, *The Road to Waterloo,* 16–22.

Smythies, R. H. Raymond. *Historical Records of the 40th (2nd Somersetshire) Regiment.* Devonport: A. H. Swiss, 1894.

Stanhope, James Hamilton. *Eyewitness to the Peninsular War and Waterloo: The Letters and Journals of Lieutenant Colonel the Honourable James Stanhope, 1803 to 1825.* Edited by Gareth Glover. Barnsley: Pen and Sword, 2010.

Steiner, E. E. "Separating the Soldier from the Citizen: Ideology and Criticism of Corporal Punishment in the British Armies, 1790–1815." *Social History* 8, no. 1 (January 1983): 19–35.

Stewart, Captain P. F. *The History of the XII Royal Lancers (Prince of Wales's).* Oxford: Oxford University Press, 1950.

Strachan, Hew. *Wellington's Legacy: The Reform of the British Army, 1830–1854.* Manchester: Manchester University Press, 1984.

Summerville, Christopher. *March of Death: Sir John Moore's Retreat to Corunna, 1808–1809.* London: Greenhill, 2003.

Surtees, William. *Twenty-Five Years in the Rifle Brigade.* London: Muller, 1833.

Tennant, R. J. "Wellington's Mules." *First Empire* 113 (July/August 2010): 7–20.

Thoumine, R. H. *Scientific Soldier: A Life of General Le Marchant, 1766–1812.* Oxford: Oxford University Press, 1968.

Tomkinson, Lt. Colonel William. *The Diary of a Cavalry Officer in the Peninsular and Waterloo Campaigns, 1809–1815.* Staplehurst: Spellmount, 1999.

Urban, Mark. *Fusiliers: How the British Army Lost America But Learned to Fight.* London: Faber and Faber, 2007.

———. *The Man Who Broke Napoleon's Codes: The Story of George Scovell.* London: Faber and Faber, 2001.

———. *Rifles: Six Years with Wellington's Legendary Sharpshooters.* London: Faber and Faber, 2003.

Ward, Harriet. *Recollections of an Old Soldier: A Biographical Sketch of the Late Colonel Tidy CB, 25th Regt.* London: Bentley, 1849.

Watson, Bruce Allen. *When Soldiers Quit: Studies in Military Disintegration.* Westport, Conn.: Praeger, 1997.

Weller, Jac. *Wellington in India.* London: Greenhill, 1993.

———. *Wellington in the Peninsula, 1808–1814.* London: Greenhill, 1992.

Wellesley, Arthur Richard, ed. *Supplementary Despatches and Memoranda of Field Marshal Arthur Duke of Wellington.* 15 vols. London: J. Murray, 1858–1872.

Wheeler, William. *The Letters of Private Wheeler, 1809–1828.* Edited by Captain B. H. Liddell Hart. London: M. Joseph, 1951.

White-Spunner, Barney. *Horse Guards.* London: Macmillan, 2006.

Willis, Clive. "Colonel George Lake and the Battle of Roliça." *Portuguese Studies* 12 (1996): 68–77.

Woodman, Richard. *The Victory of Seapower: Winning the Napoleonic War, 1806–1814.* London: Chatham, 1998.

Woolgar, C. M., ed. *Wellington Studies* 1. Southampton: Hartley Institute, University of Southampton, 1996.

Woolright, H. H. *History of the Fifty-Seventh (West Middlesex) Regiment of Foot 1755–1881.* London: Richard Bentley, 1893.

Zamoyski, Adam. *Rites of Peace: The Fall of Napoleon and the Congress of Vienna.* London: HarperCollins, 2007.

Online Sources

Arnold, James R. "A Reappraisal of Column Versus Line in the Peninsular War." Military Subjects: Organization, Strategy & Tactics. The Napoleon Series. http://www.napoleon-series.org/military/organization/maida/c_maida3.html.

Bamford, Andrew. "The British Army in the Low Countries, 1813–1814." Military Subjects: Battles & Campaigns. The Napoleon Series. http://www.napoleon-series.org/military/battles/c_lowcountries1814.html.

———. "British Forces at Cadiz. 1810–1814: Organisation, Strength, and Losses." Military Subjects: Battles & Campaigns. The Napoleon Series. http://www.napoleon-series.org/military/organization/Britain/Strength/Cadiz/c_CadizIntro.html.

———. "The Corps of Embodied Detachments, 1809." Military Subjects: Battles & Campaigns. The Napoleon Series. www.napoleon-series.org/military/battles/c_walcherendetachment.html.

———. "The Guadiana Fever Epidemic." Military Subjects: Battles & Campaigns. The Napoleon Series. http://www.napoleon-series.org/military/battles/c_Guadiana1.html.

Bois, Mark. "The Inniskillings at Waterloo." Military Subjects: Organization, Strategy & Tactics. The Napoleon Series. http://www.napoleon-series.org/military/organization/Britain/Infantry/1-27Waterloo/c_1-27Waterloo.html.

Brown, Steve. "British Regiments and the Men Who Led Them, 1793–1815." Military Subjects: Organization, Strategy & Tactics. The Napoleon Series. http://www.napoleon-series.org/military/organization/Britain/Infantry/Regiments/c_InfantryregimentsIntro.html.

Burnham, Robert. "The British Army in the Napoleonic Wars: Manpower Stretched to the Limits?" Military Subjects: Organization, Strategy & Tactics. The Napoleon Series. www.napoleon-series.org/military/organization/c_casualties.html.

———. "The British Expeditionary Force to Walcheren: 1809." Military Subjects: Battles & Campaigns. The Napoleon Series. www.napoleon-series.org/military/battles/c_walcheren.html.

————, ed. "Lionel S. Challis's 'Peninsula Roll Call.'" Research Subjects: Biographies. The Napoleon Series. http://www.napoleon-series.org/research/biographies/GreatBritain/Challis/c_ChallisIntro.html.

Burnham, Robert, and Ron McGuigan. "Not One in Ten Thousand Know Your Name: The Officers of the British 1st Battalion of Detachments in 1809." Research Subjects: Biographies. The Napoleon Series. http://www.napoleon-series.org/research/biographies/GreatBritain/Detachments/c_DetachmentsIntro.html.

Centeno, João. "General Officers in the Portuguese Army." Military Subjects: Organization, Strategy & Tactics. The Napoleon Series. www.napoleon-series.org/military/organization/portugal/c_generals.html.

Cook, John, and Robert Burnham. "Nicknames of British Units in the Napoleonic Wars." Military Subjects: Organization, Strategy & Tactics. The Napoleon Series. www.napoleon-series.org/military/organization/c_nickname.html.

Dunn, John. "The Royal Veterans." Pt. 1. *Garrison Gazette* (Spring 2007): 3–4 http://73rdregiment.tripod.com/sitebuildercontent/sitebuilderfiles/gg spring2007.pdf.

Graves, Donald. "British Military Discipline in the Napoleonic Period: Gleanings from the Inquiry into the System of Military Punishments in the Army, 1836." Military Subjects: Organization, Strategy & Tactics. The Napoleon Series. http://www.napoleon-series.org/military/organization/Britain/Miscellaneous/c_militaryjusticeinquiry.html.

————. "'Every horror was committed with impunity . . . and not a man was punished!' Reflections on British Military Law and the Atrocities at Hampton in 1813." *War of 1812 Magazine,* Issue 11, June 2009. http://www.napoleon-series.org/military/Warof1812/2009/Issue11/c_hampton.html

————. "'We Have an Immediate Opening': The Seniority of Brigade Commanders in the Peninsular Army, 1812." Military Subjects: Organization, Strategy & Tactics. The Napoleon Series. www.napoleon-series.org/military/organization/c_seniority.html.

Grodzinski, John R. "Command and Control in Upper Canada." *War of 1812 Magazine* 1 (January 2006). www.napoleon-series.org/military/Warof1812/2006/Issue1/c_ccuc.html.

————. "Much To Be Desired: The Campaign Experience of British General Officers of the War of 1812." *War of 1812 Magazine* 7 (September 2007). http://www.napoleon-series.org/military/Warof1812/2007/Issue7/c_BritishGenerals .html.

Hadaway, Stuart. "Rolica: A Most Important Affair." Military Subjects: Battles & Campaigns. The Napoleon Series. www.napoleon-series.org/military/battles/c_rolica1.html.

McGuigan, Ron. "Anglo-Allied Army in Flanders and France—1815." Military Subjects: Battles & Campaigns. The Napoleon Series. www.napoleon-series.org/military/battles/c_waterloo1.html.

————. "The British Army Stationed in British North America 1812–1815." Military Subjects: Battles & Campaigns. The Napoleon Series. http://www.napoleon-series.org/military/battles/bna/c_bna1.html.

―――. "British Generals of the Napoleonic Wars 1793–1815." Research Subjects: British Generals. The Napoleon Series. www.napoleon-series.org/research/biographies/BritishGenerals/c_Britishgenerals1.html.

Millar, C. M. H. "The Dismissal of Colonel Duncan Macdonald of the 57th Regiment." *Clan Donald Magazine Online* 10 (1984). www.clandonald.org.uk/magazine (site discontinued).

―――. "Further Notes on the Career of Colonel Duncan Macdonald of the 57th Regiment." *Clan Donald Magazine Online* 12 (1991). www.clandonald.org.uk/magazine (site discontinued).

Index

Numbered military units appear first in this index. Numbered entries and subentries are listed in numerical order. Non-numbered entries are arranged alphabetically.

1st Dragoon Guards, 106–108
1st Dragoons, 194, 277, 284
1st Foot, 11–12, 49, 315n51; Third Battalion, 14, 106, 305; Fourth battalion, 16, 88, 106, 156–157
1st Foot Guards, 11, 191, 209; First Battalion, 124, 137, 305; Second Battalion, 138–139, 161; Third Battalion, 120, 124–125, 137, 139, 305
1st Foreign Veteran Battalion, 11, 117
1st KGL Dragoons, 277, 279, 281–282
1st KGL Hussars, 196, 273, 277–279, 281, 283–284
1st KGL Light Battalion, 157, 306
1st KGL Line Battalion, 133, 140, 157
1st Life Guards, 18, 107–108, 277
1st Royal Veteran Battalion, 162–163
2nd Dragoons, 56, 315n51
2nd Foot, 49, 73–79, 94, 147–148, 155, 208, 225, 289, 305
2nd KGL Dragoons, 134, 277–279, 281
2nd KGL Hussars, 120, 273, 277–279, 281
2nd KGL Light Battalion, 157, 306
2nd KGL Line Battalion, 140, 157
2nd Life Guards, 18, 107–108, 277
3rd Dragoon Guards, 277, 284
3rd Dragoons, 73, 77, 277, 282
3rd Foot, 314n17; First Battalion, 49, 54, 80, 90, 144, 254–255
3rd Foot Guards, 53, 264; First Battalion, 137; Second Battalion, 138, 161
3rd KGL Hussars, 133, 157, 276–277

4th Dragoon Guards, 51–52, 196, 277, 280–281, 315n51
4th Dragoons, 196, 272, 277, 284
4th Foot, 171; First Battalion, 186, 227, 305; Second Battalion, 120, 173
4th Royal Veteran Battalion, 125
5th Dragoon Guards, 260, 272, 277
5th Dragoons, 18
5th Foot, 171; First Battalion, 104, 231, 305; Second Battalion, 100, 171–173
5th KGL Line Battalion, 140, 157, 224–225
6th Dragoons, 315n51
6th Foot, 136
7th Fusiliers, 58, 102, 104, 314n8; First Battalion, 46, 53, 84, 150, 177, 256; Second Battalion, 84, 100, 188, 237
7th Hussars, 107, 277
7th KGL Line Battalion, 140, 331n73
7th Royal Veteran Battalion (disbanded 1814), 125. *See also* 13th Royal Veteran Battalion (renumbered as 7th in 1815)
8th Foot, 136; Second Battalion, 100, 141; Second Battalion, 100, 141–142
9th Foot, 17, 171; First Battalion, 17, 125, 133, 305; Second Battalion, 17, 119, 125, 133, 173
9th Light Dragoons, 231, 277, 279, 281–282
10th Hussars, 62, 72, 77, 109, 277
10th Royal Veteran Battalion, 117–118
11th Foot: First Battalion, 125, 150, 207; Second Battalion, 125

11th Light Dragoons, 49, 62, 73, 272, 277, 314n19
12th Foot, 135–36
12th Light Dragoons, 61–62, 271–272, 277–278, 318n44
13th Light Dragoons, 194, 272, 277
13th Royal Veteran Battalion (renumbered as 7th in 1815), 11, 117, 231
14th Foot, 11, 58, 158, 171; Second Battalion, 158, 305; Third Battalion, 63, 88, 158, 171–173
14th Light Dragoons, 268, 277, 280–281, 284
15th Foot, 135–136; Second Battalion, 173
15th Hussars, 50, 55, 109, 277, 283–285
16th Foot, 58
16th Light Dragoons, 109–112, 268, 272, 277, 280, 284
17th Foot, 73
18th Foot, 315n51; Second Battalion, 88
18th Hussars, 196, 270, 277, 279, 284
19th Foot, 135–136, 171, 173
20th Foot, 48, 86–87, 89, 94, 96, 98, 251, 305
20th Light Dragoons, 107, 277
21st Fusiliers, 314n8, 315n51; Second Battalion, 164
22nd Foot, 135–136, 158, 171; 1st battalion, 159; Second Battalion, 158–160, 171–174
23rd Fusiliers, 58–59, 290, 296, 314n8; 1st battalion, 59, 177, 225; Second Battalion, 141–142, 153, 305
23rd Light Dragoons, 217, 268, 276–277, 297
24th Foot, Second Battalion, 145, 147
26th Foot, 315n51; 1st battalion, 114, 227, 230, 305
27th Foot, 11, 59, 100, 315n51; First Battalion, 100, 106, 177, 255–256; Second Battalion, 100; Third Battalion, 14, 88, 100, 106, 114, 209
28th Foot, 46, 83; First Battalion, 104, 120, 186, 305; Second Battalion, 100
29th Foot, 91, 94–98, 136, 144, 152, 154, 173, 217, 234
30th Foot, 290, 315n64; Second Battalion, 119–121, 147, 155–156, 164, 175, 211

31st Foot: First Battalion, 154; Second Battalion, 95, 144–147, 153–155, 167
32nd Foot; First Battalion, 305, 337n1
33rd Foot, 21, 49, 135–136, 156, 164, 167–168, 254
34th Foot, 14; First Battalion, 14; Second Battalion, 9, 14, 90
35th Foot; Second Battalion, 162–164
36th Foot, 134, 136; First Battalion, 16, 305
37th Foot, 158; First Battalion, 115, 158, 319n62; Second Battalion, 158, 162–164, 168, 172
38th Foot; First Battalion, 104, 227, 305; Second Battalion, 145
39th Foot, 47, 171; First Battalion, 104; Second Battalion, 100, 173, 236
40th Foot, 91; First Battalion, 65, 69–73, 90–94, 177, 222, 234, 247, 256–257; Second Battalion, 91–92
41st Foot, 135–136; First Battalion, 100; Second Battalion, 100, 102
42nd Highlanders, 47, 314n8, 315n47; First Battalion, 56–57, 104, 130, 305, 326n17; Second Battalion, 145, 236
43rd Light Infantry, 48, 58–59, 129, 297; First Battalion, 26, 89, 227, 256, 305; Second Battalion, 89, 91, 184, 227–228, 234–235, 305, 330n41
44th Foot; Second Battalion, 121, 144, 155–156, 163, 175, 211
45th Foot; First Battalion, 16, 150, 188, 214
46th Foot, 98
47th Foot; Second Battalion, 120–122
48th Foot, 102, 104, 136, 144; First Battalion, 63, 119; Second Battalion, 100
50th Foot, 136, 296; First Battalion, 305
51st Light Infantry, 48, 58, 63, 81, 98, 149, 305, 326n21, 331n76
52nd Light Infantry, 48, 51, 57, 129, 253; First Battalion, 26, 48, 61, 89, 106, 163, 227, 306, 337n1; Second Battalion, 12, 89, 104, 106, 162–163, 227–228, 306
53rd Foot, 135–136; Second Battalion, 78, 147–148, 155, 209, 223, 247, 251, 253

54th Foot, 156

55th Foot, 163

56th Foot, 11, 158; Second Battalion, 88;
Third Battalion, 88, 158, 162–163,
168, 172

57th Foot, 46; First Battalion, 63, 69,
80–81, 90, 119, 144

58th Foot, 49; First Battalion, 155;
Second Battalion, 113, 145, 147, 155

59th Foot; Second Battalion, 120–122,
154, 306

60th Foot; 11, 12, 48–49, 165–166, 193,
323n28; Fifth Battalion, 193, 236,
248, 308; Seventh Battalion, 165–166,
294; Eighth Battalion, 165–166

61st Foot; First Battalion, 119

62nd Foot; Second Battalion, 88,
115–116, 150–151

63rd Foot, 171–173

64th Foot, 118, 135–136

66th Foot; Second Battalion, 96,
144–147, 153, 155

68th Light Infantry, 9, 48, 94, 149, 225,
228, 230

69th Foot; Second Battalion, 54,
163–164

70th Foot, 47, 315n51

71st Highland Light Infantry, 48, 51,
67, 314n8; First Battalion, 48, 51,
63, 79–82, 142, 150, 153, 211, 306,
326n21

72nd Foot (Highlanders until 1808), 47,
51, 314n8

73rd Foot (Highlanders until 1808),
290, 314n8; First Battalion, 118;
Second Battalion, 44–45, 81–82,
156–157, 159, 291

74th Foot (Highlanders until 1808),
47, 50

75th Foot (Highlanders until 1808), 47,
314n8

76th Foot, 50, 115, 150, 306

77th Foot, 49, 114–115, 230

78th Highlanders, 46–47, 135–137,
314n8 First Battalion, 136; Second
Battalion, 48, 136, 164

79th Highlanders, 47, 136, 314n8; First
Battalion, 57, 266, 306

80th Foot, 47

81st Foot, 47; Second Battalion, 164,
306

82nd Foot, 47; First Battalion, 206, 306

83rd Foot, 99; First Battalion, 100;
Second Battalion, 88, 99–101, 114,
222, 236

84th Foot; Second Battalion, 115–116,
159, 247, 319n62

85th Light Infantry, 47–48, 50, 62, 96,
115–116, 142, 153

86th Foot, 158, 171, 315n51; First
Battalion, 159; Second Battalion,
158–60, 168, 172–173

87th Foot, 47, 59, 251, 315n51, 316n79;
Second Battalion, 59, 66–67,
122–123, 236, 250

88th Foot, 59, 102–104, 136, 207,
315n51; First Battalion, 14–15, 50,
55, 64, 102–105, 214, 250; Second
Battalion, 14–15, 102–104, 106, 114,
119

89th Foot, 82, 290; Second Battalion,
15, 82–83

90th Foot, 47–48, 315n51

91st Foot (Highlanders until 1808), 47,
314n8; First Battalion, 16, 47, 57,
306; Second Battalion, 47, 156–157,
167–168, 251, 254

92nd Highlanders, 47, 314n8; First
Battalion, 47, 306

93rd Highlanders, 314n8

94th Foot, 46, 52, 210, 315n51

95th Rifles, 11, 48, 57, 59, 129–130,
140–141, 206, 296, 314n15, 324n120;
First Battalion, 26, 63–64, 140, 162,
211, 227, 237, 251, 256, 306; Second
Battalion, 64, 66, 120–121, 140,
162, 227–229, 306; Third Battalion,
57, 88, 120, 140, 162, 227–228;
Provisional Battalion, 162–163

96th Foot, 12

97th Foot, 96, 253, 293, 308

98th Foot, 118

99th Foot, 315n51

100th Foot, 98, 315n51

101st Foot, 13, 315n51

102nd Foot, 13, 118, 166, 293, 323n28

103rd Foot, 13, 323n28

104th Foot, 11, 13, 51, 55, 98, 293,
323n28

Abercrombie, Lt. General Sir Ralph,
180–181

Acland, Major General Wroth Palmer, 23, 173, 183
Adjutant General, Post of, 5, 7, 8, 67, 154
Aitcheson, Lt. John, 53–54, 264, 266
Albuera, Battle of, 28, 46, 50, 54, 90, 95, 102, 106, 144, 146, 201, 217, 220, 235, 290, 328n78
Alcohol, 44, 50
Aldea de Ponte, Battle of, 46
Almeida, Portugal, 27–29, 74, 112, 252
Almond, Private Joseph, 256
Alten, Lt. General Karl von, 190–193, 203, 210–211
Alten, Major General Victor von, 196
American Revolutionary War, 39, 179, 297–299
Anson, Major General Sir George, 62
Anson, Major General William, 71, 148, 209
Anstruther, Brigadier General Robert, 23, 183
Antwerp, Netherlands, 25, 36, 116
Archdall, Lt. Colonel Richard, 69–73, 258
Armstrong, Paymaster William, 134
Army Corps, organization of, 179, 191–192, 201–203
Arroyomolinos, Battle of, 30
Ashworth, Colonel Charles, 192
Austrian Army, 24–25, 27, 286
Aylmer, Major General Mathew, Lord, 115–117, 150, 192, 196

Badajoz, Spain, 28; operations against (1811), 29, 103; storm of (1812), 30–31, 69, 92, 198, 211, 213, 220, 232, 235
Baird, Lt. General Sir David, 24, 184, 337n1
Baltic, Expedition to: 1808, 22, 182, 186; 1813–1814, 35–36, 44, 156–157, 161, 164, 167, 254, 304
Barossa, Battle of, 30, 47, 120, 122, 139
Bathurst, Henry, Earl of, 5, 6, 33, 37–38, 127–128, 148, 151, 158, 160–163, 169–70, 174–176, 271, 299
Battalions of Detachments: Peninsula 1809, 129–133; York's 1814 Project for, 127, 135, 170–174, 176. See also Corps of Embodied Detachments

Bayly, Major General Sir Henry, 169
Bayonne, operations against, 37
Beckwith, Major General Sidney, 63–64
"Belem Rangers," 243, 259
Bennett, Private (20th Foot), 251
Bentinck, Lt. General Lord William, 34, 188
Beresford, Lt. General Sir William Carr, 27–28, 188–189, 201, 249, 326n17
Bergen op Zoom, attack on, 36, 54, 164, 170
Bermuda, 166
Billets, 110, 240–42
Bingham, Lt. Colonel George Ridout, 78, 209
Bissett, Commissary General Sir John, 264
Blakeney, Colonel Edward, 256
Bock, Major General Baron E.O.G. von, 278
Bonaparte, Joseph, 33
Bonaparte, Napoleon. See Napoleon I, Emperor of the French
Bork, Private Henry, 282
Bowes, Brigadier General Barnard, 326n17
Bridger, Captain James, 272
Brigades: 1st Guards Brigade, 54, 124, 137–139, 235, 241, 321n27; 2nd Guards Brigade, 137, 139; 3rd Brigade (1809), 249–250; 3rd Guards Brigade, 138–139, 161; 7th Brigade (1815), 116; 12th Brigade (1815), 116–117; Brigade of Provisional Militia, 169–171, 192; Flank Brigades, 184; Household Brigade (see Household Cavalry); Hussar Brigade, 144, 196, 266, 276, 285; Light Brigade, 26, 48, 114, 190, 228; Organization of, 104, 106, 108, 176, 196, 207, 288. For brigades in peninsular army post-1809, see names of incumbent commanders
British North America, 20, 32, 38, 40, 42, 82–83, 87, 90, 96, 99–100, 104, 112, 117–118, 127, 135, 142–143, 156, 165–166, 220, 223–224, 246, 249, 255–256, 293, 295, 300, 304
Brock, Major General Isaac, 32
Brophy, Private Ody, 55
Browne, Captain Thomas, 59, 296

Browne, Lt. Colonel John, 168
Brownrigg, Lt. General Sir Robert, 5
Brumwell, Stephen (historian), 298
Brunswick-Oels Hussars, 11, 19
Brunswick-Oels Light Infantry, 11, 193
Bullocks. See Oxen, draught
Bunbury, Lt. Colonel William, 80
Bunbury, Major General Sir Henry, 6,
    178
Burgos, Spain: retreat from, 31, 56, 60,
    92, 146, 148, 201, 241, 249, 260, 266,
    282, 290; siege of, 31, 213, 231
Burgoyne, Lt. Colonel John Fox, 283
Burgoyne, Major General Montague,
    83
Burne, Colonel Robert, 195
Burrard, Lt. General Sir Harry, 23, 182,
    184
Busaço, Battle of, 27, 50, 237
Byng, Major General John, 80–81, 146,
    255

Cadiz, Spain, 7, 24, 29, 33–34, 52, 66,
    96, 102, 112, 118–122, 125, 137–140,
    154, 161, 165–166, 192, 223, 229,
    246, 251, 273, 303; siege of, 27, 29,
    245
Cadogan, Colonel Hon. Henry, 51, 63,
    79, 81–82, 189, 317n118
Calabrian Freecorps, 17
Calcraft, Major General Sir Granby,
    51–52
Calvert, Lt. General Sir Harry, 5, 58, 67,
    174, 295
Cameron, Brigadier General Alan, 129,
    188
Cameron, Lt. Colonel Alexander, 162
Cameron of Fassiefern, Lt. Colonel
    John, 47
Campbell, Lt. Colonel James, 214
Campbell, Lt. General Sir Colin, 10, 11,
    83, 125
Campbell, Major General Alexander,
    188, 195–196, 200
Canadian units, 8, 224, 253–254, 304;
    Fencibles, 32, 43, 254; Militia, 32
Canning, George, 128, 170
Capel, Colonel Hon. Edward, 120
Castlereagh, Robert, Viscount, 3, 5, 12,
    13, 21, 24, 95, 127–128, 132, 134,
    167, 188

Cathcart, Lt. General William, Earl of,
    181
Champalimaud, Coronel José, 191
Channel Islands, 15, 36. See also
    Guernsey; Jersey
Chasseurs Britanniques, 17, 192–193,
    254
Chateauguay, Battle of, 35
Chatham, Lt. General John, Earl of,
    6, 25, 92, 134, 185, 199–200, 228,
    230–231
Chippewa, Battle of, 38
Cimetière, Captain Gilbert, 144
Cintra, Convention of, 23, 182
Ciudad Rodrigo, Spain, 27, 29, 31;
    storm of, 29, 30, 190, 198, 213, 232,
    312n80
Cliford, Major Miller, 83
Clifton, Lt. Colonel Arthur, 194
Clinton, Lt. General Sir Henry, 67,
    75–77, 208–211
Clinton, Lt. General Sir William, 34
Côa, Combat on the, 56, 205
Cochrane, Lt. Colonel Hon. Basil, 134
Cocks, Captain Hon. Edward Somers,
    110, 112
Cocoa, proposed issue of, 240
Colborne, Colonel John, 29, 48, 95,
    144
Coldstream Guards: First battalion, 137;
    Second battalion, 138, 161
Cole, Lt. General Sir Galbraith Lowry,
    71, 177, 188, 201, 208–211, 215
Colley, Linda (historian), 206, 298,
    336n44
Colonial units. See Foreign units
Colours, 50, 81
Colville, Lt. General Charles, 200, 211
Commander in Chief, post of, 3, 5, 6–7,
    12, 61, 67–68, 71, 127, 133, 146, 152,
    156, 174–175, 209
Commissary in Chief, post of, 24
Commissary Services, 19–20, 189, 234,
    263
Companies of Independent Foreigners,
    166, 293–294
Conscription, 43, 286, 291
Convalescents, 243–44
Cooke, Major General George, 96,
    120–121, 161, 168
Cookson, John (historian), 47, 207

Cooper, Sergeant John Spencer, 58, 84, 237, 241, 256, 290
Coote, Lt. General Sir Eyre, 185
Copenhagen Expedition (1807), 22, 181
Corporal punishment, 52, 64–72, 74–77, 80–82, 250, 258, 260, 281–282
Corps of Embodied Detachments, 134–137, 141, 159
Corunna, Spain, 24, 266; Battle of, 24; retreat to, 55, 186, 249–252
Corunna Campaign (1808–1809), 24–25, 48, 57, 73, 86, 88, 90, 128–129, 141–142, 224, 228, 232, 235, 266, 305–306, 326n21, 337n1
Coss, Edward (historian), 238, 291, 328n27, 329n10, 330n58
Cotton, Lt. General Sir Stapleton, 133, 189, 194, 200
County titles, 47, 58
Courts-martial, 10, 62, 69, 76, 244–245, 248, 253, 280; Drumhead, 64–66; General, 53, 64; Regimental, 69, 80, 165, 260, 180–181
Cradock, Lt. General Sir John, 25, 128, 132, 187, 199, 269
Craufurd, Brigadier General James Catlin, 9, 327n59
Craufurd, Major General Robert, 26, 56, 64–65, 114, 190, 205, 208–209, 250, 256, 289–290, 297
Crowe, Lt. Charles, 209
Crysler's Farm, Battle of, 35, 83
Cumming, Colonel Henry, 62
Cunningham, Private (87th Foot), 251
Cuyler, Lt. Colonel Henry, 62

Dalhousie, Lt. General George, Earl of, 37, 201
Dalrymple, Lt. General Sir Hew, 23, 182–186
D'Arcy, Lt. John, 55
Davy, Private Henry, 253
Dawes, Troop Sergeant Major William, 283–285, 334n69
"De-Kilting," 47, 59
Depots: Army, 134–135; Foreign, 165; Regimental, 89, 96, 103, 154–155, 158, 162–63, 173, 227, 272
Desertion, 18, 65, 74, 78, 91, 131–132, 167, 218, 245–259, 307–308
Detroit, Battle of, 32

Dietz, Peter (historian), 296
Dilkes, Brigadier General William, 138
Disposable Force, 21, 90, 128
Divisions, 115, 178–201, 288, 294–295, 297–298; First Division (Hundred Days), 210; First Division (Peninsula), 54, 56, 74, 115, 124, 140, 188–193, 195, 197–198, 200–201, 205, 208–209, 212; Second Division (Hundred Days), 117, 203, 210; Second Division (Peninsula), 29, 144, 188, 190–192, 197–198, 201, 205–206, 208–209, 298; Third Division (Hundred Days), 203; Third Division (Peninsula), 102, 190, 198, 205, 207–208, 211–212, 214, 289; Third Division (Walcheren), 134; Fourth Division (Peninsula) 29, 104, 112, 148, 177, 188, 190, 195, 198, 200, 204–205, 207–208, 210, 212, 295; Fifth Division (Hundred Days), 177; Fifth Division (Peninsula), 115–116, 190, 195–196, 198–199, 205, 212–213, 290; Fifth Division (Walcheren), 185, 326n23; Sixth Division (Hundred Days), 177; Sixth Division (Peninsula), 56–57, 104, 190–191, 195–200, 205, 207–208; Seventh Division (Peninsula), 37, 185, 190–195, 197–198, 205–206, 208, 295; First Cavalry Division (Peninsula; unnumbered 1809–1811, 1813–1814), 189, 194, 206, 208, 295; Second Cavalry Division (Peninsula), 194–95, 201, 208, 278, 295; Light Division (Peninsula), 56, 140, 167, 178, 190, 192–193, 197–98, 204, 208–209, 211–212, 256, 279, 289–290, 295; Light Division (Walcheren), 185; Portuguese Division (Peninsula), 190–192, 197, 208, 327n45
Doherty, Lt. Colonel Patrick, 194
Don, Lt. General Sir George, 185
Donaldson, Sergeant Joseph, 52–53, 65, 209–210
Donellan, Lt. Colonel Charles, 63
Douglas, Captain Neil, 266
Doyle, Lt. Colonel Charles William, 67, 316n79
Doyle, Lt. General Sir John, 73–74, 78, 164, 316n79

Drill, 10, 44, 75, 165
Drummond, Lt. General Gordon, 38
Dundas, General Sir David, 5, 7,
    127–128, 132–134, 137–138, 159,
    175, 320n2
Dunlop, Major General James, 195

East Coast of Spain, operations on
    (1812–14), 19, 33–34, 100, 155, 165,
    215, 246, 269, 303
Egyptian Campaign (1801), 46, 48, 86,
    169, 180–181, 219
Egyptian Campaign (1807), 21, 181
El Bodón, action at, 53
Elting, John (historian), 242
Elvas, Portugal, 112; hospital at,
    236–237, 241
Epidemics: 1809 Peninsula epidemic,
    235–237; 1813 Gibraltar Yellow Fever
    epidemic, 223–224; 1813 Peninsula
    epidemic, 124, 191, 235, 241. See also
    Walcheren Fever
Erskine, Lt. General Sir William, 140,
    195, 201, 290
Esprit de corps, 45, 50, 53, 60, 84, 87,
    148, 211, 287, 289
Establishments, regimental, 13, 15,
    106–107 158–161

Fane, Major General Hon. Henry,
    194–195, 200
Fencible Infantry, 12, 169
Fitzgerald, Lt. Colonel John, 166
Fletcher, Ian (historian), 275, 334n69
Flogging. See Corporal punishment
Fonseca, Brigadeiro A. L. da, 191
Foot Guards, 11, 17, 26, 47, 54–56,
    137–141, 171, 185–186, 205, 209,
    258, 296–297. See also individual units
    by name
Forage, 110, 262, 268–269, 285; theft of,
    280–282
Foreigners, enlistment of, 11, 43,
    165–168, 254–255, 293–294, 308
Foreign units, 11, 17–18, 39–40, 42–43,
    165, 246, 248–249, 253, 293. See also
    individual units by name
Fortescue, Hon. Sir John William
    (historian), 301
Franck, James (Inspector of Hospitals),
    236
Frederick of Orange, Prince, 202

French Army, 17, 80, 168, 252, 270, 286
French Revolutionary Wars, 12, 19,
    179–180, 286
Frey, Silvia (historian), 296, 298
Fuengirola, Attack on (1810), 30, 82,
    120
Fuentes de Oñoro, Battle of, 28, 29, 81,
    198, 205–206, 214
Fusilier regiments, 47–48, 55, 186, 207,
    258, 314n8. See also individual units
    by name

Gaelic, 58–59
Garcia Hernandez, Battle of, 278–279
Garrison battalions, 11, 117, 224. See also
    individual units by name
General Regulations, 9, 17, 61
Gibbs, Major General Samuel, 5, 44–45,
    156–157, 160–61, 167
Gibraltar, 10, 23, 30, 33, 82–83, 96, 102,
    112, 115, 118–120, 122, 125, 158,
    166, 223, 230, 245–246, 251, 303,
    319n62
Glanders, 270–271
Gleig, Lt. George, 50, 264, 266, 283
Göhrde, Battle of, 36, 156
Gordon, Captain Alexander, 55–56
Gordon, Major General James
    Willoughby, 5
Gough, Lt. Colonel Hugh, 59, 66–67,
    250, 289
Graham, Lt. General Sir Thomas, 30,
    36–37, 40, 120–121, 138, 140, 157,
    160, 164–65, 167–168, 170, 174, 176,
    178, 199, 201, 208, 212–213, 224,
    240, 262, 264, 269, 332n15
Grant, Colonel Colquhoun, 196
Grattan, Lt. William, 14, 50, 55, 84, 206,
    209, 212, 256
Graves, Donald E. (historian), 290
Green, Private John, 9, 228, 230
Greville, Colonel Hon. Charles, 116
Guadiana Valley, 4, 27, 92, 218, 237
Guard, Lt. Colonel William, 188
Guernsey, 67, 73, 165
Gundogs, 266

Half-pay, 15, 73
Hamilton, Lt. Colonel Alexander, 164,
    290
Hampton Roads, attack on, 64, 166,
    293

Hanoverian Army, 40, 157, 181,
    203–204
Harcourt, Lt. Colonel Charles, 69
Harris, Lt. Colonel William, 44–45,
    81–82, 290
Harris, Lt. General George, 180
Harris, Private Benjamin, 65, 228–230
Hawker, Major General Samuel, 164
Hay, Captain James, 112
Hay, Lt. William, 61, 266
Hay, Major General Andrew, 71–72
Heavy cavalry, 18, 261, 269, 273, 276,
    278–279, 301. See also individual units
    by name
Helder Campaign (1799), 5, 86,
    180–181, 334n2
Henegan, Assistant Commissary
    Richard, 262–63
Highland Regiments, 47–48, 55–59, 186,
    207, 258, 297–298, 314n8, 315n47,
    315n51. See also individual units by
    name
Hill, Lt. General Rowland, Baron, 30,
    49, 187–189, 191–192, 201–202, 206
Hinüber, Major General Heinrich von,
    203
Hoghton, Major General Daniel, 66–67,
    120, 144, 192, 251, 327n45
Holtzermann, Captain Philip, 157
Honourable East India Company, 20,
    21, 43, 181
Hope, Lt. General Sir John, 37, 116,
    183, 185, 201
Horsefall, Private (7th Fusiliers), 290
Horse Guards (Army Headquarters), 4,
    19, 22, 45, 71, 74, 87, 119, 121, 125,
    132, 143, 148–149, 151, 165, 167,
    169, 175–176, 178, 182, 185, 202,
    240
Horses, 19, 260–85, 332n15
Hospitals, 8, 131, 135, 218, 228–230,
    236–237, 241, 243–244, 292, 306
Household Cavalry, 18, 107–109, 144,
    258, 276, 286, 318n42. See also
    individual units by name
Howard, Dr. Michael (medical
    historian), 220
Howard, Major General Kenneth, 192,
    201, 208
Hunger. See Malnutrition
Hussars, 18, 59, 107, 261, 273. See also
    individual units by name

India, 14, 20, 21, 32, 41, 43, 50, 73, 82,
    96, 112, 135–136, 158–159, 180–181,
    200, 220, 299
Indiscipline, 31, 44, 54–55, 60, 68
Inglis, Lt. Colonel William, 46, 63–64,
    69, 80, 189
Inspections of units, 7, 9–10, 61, 66–67,
    73–76, 79, 81, 83, 86, 96, 120,
    138–139, 149, 156–157, 161, 164,
    167–168, 230, 251, 268, 272, 280
Iremonger, Lt. Colonel William, 73–74,
    77–78
Irish Militia, 16, 57, 96, 252
Irish Rebellion (1798), 18, 169
Irish Wagon Train, 20
Italian Levy, 17

Jersey, 164, 173–174, 230
John, Major Henry, 166
Johnston, Sergeant Archibald, 56
Jones, Private Thomas, 247–248, 253

Kelly, Major Dawson, 44–45, 81, 291
Kemmis, Colonel James, 65, 258
Kempt, Major General Sir James, 212
Kennedy, Captain Arthur, 285
KGL Garrison Company, 117, 251
Kielmansegge, Major General Graf von,
    203
Kincaid, Second Lt. John, 206, 289
Kingsbury, Major John, 75, 77–78,
    316n101
King's German Legion, 11, 17, 18, 34,
    39–40, 42, 157, 181, 186, 191, 203,
    224–225, 255, 261, 294, 308, 337n1;
    artillery, 157, 274; cavalry, 18, 19,
    202, 274–285; depot, 157; infantry,
    48, 54, 133, 140–141, 182–83, 189,
    191, 193, 296. See also individual units
    by name

Lake, Lt. Colonel Hon. Gerard, 94
Lambert, Major General John, 177, 196
Lawrence, Sergeant William, 65, 91, 250,
    258
Le Brun, Private Philippe, 165
Lecor, Marechal de Campo Carlos, 191
Le Couteur, Lt. John, 51, 55
Leipzig, Battle of, 36, 157
Leith, Lt. Colonel Alexander, 145
Leith, Lt. General Sir James, 190, 208,
    212, 290, 328n90

Le Marchant, Major General John Gaspard, 196, 278

L'Estrange, Lt. Colonel Guy, 145

L'Estrange, Lt. George, 146, 207, 242

Lightburne, Stafford, 190, 327n59

Light dragoons, 18, 46, 59, 261. *See also individual units by name*

Light infantry regiments, 48, 55–57, 326n21. *See also individual units by name*

Lines of Torres Vedras, 28, 121, 237, 242, 283

Lisbon, Portugal, 7, 9, 25, 28, 99, 102–103, 112–113, 115, 117, 119–122, 131, 186, 195, 243, 251, 271, 317n110

Liverpool, Robert, Earl of, 5, 128, 145, 147, 252–253, 255, 300

Leveson-Gower, Lord Granville, 5

Long, Major General Robert Ballard, 230–231

Looting. *See* Plundering

Löw, Major General Sigismund von, 140

Lundy's Lane, Battle of, 38, 83

Macara, Lt. Colonel Robert, 57

MacDonald, Captain Alexander, 193

Macdonald, Lt. Colonel Duncan, 80–81

Mackenzie, Major General John Randoll, 188, 190

Mackenzie, Major General Kenneth, 81, 116

Mackenzie Fraser, Lt. General Alexander, 134, 183

Mackinnon, Lt. Colonel Henry, 236

Maida, Battle of, 21, 86, 181

Maine, operations in, 38, 96

Mainwaring, Lt. Colonel John Montague, 63, 81

Maitland, Lt. General Frederick, 34, 269

Malnutrition, 226–227, 235, 238–239

Manners, Lt. Colonel Lord Robert, 75–79

Manningham, Major General Coote, 66, 86

Marmont, Marshal Auguste, 29–31

Masséna, Marshal André, 27–28, 119, 121, 198, 288

Master General of the Ordnance, post of, 6

Mauritius (Isle de France), 42, 158

McGrath, Bugler Bryan, 251

McGrigor, Sir James (Inspector of Hospitals), 219–220, 222, 224, 226, 231–244, 292, 329n19

McKeen, Private William, 260, 281

Medals, 51, 301

Mediterranean Theatre, 21, 26, 29, 34, 38, 41, 106, 136, 143, 156, 220, 254

Memoirs, 49, 60, 62, 190

Messing, arrangements for, 9, 239

Military Secretary, post of, 5

Militia, 12–13, 16–17, 21, 57–59, 74, 86, 103, 124, 127, 158, 168–169, 227, 286–287, 291, 295, 298. *See also* Brigades: Brigade of Provisional Militia; Irish Militia

Monthly returns, 4, 7, 8, 9, 53, 147, 161, 173–174, 220, 232, 234–235, 243–44, 251, 264, 303, 308, 309n8

Moore, Lt. General Sir John, 22–26, 55–56, 66, 91, 94, 133, 137, 180, 182–184, 186, 188, 224, 226–227, 276, 303, 309n2, 330n41

Morris, Private Thomas, 44

Morrison, Lt. Colonel Joseph, 83, 290

Muir, Rory (historian), 151, 320n2

Mules, 19, 242, 260, 262–264, 266, 281, 332n2

Mulgrave, General Henry, Earl of, 6

Mundy, Lt. Colonel Godfrey, 77

Munro, Lt. Thomas, 130

Murray, Captain George, 110, 112

Murray, Lt. General Sir John, 34, 189

Murray-Pulteney, General Sir James, 5

Myers, Lt. Colonel Sir William, 188

Napier, Lt. Colonel George, 81–82

Napoleon I, Emperor of the French, 24, 28, 31–32, 36–37, 40, 42, 158, 160, 169, 174, 257, 293

Native American forces, 32, 35

Neglect by officers: asserted by Wellington, 31; historiography regarding, 225, 329n10

Netherlands Army, 40, 202

Netherlands Campaign (1813–1814), 16, 36–37, 42, 44, 89, 127, 156–157, 160–164, 169–175, 178, 223–224, 240, 246, 261–262, 269

New Orleans, Campaign and Battle of, 39, 220, 256

New South Wales, 20, 96, 118

New South Wales Veteran Company, 118

Niagara Campaign (1814), 38, 96, 220
Nichols, Alistair (historian), 192
Nicknames: divisional, 204–206, 328n78; regimental, 46, 49, 69, 296, 314n19
Nicol, Lt. Colonel Charles, 145
Nive, Battle of the, 37
Nivelle, Battle of the, 37, 50
Northern Italy, operations in (1814), 34

Oakes, Brigadier General Hildebrand, 180
Oman, Sir Charles (historian), 87, 108, 139, 151–152, 184, 318n42, 322n63, 328n78
Ompteda, Colonel Christian von, 203
Oporto, Portugal, 129; Campaign and Second Battle of, 26, 95, 107, 122, 130, 133, 182, 187–189, 218, 235–236
Orange, Lt. General William, Prince of, 202
Orthez, Battle of, 37, 149
Oswald, Major General John, 212
Overstretch, 22, 26, 36–37, 42–43, 99, 127, 156, 300
Oxen, draught, 262–264

Pack, Major General Dennis, 82, 200, 317n118
Paget, Lt. General Hon. Sir Edward, 184, 187–189
Paget, Lt. General Lord Henry, 24, 185, 326n23
Pakenham, Major General Sir Edward, 39, 46, 195, 200
Palmerston, Henry, Viscount, 5
Pardoning of offenders, 53, 248, 281
Patriotism, 47, 298–299
Payne, Lt. General William, 187, 189, 194, 200
Peacocke, Lt. Colonel Sir Nathanial, 79–82, 317n110, 317n118
Peacocke, Major General Warren, 7, 112–113, 115–117, 120, 317n110
Pearson, Lt. Colonel Thomas, 290
Picton, Lt. General Sir Thomas, 46, 64, 134, 190, 201, 205, 208–10, 214, 289
Plattsburg, Battle of, 38–39
Plundering, 80, 210, 240–41, 330n58
Ponsonby, Brigadier General William, 282
Ponsonby, Lt. Colonel Hon. Frederick, 62

Portuguese Army, 25, 27–30, 33, 35, 112, 181, 188–190, 192, 203–204, 250, 252, 271, 295–296, 298, 327n27. See also Divisions: Portuguese Division (Peninsula)
Portuguese Campaign (1808), 16, 22, 107, 182, 270. See also Roliça, Battle of; Vimeiro, Battle of
Portuguese Campaign (1810–1811), 27–28, 119, 121, 237–238. See also Busaço, Battle of; Lines of Torres Vedras
Prévost, Lt. General Sir George, 38–39
Prince Regent, 62, 77
Provisional Battalions, 77, 94, 100, 144–156, 172, 175, 299; First (unnumbered prior to December 1812), 95–96, 144–147, 149; Second, 77, 147–149; Third, 147–149; Fourth, 147, 149, 156; Fifth (proposed), 94, 149; under Graham in 1814, 164
Prussian Army, 36, 286
Pyrenees, 33, 35, 37, 242; battles of, 33, 204–205, 213, 273

Quartermaster General, post of, 5
Quatre Bras, Battle of, 40, 46, 50, 300
Queenston Heights, Battle of, 32
Quentin, Lt. Colonel George, 62, 72

Ramsay, Lt. Colonel Hon. James, 73–74, 78
Rations, 129, 235, 238, 240. See also Messing
Recruiting, 9, 13, 14, 45, 57–60, 74, 78, 85–86, 89, 94–96, 102–103, 142, 164, 174, 291
Regimental colonels, 3, 12, 51, 58, 316n79
Regimental histories, 297
Regiment de Dillon, 17
Regiment de Meuron, 17, 165
Regiment de Roll, 17
Regiment de Watteville, 17, 165
Rider, Private George, 53
Roliça, Battle of, 23, 91, 94–95, 182–184, 232
Ross, Captain Hew, 19
Ross, Major General Robert, 39, 86–87
Rosslyn, Lt. General James, Earl of, 185
Rottenburg, Major General Francis, Baron de, 48

Royal Artillery, 19, 262–263, 269. *See also* Royal Horse Artillery

Royal Artillery Drivers, 19

Royal Engineers, 19, 283

Royal Horse Artillery, 19, 50; 2nd Rocket Troop, 157; A Troop, 19; E Troop, 193

Royal Horse Guards (Blues), 18, 108

Royal Navy, 30, 39, 119, 286–287

Royal Sappers and Miners, 19

Royal Staff Corps, 19

Royal Veteran Battalions, 11, 117–118, 161, 224, 229. *See also individual units by name*

Royal Wagon Train, 20, 39, 264, 268, 274

Saint Pierre, Battle of, 79–80

Salamanca, Spain, 29, 31; Campaign and Battle of, 31, 69, 77–78, 121, 198, 205–206, 231

San Sebastián, siege and storm of, 33, 212–214

Schafer, Private (1st KGL Hussars), 281

Schaumann, Deputy Assistant Commissary-General August, 234, 263, 275, 279–280, 283–284

Scovell, Lt. Colonel Sir George, 290

Second-in-Command, post of, 182–185, 187, 208

Secretary at War, post of, 5, 8

Secretary of State for War and the Colonies, post of, 3, 5–6, 8, 33, 95, 114, 121, 128, 146, 148, 161, 175

Sehestedt, Battle of, 157

Seven Years' War, 12, 256, 298–299

Sewell, Lt. Colonel William, 82–83

Sherbrooke, Lt. General Sir John, 38, 74, 78, 96, 187–189, 200

Sherer, Captain Moyle, 14, 60, 84, 95, 217–218

Sicily, 17, 20, 21, 34, 100, 107, 154, 181

Sickness, 4, 8, 25, 27, 29, 30, 78, 90–92, 102, 114, 122, 131–132, 135, 141–142, 144, 147, 162, 217–244, 247, 254, 282, 290, 306. *See also* Epidemics; Hospitals; Walcheren Fever

Silveira, Marechal de Campo Francisco, Conde de Amarante, 191

Simmons, Lt. George, 237

Skerrett, Colonel John Byne, 121, 192

Sleigh, Lt. Colonel James, 62

Smith, Captain Harry, 211, 289

Soult, Marshal Jean-de-Dieu, 26, 28–30, 37, 235

South Africa, 20, 100, 181

South American Campaign (1806–1807), 22, 23, 91, 102, 181, 234

Spanish Army, 23, 25, 27, 33–35, 37, 295–296

Spencer, Lt. General Sir Brent, 22, 23, 182–183, 188, 200–201, 208, 234

Squadrons, organization of, 18, 106–107

Staff Corps of Cavalry, 19

Stanhope, Captain Hon. James, 124

Stavely, General Miles, 51

Steuart, Lt. General Sir James, 271–272

Stewart, Brigadier General Richard, 181

Stewart, Lt. General Hon. Sir William, 66, 120, 201, 209, 227, 262, 327n67

Stewart, Major General Sir Charles, 24, 130–131

St. George, Lt. Stepney, 146

St. Patrick's Fund, 51–52

Straggling, 31, 74, 218, 250, 259, 291

Stuart, Lt. General James, 180

Stuart, Lt. General Sir John, 21

Subsidies, 27, 33, 287, 302

Suchet, Marshal Louis, 34, 37

Supply problems, 27, 235, 238–241

Swabey, Lt. William, 50

Tagus, 26, 28, 92, 188

Talavera, Battle of, 16, 27, 49, 54, 56, 63, 95, 114, 122, 190, 199–200, 217, 236, 268, 276, 298

Talavera Campaign (1809), 27, 91, 95, 99, 129–130, 132–133, 187, 189–190, 195, 218, 235–236, 241, 266

Tariffa, Spain, 30, 120; siege of, 30, 59

Tarragona, attack on (1813), 34, 189

Taylor, Major General Herbert, 261–262, 264, 269

Tents, 241–242, 262

Thames, Battle of the, 35, 100

Tidy, Lt. Colonel Francis, 63

Tilson, Christopher (later Tilson-Chowne), 200–201, 249–250, 326n33, 327n68

Tomkinson, Captain William, 109, 112, 204–206, 270–271

Torrens, Major General Henry, 5, 170

Toulouse, Battle of, 37, 56, 73, 87, 90, 199, 285
Transport ships, 21, 22, 234, 270–271

Undersecretary of State for War and the Colonies, post of, 6
Unit commanders, 45, 61–85, 290
Unit Rotation, 85, 87, 112–126, 143, 155
Urban, Mark (historian), 290, 298

Vandebruck, Private Henry, 253
Vigo, retreat to, 65, 250
Villamuriel, action at, 290
Vimeiro, Battle of, 23, 73, 91, 95, 182–183, 194, 232, 296
Vitoria, Campaign and Battle of, 33, 79, 92, 197, 199, 202, 241, 264, 276, 283, 285, 298
Vivian, Major General Sir Richard Hussey, 194
Volunteers, 12, 21, 286–287, 298

Wade, Lt. Colonel Hamlet, 64
Walcheren, Expedition to (1809), 7, 24–26, 28, 31, 42, 74, 86, 88–90, 92, 94, 96, 104, 127–128, 132, 134–138, 141–142, 185–186, 190, 196, 199–200, 215, 218–219, 222–223, 225–231, 236, 245–246, 251, 303
Walcheren Fever, 25, 31, 137, 185, 226–232
Wallace, Lt. Colonel Alexander, 50, 64
Wallmoden-Gimborn, Lt. General Ludwig, Graf von, 157, 337n1
War of 1812, 20, 32, 35, 37–39, 42, 98, 100, 117, 200, 220, 224, 245, 270, 300–301, 304
War of the Austrian Succession, 179
Waterloo, Campaign and Battle of, 14, 16, 40, 45, 48, 56, 63, 73, 78–79, 91,

98, 102, 106–107, 110, 161, 200–201, 203, 210, 222–223, 263, 297, 300, 304
Watson, Bruce Allen (historian), 212–13
Wellington, Field Marshal Arthur Wellesley, Duke of, 4, 7, 16, 38, 50, 52–53, 60, 74, 77, 80–81, 104, 120, 218, 290–291, 298, 328n90; during 100 Days, 40, 117, 177, 297, 301; and desertion, 248, 252–253, 255–256; and divisional system, 178, 185–1216, 294–295; and horses, 268–274; Indian experience, 21, 25, 180–181; and manpower, 94–96, 102–103, 109, 112–119, 127–128, 132–133, 143–158, 167–176, 295–296, 299; in Peninsular War, 22–23, 25–33, 35, 37, 121, 124, 126, 128–131, 140, 182–184, 219, 223, 230, 236, 239–240, 242, 246, 263–264, 275, 279, 285, 289–293
West India Regiments, 11, 43. *See also individual units by name*
West Indies, 18, 20, 21, 41, 96, 135, 217, 220, 256, 294, 299
Wheeler, Private William, 58
Whitelocke, Lt. General Sir John, 23
Widdrington, Major General David, 230
Wilkins, Major George, 66–67
Wood, Lt. George, 206
Woodberry, Lt. George, 285
Wynn, Colonel Sir Watkin, 170, 324n120

York, Field Marshal HRH Prince Frederick, Duke of, 3, 5–6, 12, 23, 33, 35, 79–80, 86, 180, 182–186, 272, 289–290; and flogging, 67–71; and manpower, 127–128, 133, 151–156, 158–176, 299
York Light Infantry Volunteers, 18